The Economy of Ghana

The First 25 Years since Independence

M. M. Huq
Lecturer in Economics
University of Strathclyde

St. Martin's Press   New York

© M. M. Huq, 1989

All rights reserved. For information, write:
Scholarly and Reference Division
St. Martin's Press, Inc., 175 Fifth Avenue, New York, N.Y. 10010

First published in the United States of America in 1989

Printed in Hong Kong

ISBN 0-312-02107-0

Library of Congress Cataloging-in-Publication Data
Huq, M. M.
The economy of Ghana.
Bibliography: p.
Includes index.
1. Ghana—Economic conditions—1957–1979.　I. Title.
HC1060.H87　1989　330.9667′05　88-18158
ISBN 0-312-02107-0

THE ECONOMY OF GHANA

Also by M. M. Huq

CHOICE OF TECHNIQUE IN LEATHER MANUFACTURE
(*with H. Aragaw*)

MACHINE TOOL PRODUCTION IN DEVELOPING COUNTRIES
(*with C. C. Prendergast*)

To my late father, who, with no formal education in economics, knew very well the difference between consumption and saving, and sacrificed his precious consumption for my education, which he knew to be an investment.

	7.1.3	Manganese	156
	7.1.4	Bauxite	157
	7.1.5	Salt	158
7.2	Output and factor intensity		159
7.3	Recent trends		161

PART III MONEY, BANKING, TRADE AND EMPLOYMENT

8 Money and Credit — 167

    8.1 Structure of banking   167
    8.2 Money supply   173
    8.3 Credit supply   177

9 Development Banking — 181

    9.1 Need for development financing   181
    9.2 Specialised development banks   182
           9.2.1 National Investment Bank   183
           9.2.2 Agricultural Development Bank   187
           9.2.3 Rural Banks   190
    9.3 Conclusions   192

10 External Trade — 195

    10.1 Exchange rate policy   195
    10.2 Export and import policies   198
    10.3 Major exports   201
    10.4 Major imports   204
    10.5 Direction of trade   207
    10.6 Terms of trade   210
    10.7 Balance of payments   213

11 Prices and Internal Trade — 215

    11.1 Inflation   215
    11.2 Price controls   217
    11.3 Internal trade   221

12 Employment — 225

    12.1 Employment in public and private sectors   225
    12.2 Wages and salaries   230

PART IV SAVINGS, INVESTMENT AND TECHNOLOGY

13 Finance for Investment — 237

    13.1 Domestic savings — 237
        13.1.1 Public savings — 238
        13.1.2 Private savings — 244
    13.2 External capital — 246
    13.3 Conclusions — 250

14 Investment and Technology Choice — 253

    14.1 Low investment level — 253
    14.2 Public and private investment — 254
    14.3 Sectoral allocation — 256
    14.4 Rationale of investment decisions — 256
    14.5 Choice of technology — 266
        14.5.1 Technology choice in Ghana — 269
    14.6 Conclusions — 274

Postscript — 277

Appendix A: A note on data used in the study — 283

Appendix B: The *Kalabule* Economy — 309

Notes and References — 315

Bibliography — 335

Index — 345

List of Tables

1.1	Main demographic features 1960, 1970 and 1984	38
1.2	Population by regions 1960, 1970 and 1984	40
1.3	Distribution of employment by industries, 1960 and 1970	42
1.4	Occupational distribution by sex, 1960 and 1970	43
2.1	Growth in GDP and GDP per capita, 1890-1983	46
2.2	GDP and GDP per capita, 1957-84	47
2.3	Percentage distribution by uses of GDP at current prices, 1960-82	49
2.4	Percentage distribution by uses of GDP at 1970 prices, 1961-84	50
2.5	Percentage distribution of GDP by industrial origin at current prices, 1965-82	51
2.6	Percentage distribution of GDP by industrial origin at 1970 prices, 1968-84	52
2.7	Average and marginal gross domestic savings at 1970 prices, 1957-80	54
2.8	Savings and investment in selected developing countries, 1984	55
2.9	Income distribution by source and class in primary production in the Biriwa area, 1982	58
2.10	Inter-regional differences in urbanisation, 1970	59
2.11	Inter-regional distribution of student enrolments at primary, middle and secondary school levels, 1978-79	60
3.1	Physical infrastructure, 1960-80	63
3.2	Statistics on education and health, 1965-80	72
4.1	Production of important food crops, 1950-83	82
4.2	Land area under important food crops and land productivity, 1963-83	84
4.3	Distribution of farm holdings by size, 1970	85
4.4	Percentage distribution of holding size by regions, 1970	85
4.5	Average number of farms per holder by regions, 1970	86
4.6	Percentage of farms by land tenure by regions, 1970	86
4.7	Number of farms in the cultivation of major crops in three selected districts, 1982	88

List of Tables

4.8	Inter-cropping: Number of crops per farm in three selected districts, 1982	88
4.9	Hired and family labour in selected districts, 1982	90
4.10	Continuing irrigation projects and total projected irrigated area by regions, 1981	92
4.11	Fertiliser imports 1971-81	94
4.12	Loans and advances to the agricultural sector by commercial banks, NIB and ADB, 1966-81	99
4.13	Domestic livestock population by type, 1960-82	100
4.14	Livestock population by regions, 1982	102
4.15	Catch of fish from sea and Volta Lake, 1965-82	104
4.16	Number of firms, installed capacity and capacity utilisation in different sectors of the timber industry, 1982	105
4.17	Production and export of log and sawn timber, 1960-82	106
5.1	Percentage distribution of cocoa by region according to farming areas, 1970	109
5.2	Output, nominal and real producer prices of cocoa, 1959/60-1981/82	110
5.3	Cocoa production in Ghana by regions, 1975/76-1981/82	111
5.4	Comparative return to farmers on cocoa, rice and maize, 1980	113
5.5	Farmers' reaction to input price subsidy versus output price increase, 1977	115
6.1	Gross output, value added, employment, wages and salaries in medium and large-scale manufacturing, 1963-83	120
6.2	Sectoral distribution of employment, value added, gross output and fixed assets in medium and large-scale manufacturing, 1981 and 1984	122
6.3	Output, employment and capital intensity in 20 major medium and large-scale manufacturing industries, 1984	124
6.4	End-use of manufacturing output, 1968	127
6.5	Percentage distribution of value added in medium and large-scale manufacturing by regions, 1970-84	134
6.6	Employment, output, value added and capital per employee in small and large-scale sectors, 1973	135
6.7	Selected characteristics of small-scale industries, 1981	137

List of Tables

6.8	Percentage distribution of value added in medium and large-scale manufacturing by nationality and type of ownership, 1962-84	141
6.9	Manufacturing industries: estimated rate of capacity utilisation, 1978-82	143
6.10	Percentage distribution of sources of inputs for medium and large-scale manufacturing, 1970-84	144
6.11	Domestic resource costs (DRC) and effective rates of protection (ERP) in manufacturing, 1967-70	148
7.1	Employment, output, value added and fixed assets in gold, bauxite, manganese and diamond 1970, 1975, 1980 and 1982	160
7.2	Value added and capital intensity in major mining sectors, 1970 and 1980	161
8.1	Average annual growth rate of money supply, 1961-84	175
8.2	Central government budget deficit, 1961-82	176
8.3	Percentage distribution of primary commercial banks' loans and advances by sectors, 1965-82	178
9.1	Distribution of NIB Joint Venture Projects by ownership and NIB equity participation as at December 1979	184
9.2	Sectoral distribution of NIB approved loans and NIB equity investment, 1963-82	186
9.3	Sectoral distribution of ADB loans, 1979	189
9.4	Cumulative total of rural banks by regions, 1976-83	191
10.1	Official and parallel foreign exchange rates of selected currencies, 1978-83	196
10.2	Exports and imports as percentages of GDP, 1956-82	198
10.3	Effective exchange rate for exports (EER_x) and imports (EER_m), 1955-71	201
10.4	Exports by quantity, 1960-80	202
10.5	Exports by value and percentage distribution at current prices, 1960-80	203
10.6	Imports by value and percentage distribution at current prices, 1960-80	205
10.7	Composition of imports by end-use, 1957-80	206
10.8	Direction of exports, 1960-79	208
10.9	Origin of imports, 1960-79	209
10.10	Terms of trade, 1969-82	211
10.11	Balance of payments, 1970-84	212
11.1	Annual rates of inflation, 1961-84	216

11.2	Average annual inflation rates of selected items, 1960-80	217
11.3	Average annual increases in official and parallel market prices of selected commodities, 1976-84	220
11.4	Distributive trade: sales by the big seven retail stores, 1975-82	223
12.1	Recorded employment in establishments employing ten or more persons, 1957-79	226
12.2	Growth of employment by ownership in medium and large-scale establishments, 1970-80	228
12.3	Average annual wages in the medium and large-scale industrial sector, 1967-84	231
13.1	Domestic and foreign savings at 1970 prices, 1957-84	237
13.2	Public and private savings at 1970 prices, 1961-84	238
13.3	Central government current revenue, and recurrent and development expenditure, 1961-84	239
13.4	Structure of tax revenue at 1970 prices, 1961-84	240
13.5	Non-tax revenue at 1970 prices, 1961-84	241
13.6	Profit and loss account of selected state enterprises, 1965-80	243
13.7	Nominal and real interest rates, 1961-82	246
13.8	Foreign indebtedness (cumulative) by type of debt, 1976-82	249
13.9	Foreign loans and grants, 1977-82	250
14.1	Gross investment at current and 1970 prices, 1957-82	253
14.2	Public and private investment at current and 1970 prices, 1961-80	255
14.3	Allocation of investments by type of assets at current and 1970 prices, 1961-82	257
14.4	Estimate and use of shadow prices by selected organisations	261
14.5	Incremental capital-output ratios (ICOR) at 1970 prices, 1957-80	265
14.6	Characteristics of technology choice in selected projects	273
A.1	Gross domestic product, 1957-84	287
A.2	Population and GDP per capita, 1957-84	288
A.3	Expenditure components of the GDP at current prices, 1957-84	289
A.4	Expenditure components of the GDP at 1970 prices, 1957-84	290

A.5	GDP by industrial origin at current prices, 1965-84	291
A.6	GDP by industrial origin at 1970 prices, 1968-84	293
A.7	Gross and net domestic capital formation at current and 1970 prices, 1957-84	295
A.8	Implicit deflators of the components of GDP by uses, 1957-84	296
A.9	Implicit deflators of the types of assets in gross fixed capital formation, 1965-84	297
A.10	Implicit deflators of the components of GDP by industrial origin, 1968-84	298
A.11	Central government revenue and expenditure, 1961-85	300
A.12	Public, private and total savings at current and 1970 prices, 1961-84	301
A.13	Public, private and total investment at current and 1970 prices, 1961-84	302
A.14	Money and quasi-money supply, 1961-85	303
A.15	Consumer price index, 1963-84	304
A.16	Major exports of domestic produce by value at current prices, 1960-82	305
A.17	Major imports classified by main commodity groups, 1960-82	306
B.1	Percentage of people buying goods from official sources and the extent of such purchases for 1980, 1982 and 1984	312

List of Figures

14.1	Fixed factor proportions	267
B.1	Low official price and excess demand	310

Preface

The major part of the work for this book was done during my two-year stay in Ghana between 1982 and 1984, as a member of a team from the University of Strathclyde engaged in a collaborative research and training project with the Centre for Development Studies (CDS) at the University of Cape Coast. The project was funded by the European Economic Community and the Government of Ghana.

In the course of preparing lectures for Training Workshops on project appraisal and technology choice at the CDS, I discovered that no current book on Ghana was available which provided a comprehensive analysis of Ghana's economic performance since independence, based on a consistent and again comprehensive set of statistical indicators. The last work on this scale – that by Birmingham et al. – dated back to 1966. It is true that a small number of books – including a very good study by Killick (1978) which we have referred to frequently – have appeared since then, but they have remained specialised in their approach.

My stay in Ghana happened to coincide with a period of deep economic crisis. The crisis which reached serious proportions by the time Ghana celebrated 25 years of independence in 1982 was undreamt of in 1957. It was then expected, with good reasons, that the Ghanaian economy would be a 'show-piece', a model for other sub-Saharan African countries to follow. This study discusses developments in the Ghanaian economy mainly during 1957–82, hence the sub-title *The First 25 Years since Independence*.

The book seeks to identify, at least in broad terms, what went wrong. Individual chapters discuss sectoral developments, while the causes of economic decline in the round are dealt with in the introductory, Overview chapter. Many of the views expressed were presented during my stay in Ghana – in seminars at the Universities of Ghana (Legon) and Cape Coast, and also during many discussions I had in various circles in Ghana. These views – which appeared highly critical to many at that time – have now become almost conventional wisdom in Ghana. And it is heartening to note that in the recent past the economy of Ghana has shown an upward trend, following present policy directions which are generally in line with the policy measures suggested in the study.

The first draft of the book was completed in September 1984 (that is, just before I left Cape Coast for Glasgow) and at the time of writing

1982 was the last year for which comprehensive data were available. I have revised the study several times since then and in the process have incorporated recent available data so as to make the book as up-to-date as possible. However, two points need to be mentioned. First, as the study is the direct outcome of my first-hand experience gained during 1982–84, most of the comments in the book should be read as at 1984. Secondly, as will be apparent to the readers, an important objective of the study is to provide information on the sectoral developments of the economy in as much detail as possible, hoping to make it easier for others to start serious analytical work.

In the course of my research I was fortunate to be able to assemble and direct a highly competent and hard-working research team in Cape Coast consisting of Danny B. Safo, Stephen Osei-Yeboah and Bernard Buah-Nkwantabisa – all Ghanaians. Considering the serious data gaps that existed and which simply made it impossible for us to make any headway, we started by building a 'data bank'. We have depended heavily on this data bank for our analysis. Bernard was in charge of the data bank, a job which he undertook with devotion and efficiency. Danny and Stephen proved indispensable in collecting and analysing data. In Glasgow, Salifu Baba Ibrahim (a Ghanaian research student at Strathclyde University) kindly provided assistance in updating the data.

Many Ghanaians from the three Universities (at Cape Coast, Kumasi, and Legon), and also from different Government Ministries and Organisations helped extensively in my research and I am grateful to all of them. I am also indebted to the individual institutions (acknowledged separately after this Preface) which provided information, otherwise unavailable, through unpublished documents and personal interviews.

At different stages of the study I benefitted from the help of a number of friends, colleagues and others. Particular mention should be made of: in Ghana, Dr Stephen Adei (of the Ghana Investments Centre) who read the individual chapters as they took final shape and made valuable comments, and in Glasgow, Dr H. P. Kushari (of the Economics Department, Strathclyde University) who read with utmost care the whole draft and whose suggestions led to many improvements in the book. I also benefitted from the following who read parts of the earlier draft(s) and made helpful suggestions: George Adamu, Dr Edwin Amonoo, Thomas Buxton and Professor Kwesi Haizel of Cape Coast University; Lawrence Smith of Glasgow University; and Professor James Pickett, Dr Eric Rahim and Dr Douglas Strachan of

Strathclyde University. I also had the opportunity of having constant discussions with my colleague, Michael Tribe (now of Bradford University), in both Cape Coast and Glasgow. Janet Lauchlan (now of Heriot-Watt University) read the whole draft and suggested a number of improvements. Dr M. J. Abedin (now of King Saud University, Riyadh) prepared and ran a number of computer programmes, providing a set of consistent time series data. Dr Girma Zawdie was closely involved at the final stage of the study, going through the various chapters and providing useful comments. He also helped in the preparation of the index. Hayford Ofori-Atta's careful scrutiny – just before the book went to press – was very helpful.

I owe a special debt to Dr Kwesi Boakye (Acting Director of the Centre for Development Studies, University of Cape Coast, when the first draft of the book was being completed), who, besides providing constant moral support, helped me in many ways.

At Strathclyde University, Professor James Pickett, Director of the David Livingstone Institute (where I worked until August 1987) and Professor David Forsyth, Head of the Department of Economics (where I moved in September 1987) provided the much needed encouragement for the publication of the book and I am very grateful to both of them.

I would also like to thank Keith Povey for his dedicated and painstaking editing.

Typing assistance in Cape Coast was kindly provided by George Acquah, Joseph Charles Assam and Grace Blankson and, in Glasgow, by Katherine Davidson, Aliza Islam and Fiona Macdonald.

I am indebted to my wife (Kumkum) and son (Kamal) for bearing with my long periods of silence and always co-operating with me, thus making the burden of work that much more tolerable.

Needless to mention, I am alone responsible for the errors and omissions that remain despite the advice and help I have received from many individuals and organisations. I do, however, hope that the factual errors that remain are not too numerous, nor do they present a fundamentally distorted picture of the Ghana economy. There are obvious difficulties of making a wide-ranging study such as this on an individual effort, with limited resources, and I will feel delighted if I have succeeded in making a small contribution in my humble capacity.

Glasgow M. M. HUQ

Institutional Acknowledgements

Aboso Glass Factory, Aboso
Agricultural Development Bank, Accra and Kumasi
Akyeampong Ceramics, Winneba
Alajo Brick Factory, Accra
Ameen Sangari Industries Ltd, Cape Coast
Ankaful Brick Factory, Ankaful
Anwia-Nkwanta Oil Mills, Bekwai
Ashanti Goldfields Corporation, Obuasi
Bank of Ghana, Accra
Bonsaso Tyre Factory, Bonsaso
Building and Road Research Institute, UST (University of Science and Technology), Kumasi
Cannery Company Ltd, Wenchi
Central Bureau of Statistics, Accra
Centre for Development Studies, UCC, Cape Coast
Cocoa Marketing Board, Accra
Cocoa Processing Factory, Takoradi
Ghana Highway Authority, Accra
Ghana Investments Centre, Accra
Ghana Industrial Holding Corporation, Accra
Ghana National Association of Garages, Kumasi
Ghana National Manganese Corporation, Nsuta
Ghana Railways Corporation, Takoradi
ISSER, University of Ghana, Legon
Komenda Sugar Factory, Komenda
Leather and Tanning Company Ltd, Kumasi
Logs and Lumber Ltd, Kumasi
Management Development Productivity Institute, Accra
Ministry of Agriculture, Accra
Ministry of Finance and Economic Planning, Accra
Ministry of Industries, Accra
Nasia Rice Company, Tamale
National Investment Bank, Accra and Kumasi
Nsawam Canneries, Nsawam
Prices and Incomes Board, Accra

Saltpond Ceramics, Saltpond
State Enterprises Commission, Accra
State Gold Mining Corporation, Tarkwa
Technology Consultancy Centre, UST, Kumasi
Timber Marketing Board, Takoradi
Volta Aluminium Company, Tema
Volta River Authority, Akosombo

Abbreviations

ADB	Agricultural Development Bank
AFRC	Armed Forces Revolutionary Council
₵	Cedi
CBS	Central Bureau of Statistics
CDS	Centre for Development Studies
CMB	Cocoa Marketing Board
CPC	Cocoa Purchasing Company
CPP	Convention People's Party
CSIR	Council for Scientific and Industrial Research
DLI	David Livingstone Institute
ECOWAS	Economic Community of West African States
ERP	Economic Recovery Programme
GDP	Gross Domestic Product
GFCF	Gross Fixed Capital Formation
GIHOC	Ghana Industrial Holding Corporation
GIC	Ghana Investments Centre
GNMC	Ghana National Manganese Corporation
GSCA	Ghana Sample Census of Agriculture
GWSC	Ghana Water and Sewerage Corporation
ICOR	Incremental Capital-Ouput Ratio
ISSER	Institute of Statistical, Social and Economic Research
IMF	International Monetary Fund
ISIC	International Standard Industrial Classification
LDC	Less Developed Country
MBG	Merchant Bank Ghana Ltd
NIB	National Investment Bank
NLC	National Liberation Council
NRC	National Redemption Council
N₵	New Cedi
PIB	Prices and Incomes Board
PNDC	Provisional National Defence Council
SEC	State Enterprises Commission
SGMC	State Gold Mining Corporation
TCC	Technology Consultancy Centre
UCC	University of Cape Coast
UNCTAD	United Nations Conference on Trade and Development

USAID	United States Agency for International Development
VALCO	Volta Aluminium Company
VRA	Volta River Authority
WHO	World Health Organisation

Chronology of Major Events, 1957-84

1957 Bank of Ghana established on 4 March
Ghana becomes independent on 6 March
1958 Ghana pound (G£) becomes legal tender in July
Consolidated Plan (1958-59) produced
1959 Second Five-Year Development Plan (1959-64) comes into effect
Ghana becomes a Republic with Nkrumah as the President
1961 Exchange control and import licensing severely tightened in July and December
1962 Promulgation of Control of Prices Act
1964 Seven-Year Plan for National Reconstruction and Development (1963-64 to 1969-70) presented in January
1965 Decimal currency system introduced in July with Cedi as the major unit and Pesewa the minor unit
1966 Overthrow of the Nkrumah Government by the National Liberation Council in February
1967 New Cedi (N₵) introduced in February (G£1 = N₵2.00)
Currency devalued in July (from N₵0.71 to N₵1.02 per US dollar)
Imports liberalised, a process which continued for the next four and a half years
1968 Two-Year Development Plan (Mid-1968 to Mid-1970) produced in July
1969 Elected Government under Busia takes over in October
1970 One-Year Development Plan (July 1970-June 1971) produced in September
1971 Currency devalued in December (from N₵1.02 to N₵1.82 per US dollar)
1972 Busia overthrown by NRC military Government in January
Currency revalued in February (from N₵1.82 to N₵1.28 per US dollar) and exchange control imposed
1973 Foreign exchange rate adjusted to ₵1.15 per US dollar in December
1977 Five-Year Development Plan (1975-76 to 1979-80) produced in September

xxvii

1978 Currency devalued in September (from ₵1.15 to ₵2.75 per US dollar)
Acheampong replaced by Akuffo in a 'palace coup' in July
1979 AFRC Government takes over in June
Elected Government under Limann takes over in September
1981 PNDC Government takes over on 31 December
1982 Programme for Reconstruction and Development presented in December
1983 Surcharges on imports and bonuses on exports introduced in April (7.5 and 9.9 times the official exchange rate of ₵2.75 per US dollar depending on imports and exports)
Currency devalued in October (from ₵2.75 to ₵30.00 per US dollar)
1984 Currency devalued in March (from ₵30.00 to ₵35.00 per US dollar)
Currency devalued in August (from ₵35.00 to ₵38.50 per US dollar)
Currency devalued in December (from ₵38.50 to ₵50.00 per US dollar)
Minimum daily wage rate increased in December by 100 per cent (to ₵70.00, to be implemented in two stages)

Ghana: Regional Boundaries and Capitals 1983

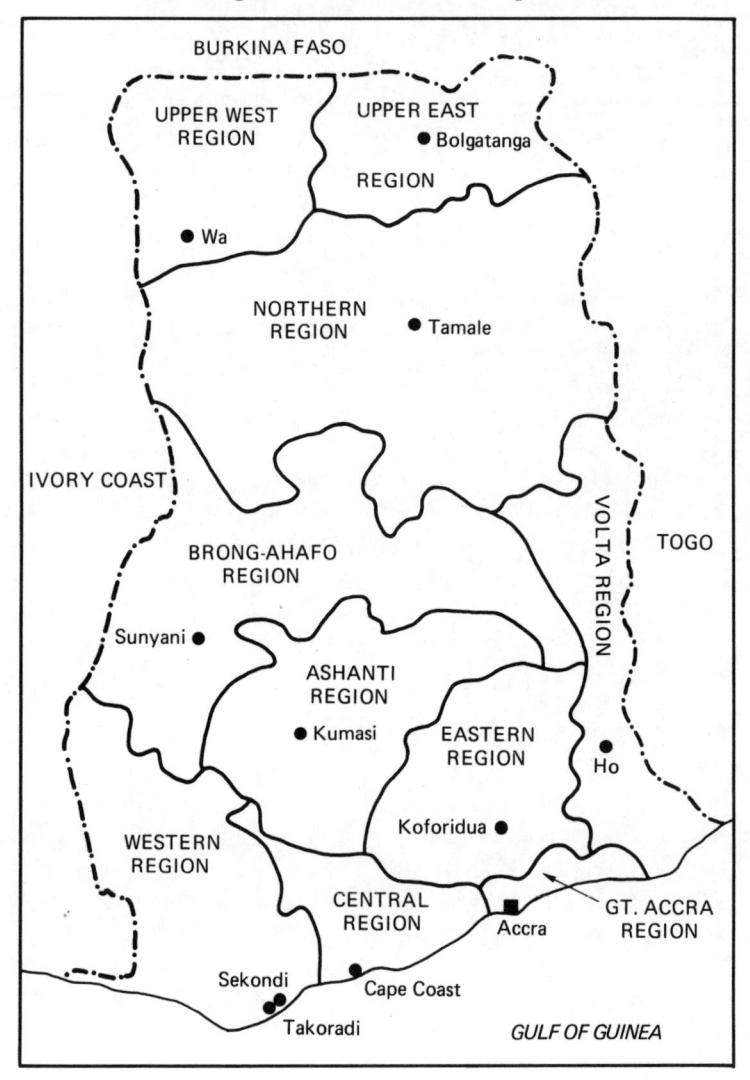

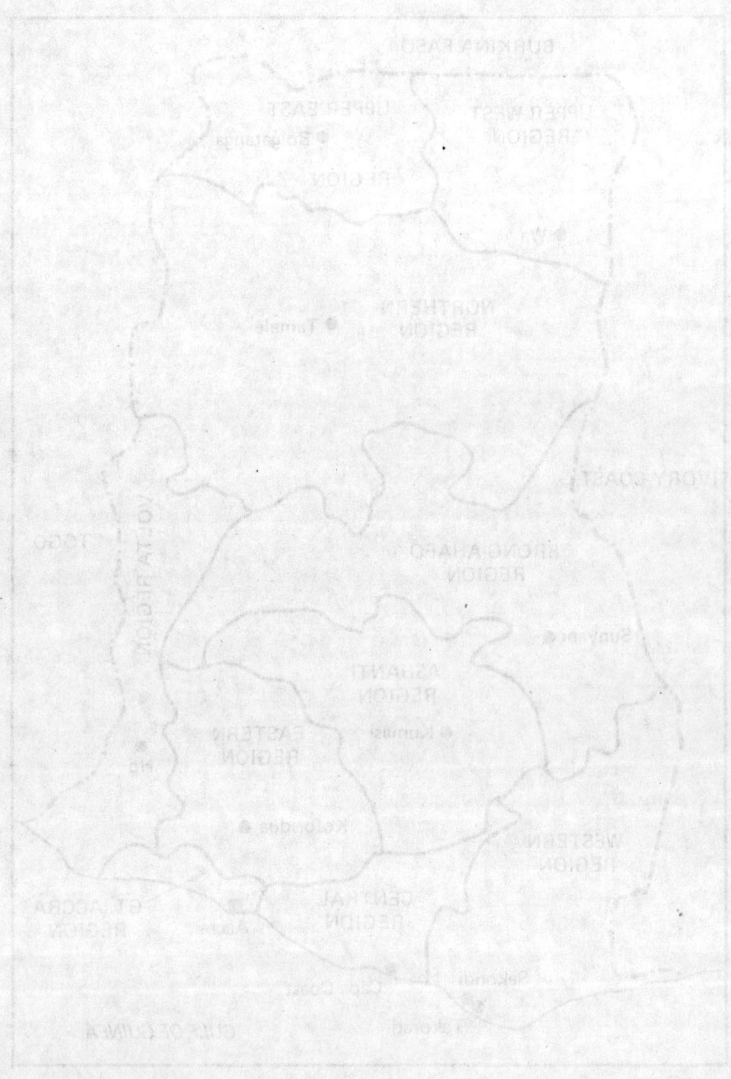

Overview

The book describes and discusses the sad story of Ghana – a nation rich in natural resources and human potential – which in the early 1980s was in a critical economic situation, the culmination of steady decline for over a decade. This is particularly unfortunate, for at the time of its independence in 1957 hopes were widely entertained that Ghana would prove to be the 'show-piece' of African economic development, a model for others to follow.

Why has this not happened? The answer to this question cannot easily be given in a short compass. Part of the difficulty is that the economic confusion of much of the post-independence period has been associated with corresponding failings in the coverage, robustness and timeliness of economic statistics. Comprehensive empirical underpinning of argument has not been possible. Moreover, weaknesses of official statistics have been compounded by the undoubtedly adverse effect of worsening economic conditions on research in universities and other institutions in Ghana. The Ghanaian economy in recent years has provided an unfavourable working environment, even to those who wished to determine the causes of economic decline.

A consequent major purpose of the present volume is to make a significant contribution on the data front. In subsequent chapters, and within a national accounting framework, a great deal of material – much of it new – is brought together, described and analysed. It is thus hoped to stimulate further and more rigorous, analysis of the Ghanaian experience.[1] In the meantime an effort is made – by focusing on the planning and management of the economy – to examine what went wrong, drawing but not exclusively relying upon the information presented subsequently.

At the outset, two points need to be mentioned. First, there are two inter-related questions which need to be answered in an analysis of the economic decline of any country. One relates to economic policies and procedures pursued by specific governments and the consequences of such policy measures. The other relates to the deep-rooted historical, social and cultural factors which have been instrumental in the adoption of certain policies leading to economic decline. Obviously, in an economic study like this our analysis will concentrate on the former.

Second, both planning and management fall under broad government policy measures and, as will be found, an attempt to describe the two under separate sections entails repetitions. Such an approach however enables us to focus more sharply on these two important areas. Before we start discussing these, let us briefly review the performance of the Ghanaian economy since independence.

THE ECONOMY SINCE INDEPENDENCE

Growth of per capita real income during the immediate pre-independence years was satisfactory and the country had a 'promising start as one of the richest, most successful and politically mature regions of black Africa',[2] having substantial sterling reserves and well-formulated development plans (see Chapter 2). The first few years of independence (1957-60) witnessed satisfactory annual average GDP growth rates of over 6 per cent. This was a good performance and quite befitting a country which in 1960 was far ahead of many other developing countries. Its per capita national income of £70 in 1960 was significantly higher than that of Egypt (£56), Nigeria (£29) or India (£25).

The average annual growth rate of GDP was lower during the 1960s, being only 2.8 per cent, and that of per capita GDP was only 0.4 per cent. Significant fluctuations occurred in GDP growth during the 1970s with four years showing negative growth rates and the average annual rate of growth over the decade as a whole being only 0.4 per cent. With annual population growth of 2.6 per cent, there was a fall in per capita GDP of 2.2 per cent per annum. In 1981 and 1982 GDP fell by 3.8 and 6.1 per cent respectively. Given the poor performance of the economy since the mid-1970s, the average Ghanaian in 1982 was much worse off than he was in 1957. At 1970 prices, GDP per capita in 1982 was ₡180, the corresponding figure in 1957 being ₡230. Moreover, absolute poverty must be much more severe as the decline in per capita income has been accompanied by a worsening of income distribution (see Chapter 12).[3]

As with GDP and GDP per capita, other economic indicators also suggest disturbing trends in the economy. The decline in the ratio of investment to GDP is particularly striking. At constant prices domestic capital formation increased from 14.1 per cent of GDP in 1957 to 22.6 per cent in 1960, but has gradually declined since then. In 1982, the corresponding figure was only 6.9 per cent (see Chapter 14).

There has also been a significant fall in public saving. From a positive figure of 4.0 per cent of GDP in 1961, public saving became negative in 1969, 1970 and 1973, and has remained so since 1975. In 1981 and 1982, central government current revenue was only about 52 and 57 per cent, respectively, of the recurrent expenditure. The negative balance and capital expenditure were largely met by budget deficits, which were in the range of 4 to 7 per cent of GDP in the early 1980s. Both tax and non-tax revenue fell sharply over the years. The contribution made by different direct and indirect taxes has been far from satisfactory. So far as non-tax revenue is concerned, the fall was drastic. At 1970 prices, it fell from ₵42 million in 1965 to about ₵7.5 million in 1982, mainly as a result of poor performances by most of the public sector undertakings.

Exceptionally poor performances were also observed in the export and import sectors. In the immediate post-independence years, at current prices, both exports and imports comprised over one-quarter each of GDP, while in the early 1980s the respective figures were less than 4 per cent (see Chapter 10). As the fixed nominal exchange rate became highly overvalued, there was naturally a major disincentive for exports. On the other hand, the trading of imported goods became so profitable that the emphasis of the economy shifted from production to trade. The system also encouraged serious corruption in the import licence allocation system.[4]

Inflation has been rapid since 1971, and particularly so since the mid-1970s (see Chapters 8 and 11). Total money supply increased from ₵697 million in 1974 to ₵6058 million in 1980, and further to ₵23 744 million in 1984. The average annual rate of inflation was 40 per cent during the 1970s, while it was about 70 per cent from 1980 to 1984.

As nominal wages have failed to keep pace with inflation, real wages have declined, particularly rapidly since the mid-1970s. Between 1970 and 1984 average real earnings of industrial workers fell to one-seventh of their 1970 value. White-collar workers, especially in the public sector, have experienced similar declines (see Chapter 12). Consequently many people have been forced to take up a subsidiary occupation (such as farming, trading, etc.) often to the neglect of the main occupation, if it was found to be a secure (for example, a government or semi-government, job).[5] Of course, some people such as traders and those actively engaged in *kalabule* or illegal practices were quite well off, but their number was unlikely to be very high.

With the decline in real wages, professional people and others with talent and skill began to leave the country to take up employment abroad. This has drastically reduced the ability of the government to administer the economy, thus making it extremely difficult to implement the recovery programme.

Over the years there has been a rapid increase in public sector employment, mainly in the form of overmanning in state enterprises, government corporations, local and central government services, and so on. In a state of falling output, the growth in employment has resulted in a significant fall in productivity, besides causing a serious strain on government expenditure (see Chapter 12). There has also been a rapid growth in employment in the low-productivity informal sector, caused by large-scale rural-urban migration. According to an estimate by the World Bank, 32 per cent of the population moved out of the rural areas during 1960-70.[6] The same source estimates that 18 per cent of the total labour force remained 'unemployed'. The open urban unemployment problem was mitigated to some extent by international migration, but a reverse migration – such as that following the expulsion of over one million Ghanaians from Nigeria in early 1983 – may again accentuate the problem.

The fall in the internal value of the currency has been so rapid that by 1981 one cedi commanded only one-tenth of its 1971 purchasing power. There has naturally been a corresponding fall in the external value of the cedi, although this was not reflected in the official exchange rate, which remained fixed at ₵1.15 per US dollar from December 1973 to September 1978 and thereafter at ₵2.75 up to October 1983 (see Chapter 10). As the official supply of foreign exchange failed to satisfy its demand, there developed a black or 'parallel' market in hard currencies, the rates for which started rising as the imbalance in supply and demand became wider and wider. The parallel market rate of the US dollar, which was reported as being three times the official rate in September 1978, went up fifteen times in October 1981 and further to about twenty-six times in March 1983. Although following a series of devaluations the foreign exchange rate has been adjusted a number of times since April 1983, the official rate of the dollar as at September 1984 was still much below the parallel rate.

Since the early 1960s price controls in some form or another have remained in force. The NRC (National Redemption Council) government, which took over in January 1972, extended considerably the list of goods covered under the price control measure. The system has also been used by subsequent governments, though by 1984 the official prices were drastically raised, thus substantially reducing the gap between official and parallel market prices (see Chapter 11). It was hoped that by controlling prices the government would be able to check inflation, reduce monopoly rent and safeguard the interests of the poor. While none of these objectives has been achieved, the adverse effects of price controls, in the form of a decline in the production of manufactured goods due to reduced profitability, have been substantial, thus further adding to inflationary pressures. There have also developed flourishing illegal activities commonly known as *kalabule* trading, directly encouraged by the existence of excess demand at controlled prices. A rough and ready estimate suggests that the size of the *kalabule* or black economy in 1981 could have been as large as two-fifths of the GDP (see Appendix B).

In the early 1980s, the economic situation reached a critical state. Because of a fall in government development expenditure there was 'a marked deterioration in critical infrastructure, including roads, railways, electricity and telecommunications. The decline in infrastructure services, particularly transport, reduced the ability of the country to move export products to ports for shipment, which further reduced export earnings, taxes, etc. The distortion induced by the overvalued exchange rate, and the shortage of imported foodstuffs, induced farmers to move out of export to produce for the domestic market. A serious drought in recent years, however, combined with shortages of fertiliser and other inputs has resulted in declining output of domestic food production as well'.[7]

The economy, already suffering as a result of various bad policy decisions, was adversely affected by 'the rapid rise in petroleum prices, declining demand for exports, and weakening commodity prices. The situation was exacerbated by the return of over one million Ghanaians from Nigeria, drought and the outbreak of bush fires which destroyed part of the cocoa tree stock'.[8] On top of all this, there was little help from the aid donors who drastically reduced their support as they were 'discouraged by the deterioration in the economy, political instability and poor policy performance', thus further worsening the balance of payments.[9]

The public services, which at the time of independence were so competently run, almost disintegrated in the early 1980s. People had little faith in postal services. Telephone services hardly worked outside Accra and even within Accra one was lucky to have one's telephone working. The electricity supply, once the best in Africa, had suffered severely. Water services had deteriorated so that there were frequent stoppages.

Fortunately, by 1985 there were some signs of recovery. Although it is a bit too early to pass any serious judgement on whether the downhill slide of the economy has been stopped, one cannot ignore the feeling of optimism among the people. One keen observer of Ghana found during his visit in April 1985 that his 'Ghanaian friends viewed life's knocks from a different perspective ... they were encouraged by the marked improvement since last year. There is a spirit of hope about, as round a sick-bed when you feel that the depth of crisis has passed, and that the patient, though very ill, will nevertheless pull through'.[10]

It is difficult to imagine that a country like Ghana should have deteriorated so badly and should have reached such a critical state. The country did not have any war, but 'some might say that its governments made up for it'.[11]

In the following two sections I will present, mainly in a descriptive form, a discussion of planning and economic management in Ghana in a chronological order. This is followed by an analysis in the form of a general review of economic planning and management, which highlights the causes of Ghana's economic decline.

PLANNING

Development planning in Ghana has a long history, dating back to 1919 when a *Ten-Year Development Plan* was launched.[12] The plan, initiated under the colonial governorship of Gordon Guggisberg, was scheduled for implementation between 1920 and 1930. The economic package, which had a considerable degree of survey and research work put into it, enabled the country to 'build a relatively advanced physical and social infrastructure'.[13] Between 1920 and 1927, during which period the plan was implemented, total investment was £12.4 million, about 50 per cent of the total planned expenditure. By 1927 about 333 Km of new railway lines had been constructed. Other physical developments included the construction of new roads, as well as the development of harbour and water supply systems. In social services, 19 new hospitals, including Korle Bu (now Teaching) Hospital, came into being and Achimota Secondary School (then Prince of Wales College) was established.

The next significant attempt at planning was the drafting of another *Ten-Year Development Plan* for the period 1946–56. This plan, one among many drawn up by the Colonial Office in London for a number of British colonies, had very little impact. It is pointed out by Niculescu that after the launching of the plan various government departments 'continued their work without being affected by the almost theoretical existence of a plan'.[14]

The year 1951 saw the launching of yet another *Ten-Year Development Plan*. The emphasis, according to the plan, was on 'economic and productive services', but an analysis of the investment allocation in the plan shows that as much as 68.4 per cent of the total planned investment of about £74 million was in the form of infrastructure development, economic and productive services taking up only 16.9 per cent and the remaining 14.8 per cent covering common services and general administration.[15] A likely reason, according to Omaboe, was that 'economic and productive services were given a wider and a more general interpretation than they now have'.[16]

The self-government campaign, which had gained ground by the 1950s, culminated in the election of the first African-majority government in 1951. This was shortly after the introduction of the Ten-Year Plan, and the CPP Government under Nkrumah decided to implement the plan in five years and with considerable changes, although the basic structure remained intact. The implementation period thus became 1951–56.

Independence came to Ghana in 1957, and the government in power wanted a comprehensive development programme for rapid economic growth. As the preparation of such a plan required time, a *Consolidated Plan*, intended to be an interim one, was introduced to cover the period 1958–59. During the time of its implementation, work on the *Second Five Year Plan* also proceeded. The plan was scheduled to cover the period 1959–64, but it was dropped in 1961.[17]

The *Seven-Year Plan for National Reconstruction and Development* (1963/64–1969/70) was launched in 1964 with a total gross investment of G£1016.5 million (net £876.3 million), 37.3 per cent in the directly productive sectors and the rest in social services and infrastructure.[18] An important aim of the plan was to build a 'socialist state' so as to achieve a rapid rate of economic growth.[19] According to the planners: 'Government's participation in the economy must be on such a scale as to enable her to implement her socialist policies with respect to the distribution and utilization of the national income'.[20] Long-term objectives of the plan included the achievement of full employment and a complete diversification of the Ghanaian economy from the primary export-oriented type.

In his comments on the plan Killick, who has studied it in depth, observed that

> It was sensible and specific about institutional arrangements for plan implementation, explicit about the rather stringent budgetary implications of the government's ambitions, and laid down sensible principles for the acceptance of external finance.[21]

The 7YP (Seven-Year Plan) has been widely praised as well as criticised by many writers including Omaboe, Killick, Bissue and Ewusi.[22] The defects of the plan were however of 'little consequence, for while the plan remained officially in operation it was never actually implemented'.[23] The Minister of Finance was not willing to pay adequate attention to the plan while making the annual budget. Nor were the other Ministers willing to accept the plan provision that 'all projects and contracts had first to be subjected to careful economic and financial screening by the Planning Commission and the Ministry of Finance before they could even be considered by the Cabinet'.[24] Indeed, the politicians who were always ahead of the planners conceived or considered almost all projects before referring them to the planners. Omaboe, in referring to the role of politicians, says:

> In many cases (relating to projects under the plan) some form of commitment is entered into before the civil servants are called in and they are therefore handicapped in the application of their skill and experience. There is little room for manoeuvre once commitments are made.[25]

The following views of Rimmer about the way projects were conceived and implemented are basically correct:

> New projects appeared which had never been envisaged in the Plan but were now being pushed by the contractors willing to pay commissions to the persons who accepted them. Projects were begun without feasibility studies and without competitive tendering.[26]

Related to this was the lack of 'distinction between the functions of the Cabinet and the responsibility of the planning agencies'.[27]

According to Omaboe, 'There are instances where decisions are taken by the Cabinet which have economic and financial implications and which have not been properly assessed in terms of the existing development plans. As the Cabinet is the supreme governmental body in the country it can initiate and take decisions in any field. Moreover, its decisions override those of all other agencies to which it has delegated some of its powers'.[28]

A likely cause of estrangement between the planners and politicians is the absence of Ministers from membership of the Planning Commission, but Nkrumah, as its Chairman, was not a good example of a Minister supporting the plan. He did not regard himself as bound by it.[29] Thus, as observed by Killick, Nkrumah's 'belief in planning as an allocative device appears to have lacked much consistency'.[30]

It is therefore no exaggeration to say that the plan existed on paper only, being ignored in different quarters and with various Ministers acting as if there were no plan guidelines. Then came the February 1966 coup when the 7YP was formally dropped.

The coup was followed by two successive governments, under National Liberation Council (February 1965–October 1969) and Busia (October 1969–January 1972), aiming to pursue liberal policies. The period saw the implementation of a *Two-Year Development Plan* for the mid-1968 to mid-1970 period, and a *One-Year Plan* covering the period of mid-1970 to mid-1971. A medium term plan was almost ready when the military coup under NRC (National Redemption Council) took place in January 1972.

The two-year and the one-year plans obviously were not prepared with a long-term perspective. These plans, however, were prepared against a different ideological background than was the case with the 7YP. The Two-Year Plan (2YP), which was particularly critical of the previous government's industrial policy, declared its belief in private enterprise and proposed to expand investment in this sector. In a Foreword to the 2YP, the government declared that the Ghanaian economy was essentially a private enterprise one; therefore the government intended to rely on private enterprise whenever this could lead to the desired objectives.[31] As a method of ensuring higher efficiency in production, major cuts were proposed in public sector spending, leaving only economically sound projects, while in the private sector a continuing expansion of investment was projected.

The *One-Year Plan* aimed at consolidating the strategy of the preceding 2YP. The government remained committed to private enterprise, because of the high level of efficiency it associated with this sector. The planners again emphasised that the government would continue to promote the growth of the private sector and would not establish state-owned factories for producing goods which private enterprise could successfully manufacture.[32]

An important development which took place during this period was the strengthening of the planning agency with Ghanaian and foreign experts. 'But', according to Killick who was attached to the planning agency at that time, 'all this meant little unless there was a parallel improvement in government commitment to make planning work and, so far as the Busia period was concerned, it seemed ... that the position had improved little upon the Nkrumah era'.[33] It is therefore not difficult to realise the frustration felt by the people working within the planning agency at that time:

> If conditions are such that development plans cannot or will not be implemented it would be far better not to bother with plan-writing at all and to use the manpower thereby released for more productive activities.[34]

The NRC government which took over in January 1972 initially managed the economy without the guidance of a formal development plan. A document entitled *Guidelines for the Five-Year Plan 1975–80* was published in January 1975, but it was not until 1977 that the *Five-Year Development Plan* (1975/76–1979/80) was launched. The planned investment of ₵2192.3 million was arrived at from estimates of requirements of various sectors of the economy, the implied incremental capital output ratio being about 3:1.[35] Like the 7YP, this Five-Year Plan (5YP) envisaged a greater state participation in direct production. The plan aimed to build an independent economy through a policy of self-reliance.[36] The planners believed that the manufacturing industries made a significant impact on the economy, following the adoption of the import-substitution industrialisation strategy.[37] According to the plan, any growth rate achieved in the secondary sector was accounted for largely by the large-scale industries; therefore, 'industrial policies and programmes have been formulated to stimulate an increased investment in manufacturing activities'.[38] The average increase in GDP was expected to be 5.5 per cent per annum during the plan period.

The 5YP was a well-prepared document and followed the comprehensive approach adopted in the 7YP. Thus it concentrated on the macro-economic variables and, as before, micro aspects such as project appraisal remained weak. And as in the 7YP, the 5YP suffered from non-implementation (its causes and effects are discussed later). Curiously enough, following the publication of the *Guidelines* in January 1975, the document was used as a framework by the relevant Ministries and investing agencies, mainly to work towards achieving the food self-sufficiency objective of the government. But a year after the publication of the 5YP the government itself was not making any serious effort to implement it.[39] Indeed, government actions in the form of significant rises in salaries and wages and large budgetary deficits incurred year after year went directly against the plan's resource balance objectives.

The period from 1978 to late 1979 was characterised by a series of military uprisings until the Limann Government was voted into office in September 1979. A five-year plan, entitled *Government's Economic Programme* 1981/82–1985/86 was prepared by the Limann Government, but the document remained unpublished presumably because of the military takeover in December 1981.

In December 1982 the PNDC (Provisional National Defence Council) Government proposed a recovery programme which has been formally put forward in a two-volume document, released in August 1984.[40] Set within a four-year time span, the first year (1983) was devoted to stabilisation and consolidation, thus preparing for the launch of a three-year medium term plan (1984–86). The sectoral allocation was conceived under two alternative scenarios – Base case (₵3220 million) and High case (₵3920 million). According to the government, the size of the economic recovery programme is only one-quarter of the requirements identified.[41] As much as 62 per cent of the total fund is allocated to physical infrastructure, fuel and power. The production sectors, including export-oriented activities, are provided with 30 to 32 per cent of the total fund. The rest, less than 5 per cent, is allocated to social services.

Below we list three of the important policy measures outlined in the *Economic Recovery Programme* 1984–86 (ERP). First, in an attempt to find a realistic foreign exchange rate it is 'being envisaged that the effective exchange rates will be adjusted periodically so that the real purchasing power of the exchange rate in terms of currencies of Ghana's major trading partners is maintained'.[42] Second, fiscal policy has been formulated 'with a view to ensuring financial discipline and eliminating the traditionally high deficit' in the government budgets.[43] Third, the new commodity pricing policy will be based on production costs together with appropriate incentive margins, so as to 'tackle production bottlenecks, raise productivity and production, and encourage responsible financial management'.[44]

The ERP is not a comprehensive plan. It is mainly sectoral in its approach, aimed at implementing a programme geared towards the recovery and rehabilitation of the country. The programme appears to be based on a realistic approach combining a direct government role (in areas where private sector investment is not likely to be forthcoming) with active private sector participation in a rational pricing framework. At the time of writing this book the ERP was being seriously implemented by the government.[45]

ECONOMIC MANAGEMENT

Managing an economy like Ghana's, with its heavy dependence on one particular export commodity which is subject to sharp price fluctuations, is not an easy job.[46] Moreover, the knowledge of the policy makers about the short-run behavioural characteristics of the economy has remained poor, thus making it difficult for them to respond in an effective manner to any given policy. On top of all this, taking hard economic decisions demands political courage which will not be easily forthcoming unless the government feels itself strong enough to carry the people with it.

The period from independence in 1957 to 1960 was a normal one in the sense that the previous practice of conservative monetary and fiscal management was more or less maintained (see Chapter 8). Moreover, as observed by Green: 'Export promotion was not related to preferential market creation. Similarly, while grants were welcomed, foreign finance was not seen as crucial for the public sector'.[47] The *Whitman Report* of 1959 found the country politically stable, the government having embarked 'upon a vigorous campaign designed to encourage foreign investment ... free from restrictive exchange and other controls'.[48]

With the turn of the decade however things started moving quickly, particularly with the launch of the *Second Five Year Plan* in 1959 at which time independent Ghana entered 'the arena of conscious efforts for economic development. The development efforts gained momentum in 1960 when government capital expenditure was increased substantially'.[49] The period since 1960 can conveniently be discussed under four distinct sub-periods, as follows.

(a) 1960–65: Period of monetary expansion and deficit financing

This was the latter part of the Nkrumah era, during which the government was pursuing a policy of development at breakneck speed. Naseem Ahmad, who studied the period very closely, observed that the year 1960 'marked the beginning of an uninterrupted series of sizeable budget deficits as well as the use of money creation as an instrument of financing these deficits'.[50]

The overall budget deficit was 7.0 per cent of GDP in 1961. The figure increased to 9.4 per cent in 1962 and further to 9.9 per cent in 1963. After falling slightly to 6.7 per cent in 1964, it reached double figures in 1965 – 10.9 per cent. At current prices, while government revenue increased by only 42 per cent from 1961 to 1965, the increase in government expenditure was 63 per cent, recurrent expenditure increasing by 53 per cent and development expenditure by 79 per cent.

There was a phenomenal increase in money supply during the period, at an average annual rate of 12 per cent (13 per cent including 'quasi money' consisting of time and savings deposits of commercial banks) from 1961 to 1965. The average real GDP growth rate was only 3.2 per cent per annum during the period. The increase in government consumption at constant prices was at an average annual rate of 15 per cent from 1960 to 1965, while the increase in gross investment was only 3.4 per cent.

According to Killick, following the import substitution industrialisation strategy of the first half of the 1960s, Ghana was witnessing the characteristic symptoms of 'structural inflation'

> with the domestic structure of industrial production proving too inflexible to accommodate major new demands being made on it as a result of import restrictions. It was not merely that industry could not catch up quickly enough with demand: the industrialisation was highly inefficient, fostering the emergence of high-cost producers charging prices well in excess of the imports they were replacing.[51]

During this period, the internal public debt increased from ₵76.5 million to ₵407.1 million (a five-fold increase), while the increase in external debt was from ₵12.7 million to ₵378.4 million (a thirty-fold increase), much of it in the form of suppliers' credit at high interest rates. So far as external financing was concerned, 'the heavy reliance on suppliers' credits (also) violated the principles of the 7YP that financing should not be tied to specific projects and – much more important – that loans should be long-term and at low interest rates'.[52] The following comments by Dowse (who studied Ghana up to the mid-1960s) succinctly summarise the situation.

> Signs of obvious industrial and commercial mismanagement were everywhere ... (By 1965) some £40 million had been invested in 32 state enterprises, only two of which showed profit... Government expenditure raced ahead of revenue and most of the deficit was financed by short-term producers [sic], credits over which the Planning Commission had scant control or knowledge.[53]

In brief, Nkrumah's so-called socialist régime became over-dependent on suppliers' credit, and 'wasted its capital resources on parastatal enterprises which were nearly all unprofitable, often inherently unviable, and largely devoid of stimulating linkages with other sectors of the economy'.[54] These enterprises were also grossly over-manned, following the attempt by the government to provide (preferably white-collar) jobs for educated and semi-educated political clients. Nkrumah, for reasons that are understandable in the light of the part he played in the political struggle for Ghanaian independence, believed implicitly in the primacy of politics over economics. Increasingly he tended not so much to seek to manage the economy as – in an important sense – to ignore it. Economic management was reduced – indeed, perhaps, equated – to political will.

This conscious and/or unconscious downgrading of the economic element in development meant that resource allocation was not taken seriously. It was not essentially that Nkrumah preferred planning to the market – though he thought he did. It was that he really did not have any serious investment criteria of any kind. He no more understood the technocratic requirements of a command economy than he did the workings of the market mechanism. Jeffries has summed up the mismanagement during the period as follows.

> The milking of cocoa producers to finance such 'development' projects in the absence of foreign investment served as a disincentive to production which, in the long run, was a far more important cause of declining revenues than the fluctuations in world market prices. This in turn led to a growing shortage of foreign exchange and thus of essential imported inputs for both agriculture and industry. Given its inward-looking orientation, however, Nkrumah's

government refused to devalue the national currency and resorted instead to a system of import controls. Although intended to give preference to essential capital or intermediate goods, this system was mismanaged by corrupt bureaucrats (as any such system is more than likely to be in the Ghanaian context) so as to facilitate the establishment of virtual monopolies in imported consumer items. Hence the alarming and, as it turned out, portentous increase in administrative corruption and the rate of inflation in 1964-66.[55]

(b) 1966-71: Attempts towards stabilisation

The National Liberation Council (NLC) which ousted Nkrumah, and the subsequent government under Busia took serious measures to curb the runaway growth of public consumption. At 1970 prices, government consumption increased from ₵151 million in 1959-60 to ₵349 million in 1965-66, while over the next six years – following a course of ups and downs – the figure declined to ₵275 million in 1971-72. At constant prices, government consumption comprised only 12 per cent of GDP in 1971-72, as against 17 per cent in 1965-66. The period also saw a significant drop in the rate of inflation (see Table 11.1), thanks to better monetary and fiscal management.

As shown in the preceding section, both the NLC and the Busia governments placed emphasis on private sector growth, which was expected to be associated with higher efficiency. Some writers are, however, critical of the NLC and Busia governments for not taking adequate measures towards fiscal and monetary management. Following the IMF advice the central government's share in fixed capital formation was drastically reduced, but the advice to increase taxation was not followed. 'Instead', according to Esseks, 'in both the 1966-67 and 1967-68 budgets, the NLC reduced import duties on basic consumer items and raised the exemption limit for taxable personal income'.[56] According to Jeffries, the 'practical moves in the direction of liberalization (by the NLC government) were extremely limited. In effect, it simply attempted to improve the administration of import programming. The Busia regime, too, though more genuinely committed to the abolition of import controls, was unprepared to take adequate measures, such as tax increases, to contain the demand for imports to levels within the country's import capacity ... until the major balance of payments crisis of 1971'.[57]

The December 1971 devaluation (by the Busia government) which reduced the value of cedi by 78 per cent in terms of the US dollar was, however, a major attempt to correct the serious distortions in the foreign exchange rate.[58] But the jump was too big and the move too sudden, and was followed by the January 1972 military coup.

(c) 1972–83: Complete mismanagement

Rado, who has been keenly following the developments in Ghana, has described the period from the Acheampong takeover in January 1972 to the first two years of the PNDC rule as a long period of 'acceleration towards the abyss'.

> These eleven years are Ghana's nightmare : . . . This period saw the recurrence of all the faults of the Nkrumah period of economic management, only magnified manifold.[59]

The economic policies of the successive governments under NRC, Limann and the early PNDC led to a fall in real per capita income of about 40 per cent, and a sharp decline in exports and imports (see Chapter 10), savings (see Chapter 13) and investment (see Chapter 14). All the directly productive sectors of the economy showed a significant reduction in output (see Chapters 4 to 7).

In retrospect, Acheampong's overthrow of the Busia government was probably the 'greatest single disaster in Ghana's history'.[60] As observed by Jeffries:

> Among the first actions of the new government was a revaluation which undid about two-thirds of the effect of the December 1971 devaluation. In the face of such a severe balance of payments crisis, this left the NRC no alternative but to return to comprehensive import controls which were maladministered even more thoroughly than under Nkrumah. In addition, the attempted justification of Busia's overthrow on the grounds of unacceptability of devaluation placed a political imperative on the NRC regime not to devalue even as the foreign exchange crisis deepened ... Despite nominal attempts to enforce price controls, it was hardly surprising that the combination of import controls with chronic excess demand, fuelled by a phenomenal increase in government borrowing from the banking sector, resulted in most imported goods being sold ... at ever higher prices.[61]

During the two-year period (1979–81) when the Limann Government was in power there were hardly any major policy decisions. This government 'pursued essentially the same policies, refusing to devalue, to reduce its borrowing sharply or, until shortly before its demise, to raise significantly the producer price for cocoa, with the result that it simply deepened the social and economic crisis'.[62]

The early period of the PNDC rule, that is from the overthrow of the Limann Government in December 1981 to the April 1983 budget, has been described as 'a revolutionary-socialist-ideological phase ... during which the economy continued to plummet'.[63] As in the Acheampong and the Limann periods, policies such as price controls, deficit financing and the maintenance of a highly over-valued domestic currency were pursued with vigour, and the economy continued its rapid decline.

(d) 1983 to present: Moderate pragmatic phase

Following the April 1983 budget, the PNDC government has reversed many of its earlier policies. The government has committed itself to reducing significantly the budget deficit, achieving a realistic foreign exchange rate and encouraging production and exports by providing price incentives. Official controlled prices, though still in operation, have been substantially raised through adjustments 'at regular intervals to provide for full coverage of production costs and a reasonable profit margin',[64] thus narrowing the gap with prices prevailing in the parallel market. Moreover, the list of goods under official price control has been substantially reduced, to only 17 manufactured and imported products.[65]

The government is also committed to better monetary management. The annual increase in money supply, though still very high, was down to 22 per cent in 1984 from 41 per cent in 1983.[66] The money supply was 13 per cent of GDP in 1983 and 12 per cent in 1984, as against 29 and 23 per cent in 1977 and 1979 respectively.

In an attempt to solve the problem of overmanning in the public sector, 31 700 persons have already been scheduled for redeployment of which 26 200 are from state enterprises.[67] Attempts are also being made to make the state enterprises stand on their own feet.

The change in policy measures has, however, been helped through the availability of external assistance and advice from the IMF and the World Bank.[68]

REVIEW OF PLANNING AND MANAGEMENT

From the preceding sections it is not difficult to see that the serious economic decline suffered by the country was an inevitable outcome of its poor, and at times highly unsatisfactory, planning and management efforts.

In the field of planning, one observes a difference in approach between plans prepared during the pre-independence period and those prepared after 1957. The plans prepared during the colonial period were basically of a 'shopping-list' type[69] which focused attention mainly on public investment in infrastructural projects, while those prepared in the 1960s and 1970s, especially the 7YP (1963/64-1969/70) and the 5YP (1975/76-1979/80) were based on a comprehensive planning approach, using macro-economic variables, seeking consistency and sectoral balance and involving detailed forecasting. But in development planning the use of sophisticated techniques and planning models is only a means to an end. However good a plan may be on paper, it is of little use unless it is implemented. This is where Ghana has largely failed.

In Ghana, development plans have a very poor record of implementation. Despite this, successive Ghanaian governments have not lost their faith in the preparation of such plans. It is therefore not surprising that many people feel concerned about the waste of manpower and resources in formulating plans and programmes that are not likely to be implemented.[70] The main reasons and consequences of the non-implementation of development plans are discussed below.

As already mentioned, during the time the 7YP remained operational it was generally not binding on operating ministries and agencies. Firstly, without due regard to constraints imposed by the plan, funds were committed for unplanned projects. Secondly, against the wishes of the planners, Ministers entered into contracts for major projects to be supplied on relatively short-term suppliers' credits.[71]

In Ghana, political support for the development plans have remained poor. Moreover, the plans have not been properly linked with the annual budgeting and evaluation processes. Ghanaian experience with plan implementation shows that there are two sets of forces which have worked against successful implementation.

- The first set of forces may be classified as beyond the control of those charged with plan implementation. They include inadequate political commitment, political instability, natural disasters and severe unanticipated adverse movements in the terms of trade.
- The second set of forces are relatively more within the ambit of planners and include inadequate involvement of the agencies

expected to implement the plan in the process of plan preparation, inadequate institutions, shortage of skilled manpower and inadequate policy instruments and projects.[72]

Ironically, the 5YP which identified the above factors itself suffered the fate of non-implementation within a year of its launching, although the government which prepared the plan was in power for another two years. The main reason for its non-implementation can even be described by using its own words as 'the reluctance of policy makers to adhere to the discipline required to sustain successful development planning'.[73]

The successful implementation of a development plan is obviously not easy unless the various ministries and investing agencies concerned are willing to submit to the required discipline. The observation by Arthur Lewis that most West African Ministers consider themselves to be above the law, thus allowing them to make arbitrary decisions,[74] has been particularly relevant to Ghana. Moreover, as the heads of successive governments have often preferred not to submit to plan disciplines (Nkrumah included), there is hardly any hope of plan implementation. Nkrumah went ahead with his vast domestic and foreign borrowing and Acheampong 'awarded several large wage and salary increases', although such measures went against the plans they themselves were instrumental in preparing.

Obviously, the task of implementing a plan becomes particularly hard when a small minority interest group,[75] with controlling power, finds its self-interest threatened by a proper implementation of such plans, which in effect aim to channel income into the productive sectors and alleviate poverty through improved distribution of income. This small minority has been used to maintaining a high standard of living often acquired through enormous monopoly profits, large illegal income from corrupt practices and the maintenance (up to the late 1960s) of a European salary level which is difficult to justify in a poor country.[76] The Ghanaian élite, used to conspicuous good living, was not prepared to undergo the initial sacrifices required for any development programme to succeed. The Ghanaian 'privileged urban class' and their ostentatious luxury has been severely attacked by many writers.[77] Food imports in 1962 were double those in 1952 and 'only 4.5 per cent of these imports consisted of low income foods'.[78] The comfortable living of the Ghanaian élite and the tension it created has been described by Dowse in the following manner.

Two large houses, gaming casinos, expensive American cars, private swimming pools, free trips abroad, expensive marriages, imported luxuries, expensive girl friends, obvious parties – all of which flourished in Ghana – cannot but create envy and tension.[79]

The government is equally to be blamed as it allowed itself similar ostentatious luxury 'by spending scarce resources on expensive embassies, allowing regular supplementary allocations to government departments, purchasing the latest aircraft, tolerating inefficient state corporations, building unproductive palaces and prestige conference edifices ... Moreover, its premature concentration on social services, industrialisation and public building directed resources and attention away from the agricultural rejuvenation essential to underpin other forms of development'.[80]

Nkrumah developed 'a distrust of markets in allocating resources' and the 7YP (1963/64–1969/70) was prepared in this mould.[81] The 5YP (1975/76–1979/80), prepared under the NRC government, argued for investment in projects with 'low private but high social profitability'.[82] The allocation of resources on the basis of social rather than private profitability is, of course, a laudable objective. But its application is not very easy. It demands an estimation of social rates of return for different sectors and sub-sectors of the economy, using shadow prices preferably determined by a central planning authority based on relevant demand and supply data and also on the objectives of the society. In Ghana there have been no central guidelines for the use of a set of shadow prices, and the absence of such guidelines has made the allocation of resources through administrative measures that much more difficult, especially when official as well as open market prices remained highly distorted (see Chapter 14).

A basic tenet guiding economic management in Ghana has been the faith of successive governments, barring one or two exceptions, in administered allocative mechanisms rather than in market mechanisms (see Chapter 10). Policy measures which characterised the mismanagement of the economy include the sharp decline in real producer prices of exports (see Chapters 5 and 7); a bias against export promotion and wide inter-industry variations in degrees of import protection (see Chapters 6 and 10); negative real interest rates (see Chapter 12); prices charged for public utilities being too low to cover even a fraction of their costs (see Chapter 3); negligible or negative contributions by state enterprises (see Chapter 13); unrealistically low controlled prices of imported goods and of locally manufactured goods (see Chapter

11); expanding public employment at a time of declining output and declining real wages and salaries of fixed income earners (see Chapter 12); and financing government expenditure through large budgetary deficits (see Chapter 13).

The above are obviously some of the consequences of the non-implementation of plans. As observed in a recent country-study on Ghana conducted by the World Bank, 'The situation was compounded by a succession of politically unstable governments who failed to take corrective action in time. The resulting drift and economic mismanagement culminated in severe distortions in the economy'.[83]

Price distortions represent a situation in which goods and services, as well as capital and labour, are priced in such a way that they fail to reflect correctly their scarcity value. There is a close inverse relationship between price distortions and economic growth in less developed countries (LDCs). In general, as shown by a World Bank study, the higher the distortions the lower the economic growth.[84] Of the 31 LDCs examined in the World Bank study, Ghana came out as the country with the highest level of distortions, and a very low annual GDP growth rate during the 1970s (−0.1 per cent GDP growth for Ghana against the overall average of 5.0 per cent for the selected developing countries).

In Ghana, the over-riding cause of the wide divergence between the official and open market prices has been the maintenance of an over-valued exchange rate and the absence of accompanying measures in the form of taxes and subsidies to reflect in prices the actual scarcity. Among the different agencies which have been involved in the fixing and implementation of controlled prices, the most important one is the Prices and Incomes Board (PIB), established in October 1972. The controlled prices fixed by the PIB have failed to take into account the scarcity value of the goods reflected in open market prices. Another agency controlling prices is the Cocoa Marketing Board (CMB) which, by fixing prices of cocoa beans at a very low level, based on the over-valued exchange rate, removed any incentive for the expansion or even maintenance of cocoa production.

Official estimates of the differences between the import prices of a number of agricultural products and the wholesale prices of the domestically produced equivalents showed that the landed import prices in 1982 were less than 10 per cent of the farm gate prices of domestically produced sugar, palm oil and rice; for maize it was 10.1 per cent.[85] These massive differences can partly be accounted for by high local costs of production, but they are primarily due to the over-valuation of the cedi.

The consequences of the over-valuation of the cedi have been disastrous, to say the least. As the over-valuation of the cedi increased rapidly, the divergence between the official and parallel market prices became ever wider. A stage was reached when it became significantly cheaper to import almost anything rather than to produce at home, and also more profitable to sell in the local market (and even more so to smuggle outside the country) than to export through official channels. By the late 1970s the export sector had virtually collapsed.

The resulting scarcity of foreign exchange had severe repercussions on the production capacity of the country. For example, there developed an acute shortage of spare parts for essential repair and maintenance. Moreover, the over-valued exchange rate and the resultant encouragement of imports and discouragement of exports directly contributed to a serious misallocation of resources by substantially distorting comparative advantages in sectors such as cocoa and timber. (The illustration of this point in the context of cocoa is given in Chapter 5.) The low producer price of cocoa also encouraged the smuggling of the commodity into the neighbouring countries, where the price was substantially higher when converted using the parallel market exchange rate.

In other areas of the economy also, the insistence of the government on implementing a pricing policy which bore no relation to the market realities of the situation provided disincentives to production. For example, it was much more profitable for a fruit processing manufacturer to obtain sugar at the low official price and sell at least part of it in the parallel market at a premium of over 500 per cent, than to produce jam and sell it at the extremely low official price. Similarly, it became more profitable to sell locally rather than to export. By exporting, say, one pineapple an exporter would earn ₡1.38 (US$0.50 f.o.b. price) plus a small bonus in 1981, while the same product could be sold in the local market at two to three times the cedi value of the export price. It is therefore not surprising that there was hardly any export of pineapples from Ghana while the neighbouring Ivory Coast had developed a considerable export trade in the same product.

Given the sharp fall in export revenue, one is tempted to ask whether it is due to a decline in the terms of trade. Ghana's terms of trade declined severely in the early 1970s; the ratio of the export prices index to the import prices index went down from 100 in 1968 to 71.7 in 1972 (see Chapter 10). But the terms of trade recovered quickly, following an export price boom, although Ghana failed to take advantage of it because of a decline in exports. Indeed, there was a drastic reduction in the export of most of the major export items; the amount of gold exported in 1980 was almost half of the quantity exported in 1975, and the export of cocoa declined by over one-third during the same period. The decline in the amount of timber exports during the same period was as much as two-thirds. The over-valuation of the cedi and the conversion of foreign exchange earnings of exports at the official exchange rate meant that the export sector became increasingly less lucrative, thus ultimately causing a sharp decline in the volume of exports.

It is true that a distorted price structure does not of itself necessarily lead to as serious a misallocation of resources as occurred in Ghana. First, it is possible, as has been demonstrated by the centrally planned economies, to by-pass price signals through rationing and direct allocation of materials and consumer goods. Secondly, investment allocation can be decided on the basis of alternative prices known as accounting or shadow prices (these prices being set as approximations to the true value of resources from the society's point of view). Thirdly, a society might decide to maintain a distorted price structure if it aids the achievement of objectives such as rapid industrialisation, increasing the savings ratio or raising the income of the poorer sections or areas of the community.

In most developing countries, governments are not able to offset adequately the effects of a distorted price-structure through direct rationing and allocation of materials and consumer goods. In Ghana, the mechanism for an efficient administration of an alternative set of shadow or accounting prices was found to be non-existent. At the time of writing (mid-1984), the country did not even have an effective planning body to determine a set of shadow prices. Nor did the country have the expertise or the administrative machinery necessary to apply consistently and administer effectively such an alternative set of prices.

In the absence of central guidelines for the use of a set of shadow prices, different investing agencies were using different criteria in appraising projects (see Chapter 13). A number of investing agencies did not use shadow prices in their appraisal of investment projects, although market prices remained highly distorted. Only the National Investment Bank (NIB) and the Ghana Investments Centre (GIC) were found to use shadow prices which were based on partial measures, but the shadow price ratios used by the NIB differed from those applied by the GIC. The dangers of applying such an uncoordinated and haphazard approach can hardly be over-emphasised.

It is, therefore, apparent that in Ghana different policy measures worked to maintain a highly distorted price structure, and no substantial attempts have been made to estimate and use shadow prices systematically so as to avoid serious misallocation of resources. For an application of an alternative set of prices which are significantly different from market prices it is essential that:

(i) internally, taxes and subsidies are strictly administered so that the allocation of resources is reasonably efficient in sectors outside the direct control of the government; and
(ii) externally, goods and services are not diverted through smuggling.

In view of the serious administrative problems, partly caused by a decline in the efficiency of administration and exacerbated by the fact that the country is surrounded by hard currency francophone zones with which it shares about one thousand miles of frontiers, neither of the above conditions is likely to be fulfilled in the case of Ghana.

It is, of course, wrong to put all the blame on government policies which have caused serious price distortions, supported an inefficient and over-manned public sector and maintained large budgetary deficits. Such a stand (emphasising mainly incorrect internal policy measures) was taken by the World Bank in its *Agenda for Action* for Sub-Saharan Africa,[86] and has rightly been criticised.[87]

In the case of Ghana, a number of external factors over which it had no control made matters worse. First, partly because of the oil price increase of the 1970s and partly as a result of the fall in the international prices of Ghana's major exports (cocoa, gold and manganese), imports of various items including essential spares had to be sharply curtailed, thus causing serious damage to the economy.[88] Secondly, the severe prolonged drought of the early 1980s created a serious food shortage and also caused significant losses in cocoa plantations because of severe bush fires. Thirdly, in 1983 over one million Ghanaians were expelled from Nigeria, putting a heavy strain on the food and employment situation.

Besides the above direct exogenous factors, there is an indirect factor – external dependency. One commentator has observed, without making any serious attempt to substantiate the assertion, that 'Ghana's stagnation and its inability to develop is due to its external dependency'.[89] In the literature on the dependency thesis, a central theme is the supposed excess of capital outflows (through the repatriation of profit by foreign companies) over capital inflows. According to Ann Seidman, who closely examined the period from 1950 to 1965 in Ghana, during the colonial time and the years immediately after independence, Ghana's external dependency was very high, with foreign private firms largely controlling export and import trades, banking, shipping and insurance. They were instrumental in causing 'leakages' of potentially investible surpluses out of the country.[90]

However, participation by the government in such institutions as the Ghana National Trading Corporation, the Black Star Line and the State Insurance Corporation has helped to reduce the domination of the economy by large foreign firms. But the transfer, in a way, led the country from the frying pan into the fire of inefficient management and overmanning (see Chapter 12), with the resultant low labour productivity and recurrent budget deficits to meet losses in operation incurred by many of these institutions.

A further aspect of external dependence is the debt service ratio, defined as principal and interest payments as a percentage of total export earnings. For an economy such as Ghana's, a reasonable debt service ratio would be about 15 per cent of export earnings, but the actual figure in 1966-67 (at the height of Ghana's much-publicised foreign indebtedness) was of the order of 50 per cent, a result of new foreign liabilities which were running at N₵155 million per annum three years previously.[91] However, following the rescheduling agreements carried out in 1966, 1968 and 1970, Ghana was able to obtain short-term debt relief, although the long-term debt burden on the economy obviously remained as the repayment of principal was deferred to 1984 and after (see Chapter 13).

The 1970 figure for the debt service ratio as reported in the *World Development Report* was only 5 per cent of export earnings.[92] The corresponding figure for 1983 reveals a substantial increase (14.2 per cent), but this is still much lower than that in the mid-1960s. As estimated by the World Bank, the weighted average debt service ratio for the low income countries was 12.8 per cent for 1970 and 14.4 per cent for 1983. Thus, on the basis of the above figures it does not appear that in the recent past Ghana's foreign indebtedness has been as severe a problem as it was in the mid-1960s. However, as estimated by the Government of Ghana, debt service payments in 1984 are as high as $251 million or 43 per cent of total exports, largely accounted for by old debts and oil credits. The corresponding figure for 1985 is even higher – $423 million.[93]

FURTHER REASONS FOR INCORRECT POLICIES

In the case of Ghana, the important point is that mistaken policies were not only adopted but they were pursued repeatedly. Was it because of the shortsightedness and incompetence of the leaders, or because of their avarice? Or was it due to lack of respect for professional advice (assuming that such advice was available)? Or were there other deep-rooted reasons?

Ghana has been unfortunate in that it has tolerated too long the avarice of its leaders. As early as 1961 Nkrumah was complaining about the involvement of some of his party members in bribery and corruption.[94] But things did not improve. If anything, by the mid-1960s, the 'public administrative machine had been rusted by corruption and (the government) was no longer capable of putting the new policies (of the 7YP) into practice'.[95]

The misrule during a large part of the 1970s has been summed up by Jeffries as follows:

> Acheampong and his henchmen encouraged smuggling by fairly openly engaging in it themselves. They not only condoned but actually ordered the allocation of import licences and scarce foreign exchange to girl-friends and cronies intent either on making monopoly profits in the domestic black market or else on diverting already scarce imported items into neighbouring territories. (It became a common, if heavily ironic, joke during this period that the most important qualification for obtaining an import licence was the possession of a beautiful body and a coiffured wig.) Similar charges can no doubt validly be levelled against leading members of the PNP during Limann's administration.[96]

The political instability, and the resultant difficulties faced by successive governments in taking necessary though unpopular policy decisions in time, has often been put forward by many observers as a reason for the continuation of mistaken policy measures. Indeed, the indecisive rule of the Limann Government (1979–81) is partly attributed to the fear that a realistic pricing policy – based on devaluation and the removal of price controls – would have made the government highly unpopular, thus forcing it out of office (as had happened with Busia).

But one must admit that there is something more than political instability to consider. Another factor is the general attitude towards employment of the various Ghanaian governments since independence. As pointed out by the Council of State in its 1981 Report:

> Broadly speaking successive Governments of Ghana have behaved as if they were solely responsible for the employment of all persons

and that such employment does not have to be related to performances.[97]

The Report also identified two fundamental factors which underlie all the economic problems of Ghana:

(i) The attitude and policies of successive Governments of Ghana which not merely maintained an overgrown, inefficient and undisciplined public sector but also discouraged the development of an effective private sector; and
(ii) A faulty price-incentive system which does not reward productive activity commensurately.[98]

It appears, as mentioned before, that with the April 1983 Budget the PNDC government has made a decisive move from the earlier 'revolutionary-socialist-ideological phase' to a 'moderate pragmatic phase', and that they have heeded the professional advice offered by the World Bank and the IMF.

Considering the importance of professional advice to the running of an economy, one needs to ask whether the previous governments of Ghana also received such advice and, if so, how they reacted to it. Arthur Lewis worked for two years as Nkrumah's economic adviser and drew up the *Second Five Year Plan* (1959-64). It is now on record that Lewis found Nkrumah a difficult man to work with.[99] Nkrumah also received professional advice from other economists, including Nicholas Kaldor. Emil Rado, based on his experience of working in Ghana (four years each side of independence), categorises Nkrumah as 'a flawed visionary':

> Having been instrumental in bringing Ghana to independence, his next aim was to make her a modern, industrial state, at breakneck speed ... He did not believe his advisors that resources (especially foreign exchange) were limited. There were repeated warnings from economists like Seers and Ross, Birmingham, Killick & Omaboe that the economy was highly inflation-prone, because of a high marginal propensity to spend on food, and a low elasticity of food-supply. But prosaic matters, like improving the cultivation methods of 3-4 million small farmers and devising workable schemes of agricultural credit, did not interest Nkrumah. His only gesture of agricultural innovation was the establishment of a few state farms, which, predictably, lost money. His heart was in the

dramatic and flamboyant: the Volta Project, the 600 factories,[100] skyscraper hotels and conference halls. When things started to go wrong, and reality caught up with him, he disbelieved reality.[101]

Nkrumah also ignored the advice on fiscal management offered by the IMF and the World Bank.[102] His over-optimism about the future of the economy and desire for economies of scale resulted in an under-utilised and high cost industrial structure, wholesale and arbitrary protection, biases towards capital-intensive ventures having few linkages with other sectors, poor quality investment decisions and sub-standard performance of state enterprises.[103]

Acheampong who ruled Ghana during 1972–78 did not care to listen to the professionals and his 'attempt to run the economy', in the words of an expert on Ghanaian developments, 'reminds one of a drunk on the motorway driving against the flow of traffic: the crash was only a question of time'.[104] Limann (1979–81) remained equally unresponsive to professional advice. Ironically, the advice contained in the 1981 Report prepared by the Council of State during Limann's rule is precisely what has now been put forward by the IMF and the World Bank to the PNDC which ousted the Limann government.[105]

The Council's Report, written by Ghanaian experts, contains adequate warnings and suggestions, and the adherence to such sound professional advice by the present government appears to be bearing fruit and the economy has started picking up. There was also a period of about six years between 1966 and 1972 when good professional advice was largely heeded, with beneficial results. It should, however, be added that the government during this time, particularly the Busia government, was composed largely of intellectuals.[106]

In Ghana, as in many other African countries, there has generally not been much respect for advice offered by local professionals. The 1981 Report by the Council of State is a case in point. The warnings and suggestions so clearly contained in the report were largely ignored until the same advice was put forward by the IMF and the World Bank. It is true that governments in most developing countries are generally not very keen on seeking advice from international bodies like the IMF, rather such advice is offered to them in the form of conditions of foreign aid. Such conditionality can, however, prove useful, as observed by Gordon and Parker while reviewing the World Bank's *Agenda for Action*, 'to those African governments who see the need for reform but fear its political consequences. These leaders might be able to use conditionality to deflect the political heat arising from tough decisions away from themselves and on to the 'foreign devils' from the World Bank and the IMF'.[107]

A HOPE FOR THE FUTURE

Given that the continued economic decline has severely affected various sectors of the economy, any recovery programme is likely to be lengthy and difficult. Moreover, the success of such a recovery programme will largely depend on both internal economic management and external economic forces, as already identified, becoming conducive to economic development. For an individual country like Ghana it is obviously too much to expect that it can exert a significant pressure to change the external factors to its favour. It is mainly in the internal policy areas that the government of Ghana can hope to play an immediate role.[108] Keeping this in mind we concentrate below on internal economic management.

The most obvious conclusion is that a country such as Ghana must avoid serious price disincentives if it wants to raise production, encourage exports, raise the investment ratio and channel investment in desirable directions. In the absence of an executive machinery capable of operating an alternative set of prices (for planning purposes) significantly different from parallel market or officially controlled prices, there are bound to be serious distortions caused by the system of administrative resource allocation.

The emphasis on establishing realistic prices needs to be viewed as pro-development, and not as a reflection of any ideological predilection. In this respect, two points are in order. First, the emphasis on price signals should not be considered as something which goes against the socialistic aspirations held by many developing countries, including Ghana. The following observation by Gordon and Parker is highly pertinent in this context.

> The use of market and price mechanisms within an overall socialist framework has an honourable heritage in socialist economic thought going back to Oscar Lange and running through the Liberman reforms in the USSR and current trends in Hungary, the People's Republic of China and other socialist states.[109]

Secondly, by emphasising the importance of realistic pricing policies, in no way do we imply that this measure alone is adequate for rapid economic development. Indeed, a realistic price structure without a full government commitment to economic development is not likely to be effective. Such a commitment does not necessarily have to take the form of an administered allocative system with extensive controls over prices. South Korea is an example where the government was committed to economic development, pursued comprehensive economic planning and at the same time allowed the price mechanism a big role in resource allocation.[110] In Africa, Kenya provides a good example of a country which has systematically followed comprehensive planning but has not wantonly interfered with prices.[111] Indeed, there has been a recognition of the importance of 'getting prices, if not 'right' then at least 'better'.'[112]

Economic planning based on price incentives in a framework of realistic prices can be directed to achieve, among other things, an increase in the ratios of savings and investment to GDP, development of the sectors with higher potential, an improvement of the basic infrastructure and faster economic growth of the poorer areas and sections of the community. In a situation where the level of savings is too low to achieve the targeted rate of growth, as in Ghana, there is obviously a need to encourage the growth of the sectors which have a marginal rate of savings higher than the average rate. In Ghana, as in many other developing countries, there is also the need to encourage the export sector, thus enabling the economy to alleviate the problem of foreign exchange constraint.

If Ghana does not have the administrative capacity to follow the above type of planning, it is simply wishful thinking that it can hope to allocate resources through administered allocative mechanisms, instead of through the price mechanism, as would follow from an alternative planning strategy. Indeed, even countries like India and Malaysia, which have better administrative structures than Ghana, have generally followed a type of planning which allows the price mechanism an important role.

As mentioned earlier, the 1983 Budget took a big step towards correcting the over-valuation of the cedi. Attempts were also made to reduce the wide divergence between the official and parallel market prices of goods and services and thus to establish a more realistic official price structure. Measures have also been taken to remove most price controls, to phase out subsidies on public services such as telephone, electricity, water and postal charges and to reduce subsidies on education and health.

The move towards a realistic pricing structure needs also be accompanied by other essential internal reforms. As already seen, the

Economic Recovery Programme (1984-86) has started taking measures in this respect (for example, to reduce deficit financing, to carry out redeployment exercises so as to attack the problem of overmanning in the public sector and the like). Negative public savings needs to be reversed and, towards this end, there is need for a significant improvement in the efficiency of the public sector undertakings, aimed at reducing their costs and raising their revenue. Such a stance should not be interpreted as an anti-public sector one. While a number of public undertakings (for instance, Ghana Commercial Bank) have played a very useful role in the Ghanaian economy, many others have continued to be a heavy drain on the government exchequer, thus reducing government investment in other vital areas. The same reasoning can be advanced against the premature adoption of a social welfare programme which the government cannot hope to support without reducing government investment in other essential areas, including even the construction of schools and hospitals.

A realistic pricing structure is expected to help greatly in raising exports. As already shown, the inability of the economy to raise export earnings forced the government to reduce imports even of essential items, such as spare parts and equipment which were much needed for economic recovery. The import policy will, of course, have to be rationalised. The import-substitution industrialisation strategy needs to be very carefully examined since in the past it played a highly unsatisfactory role (see Chapter 6).

In conclusion, though one particular cause appears to stand out more prominently among the rest (that is, highly distorted price structure), it would be wrong to single out any one factor as the sole cause of Ghana's economic decline which, it should be admitted, is the result of many factors inter-linked with one another. Moreover, some of the accumulated consequences of the initial decline, such as the fall in investment, migration of skilled and managerial personnel and the run-down condition of infrastructural facilities, themselves reinforce the economic decline. Given the severity of the problem, it would therefore be misleading to suggest that there are easy and quick remedies to Ghana's economic ills.

The policy framework initiated by the government in the 1983 Budget has been followed by two important developments. The first has been the launching of an *Economic Recovery Programme* (ERP) in the form of a Three-Year Medium Term Plan, 1984-86. Some of the policy directions of the ERP have already been mentioned. The main thrust of the policy package of economic reforms is to re-align, through periodic devaluations and adjustments of commodity prices, relative prices in favour of the productive sectors (particularly cocoa, timber and minerals), improve the financial position of the public sector and encourage private investment.[113] The ERP is indeed a very important package for the economic recovery of Ghana, and there is much truth in the government claim that it 'constitutes the first serious attempt in two decades at addressing issues relating to the proper management of macro-economic and structural adjustment policies'.[114] The ERP, largely a product of the World Bank-IMF conditionality, has so far received generous support from the donors.

The second important development is the positive growth in GDP per capita in the recent past. Preliminary data available for 1985 and 1986 show that the GDP per capita growth rate of above 5 per cent, achieved for the first time in 1984 after a prolonged period of decline, has been maintained.[115] The years 1984 and 1985 also witnessed significant growth in exports, at 28.9 and 7.8 per cent respectively. Higher exports and the availability of foreign assistance have enabled the government to increase imports, by 23.2 and 18.0 per cent in 1984 and 1985, respectively.

The present policy framework initiated by the government since the 1983 Budget should be able to reverse the trend of economic decline, provided that it is backed up by the required goodwill and co-operation from the different parties, both domestic and foreign, involved. Until now the government has been largely successful in commanding support from the people, especially from the professional groups, as the signs of recovery have been marked. However, a sudden change in government policies with false promises of recovery without sacrifices and/or a drastic fall in export earnings because of a fall in Ghanaian export prices can easily halt the process of recovery. Moreover, it needs to be seen whether the international community is willing to continue its support (with grants and concessionary aid) for the government's recovery programme[116] and, further, whether the government can also remain fully committed to the much needed, though often unpopular, policy measures.

Having analysed the economic decline of Ghana, we now proceed in the following chapters to examine in detail the basic facts and figures of the Ghanaian economy.

Part I
Socio-Economic Structure

1 Land and People

1.1 THE LAND

Until her political independence in 1957 Ghana was known as the Gold Coast. The change of name was effected mainly on account of the belief that the Akan ethnic group, which constitutes about half of the total population,[1] migrated to its present location from the old Ghana Empire when it collapsed in the 13th century. The new name was also expected to serve as a mark of national identity and hence as a morale booster to the then emergent liberation movement in Africa.[2]

The first Europeans to have had contact with the area now constituting Ghana were the Portuguese. They first reached the area in 1471. By 1482 they had started constructing settlements or trading posts, lodges, forts and castles on the coast. Other European nationals were attracted by the famed gold, ivory, slaves and other resources of this part of Africa. After a prolonged and intense struggle among the European nations who considered one another as interlopers, and much resistance from some of the indigenous kingdoms, especially the Ashanti kingdom, the whole area now known as Ghana was eventually colonised by the British in 1901. Part of German Togoland was added to the British possessions at the end of World War I.

Ghana was the first black African country to achieve independence from a major European power. The Convention People's Party, led by the dynamic, revolutionary and socialist-inclined Nkrumah, which had spearheaded the struggle for independence since 1951, ruled the country until February 1966 when it was overthrown in a military coup. The National Liberation Council, a military-cum-police junta, then assumed the reins of power until October 1969, when it handed over to another civilian administration headed by Busia, a pro-western, capitalist-oriented intellectual. This government was toppled in another *coup d'etat* in January 1972. There were successive military governments until September 1979, when a civilian government was sworn in. On 31 December 1981 that government too was overthrown by the military.

The country is bordered on the east by Togo, on the north and northwest by Burkina Faso and on the west by the Ivory Coast. The south of the country, which faces the Gulf of Guinea, has 554 km of Atlantic coastline. All three of the countries bordering Ghana are former French colonies and belong to the CFA monetary zone.

Ghana extends inland for 675 km and lies approximately between latitudes 4°45' and 11°11' North; and longitudes 1°11' East and 3°15' West.[3] Thus it has close contacts with both the Equator and the Greenwich Meridian. The latter passes through the port of Tema. Ghana covers an area of 238 538 square km (92 000 square miles). Until 1983, the country was divided into nine administrative regions (Ashanti, Brong-Ahafo, Central, Eastern, Greater Accra, Northern, Upper, Volta and Western); in 1983 the Upper Region was divided into Upper East and Upper West.[4]

A large part of the country is drained by the Volta River and its tributaries. That river system has its sources in Burkina Faso and empties into the Gulf of Guinea in the south. Dams have been constructed on the river at Akosombo and Kpong for the generation of hydro-electricity. The Akosombo dam, commissioned in 1966, has resulted in the formation of one of the world's largest artificial lakes – approximately 9100 square km in area and 650 km long.

Ghana's proximity to the Equator has a great influence on its climate, especially accounting for the relatively high temperature throughout the country all the year round. The mean monthly temperature never falls below 24.5°C. The country has two main seasons – the rainy and the dry – with rainfall generally decreasing from Axim in the extreme southwest (with more than 200 centimetres per year) to Bawku in the north-eastern corner (with 100 centimetres per year), and the capital city of Accra in the southeast with a little less than 75 centimetres.

The south has two main rainy seasons: the major one from March to July and the minor season in September and October. The north, by contrast, has only one wet season from May to October. Two major wind systems affect the climate. These are the northeast trade winds (or harmattan) which blow from the Sahara Desert and which are hot and dry; and the southwest (or monsoon) winds blowing from the South Atlantic which are cool and rain-laden.

In the tropics, rainfall is a major determinant of vegetation. The heavy rains in the south help produce a luxuriant vegetation in the southern forest zone; and the lighter rains, as one moves inland, finally result in the much sparser northern savanna and brush. The latter vegetation also prevails in the Accra Plains, on account of its comparatively meagre rainfall. The south, therefore, supports the growing of cash crops like cocoa, rubber, coffee, oil palm, kola nut and coconut, and foodcrops such as maize, cassava and plantain. The north, on the other hand, is particularly suited to cereal cultivation and cattle rearing. Tsetse flies preclude extensive cattle rearing in southern Ghana, except in the Accra Plains.

The south is also the source of the country's timber and mineral wealth. Minerals such as gold, diamond, manganese, bauxite and a small quantity of petroleum are currently being exploited, while rich deposits of iron, limestone and other minerals are known to exist (see Chapter 7).

1.2 POPULATION AND LABOUR FORCE

Ghana's experience of the administration of censuses predates colonial times, with chiefs sometimes counting their people by requiring each of them to deposit a grain or a shell or some other easily countable object into a receptacle. In 1891, however, the British Colonial Government conducted a census of the whole Gold Coast Colony. The subsequent 1901 census was extended to cover Ashanti and parts of the then Northern Territories. Censuses were also conducted in 1911, 1921, 1931, 1948, 1960, 1970 and, more recently, in 1984.[5]

For a number of reasons it is difficult to compare the population figures deriving from these various censuses. The earlier ones were invariably incomplete in their enumeration. In the first two national censuses, for instance, the traditional method of counting described above was widely used. Only the 1960, 1970 and 1984 censuses would qualify to be treated as properly conducted. The earlier ones were low-budget undertakings that did not elicit most of the vital statistics of interest to, among others, demographers and planners.

As the final report of the 1984 census was not ready at the time of writing, most of the data used here are based on the 1960 and 1970 censuses, although we have also incorporated data as available from the preliminary results of the 1984 census.

As is apparent from Table 1.1, urbanisation in the country is proceeding at a fast pace. This is a result of more people moving into towns and cities and also due to more centres being upgraded to urban status on the attainment of the 5000 population mark.[6] The average annual growth rate of urban population for the whole country between 1970 and 1984 was 3.2 per cent, while for the rural population the rate of growth was 2.3 per cent. Available data from the 1984 census, however, indicate a lower rate of population growth than found in

Table 1.1 Main demographic features 1960, 1970 and 1984

Features	1960	1970	1984
Population			
Total Population ('000)	6727	8559	12 206
Urban	1561 (23%)	2473 (29%)	3 825 (31%)
Rural	5165 (77%)	6086 (71%)	8 380 (69%)
Annual Growth Rate of total population (%)		2.4	2.6
Density (per sq km)	28	36	51
Labour Force			
Total Labour Force ('000)	2723	3332	
Average Annual Increase (%)		2.1	
Total Employed ('000)	2573	3137	
Male	1574 (61%)	1718 (55%)	
Female	999 (39%)	1419 (45%)	
Age Distribution ('000)			
Under 14	2999 (45%)	4016 (47%)	
65+	213 (3%)	311 (4%)	
15 to 65	3518 (52%)	4232 (50%)	
Economic Activity (Aged 15+) ('000)			
Employed	2559 (69%)	3133 (69%)	
Unemployed	164 (4%)	199 (4%)	
Homemakers	688 (19%)	625 (14%)	
Others	319 (9%)	586 (13%)	

Notes: Figures in brackets show percentage distribution of population. Because of rounding off the total may vary slightly.
Sources: CBS, *Population Census of Ghana*, 1960, 1970 and 1984.

the 1970 census for the three major urban centres of the country. The population of Accra City grew from 636 000 in 1970 to 965 000 in 1984, that of Kumasi from 345 000 to 489 000, Tema from 102 000 to 191 000 and Secondi-Takoradi from 161 000 to 175 000 over the same period.[7]

The annual growth of population for the whole country, at 2.4 to 2.6 per cent, though comparatively low by developing countries' standards, is still high enough to keep the rate of growth of per capita income down.

The crude birth rate has been fairly constant over a long time span, although due to advances in medical care and the improving standard of living the death rate has been steadily declining. The net rate of migration, though positive and very high prior to 1970 on account of Ghana's comparatively better economic conditions *vis-à-vis* other West African countries (except perhaps the Ivory Coast), is now negative. At the end of 1969 Ghana instituted the Aliens Compliance Order by which thousands of aliens without resident permits – mostly nationals of neighbouring countries such as Nigeria and Burkina Faso – were deported. Since the mid-1970s the Ghanaian economy itself has been on the decline, with the result that thousands of Ghanaians have emigrated, mostly to Nigeria. Although in 1983 one million of them were repatriated from Nigeria, there is quite a substantial number of Ghanaians still resident in that country. An educated guess would put the number at over a million. Also, in the mid-1980s the number of foreigners leaving the country far outstripped the number entering.

In both the 1960 and 1970 censuses the unemployment rate is registered at 6 per cent of the labour force. For men the rate of unemployment increased from 6.5 per cent in 1960 to 7.6 per cent in 1970, while for women there was a decline from 5.2 to 3.9 per cent over the same period. The unemployment problem is particularly acute for both males and females in the 15–24 age group. The rate of unemployment for this group increased from 14.4 per cent in 1960 to 17.7 per cent in 1970. Considering that the figures for unemployment in the 1960 and 1970 censuses did not attempt to take into account the underemployment and disguised unemployment, which prevailed in sectors such as agriculture and trading, one can safely conclude that the unemployment rates should be significantly higher than those shown in the censuses.[8]

The number of those above 65 years of age is not particularly large. On the other hand, the large and increasing number of those under 14 implies a growing demand for food, clothing, shelter, health and educational facilities. This is a problem resulting from the transition from a subsistence to a monetised economy. Economic growth usually brings forth increases in the rate of urbanisation, and makes more and more of the young dependent on the working population for their education, health and maintenance. Also, mainly because of limited investment resources in the modern sector, not all in the working age population can be absorbed, with the result that open unemployment becomes a growing social and economic problem of an economy in transition.

Economic development involves spatial redistribution of the population. Some regions of a country grow faster population-wise than others. Again, there is a shift of population from rural localities to

Table 1.2 Population by regions 1960, 1970 and 1984

Region	1960 '000	1960 % Distribution	1970 '000	1970 % Distribution	1984 '000	1984 % Distribution
Western	626	9.3	770	9.0	1117	9.2
Central	751	11.2	890	10.4	1146	9.4
Greater Accra	492	7.3	852	10.0	1420	11.6
Eastern	1095	16.3	1262	14.7	1679	13.8
Volta	777	11.6	947	11.1	1201	9.8
Ashanti	1109	16.5	1482	17.3	2090	17.1
Brong-Ahafo	588	8.7	767	9.0	1179	9.7
Northern	532	7.9	728	8.5	1163	9.5
Upper	757	11.3	861	10.1	1211	9.9
All Regions	6727	100	8559	100	12 206	100

Sources: CBS, *Population Census 1960 and 1970,* and *1984 Population Census of Ghana: Preliminary Report.*

towns and cities.[9] Table 1.2 shows population by regions in the last three censuses. Greater Accra, which is the location of the country's capital and contains the new industrial and port town of Tema, grew at an average annual rate of 5.6 per cent between 1960 and 1970; its annual growth rate of 3.3 per cent between 1960 and 1984, although higher than the national rate, was much lower compared to that of the 1960-70 period mainly because of the decline in industrial activities during the latter half of this period.

The Northern Region also grew faster (annual average growth rate of 3.1 per cent between 1960 and 1970, and 3.4 per cent between 1970 and 1980) than the average for all regions (2.4 per cent and 2.6 per cent, respectively). This is probably mainly due to the intensification of agricultural activities in that region. The same reason may account for the higher than average growth of the Brong-Ahafo Region. As in Greater Accra, the Ashanti Region which contains the city of Kumasi, the second largest in the country, experienced more rapid population growth between 1960 and 1970 than in the 1970-84 period. It is possible that the Eastern, Central and Volta Regions, being adjacent to the capital city, are losing population to Accra. This may account for their lower than average growth in population. The Western Region too has lost its premier position as the most industrialised region and the only one in the past with modern port facilities. The Accra Region is now the most industrialised, and competes keenly with the Western Region in the provision of port facilities.

Table 1.3 shows the occupational distribution of the labour force. In 1960, 61.5 per cent of the labour force was engaged in agriculture, 15.6 per cent in industry and 22.9 per cent in services (including commerce and transport). By 1970 the respective figures were 57.0, 15.8 and 27.2 per cent.

Thus while agriculture's proportionate share of the labour force fell, the other two sectors gained. This trend is in keeping with the universal behaviour of these three broad sectors in the course of economic development. In poor societies, nearly the whole of the adult working population is engaged in agriculture just to produce enough to feed the community. However, as productivity in agriculture increases, so that one man can feed some other members of the society in addition to himself, part of the agricultural labour force can be released, without much danger to food supplies, to secondary and tertiary industries.

The numbers employed in construction and in utilities declined absolutely from 1960 to 1970. This is more difficult to explain. Perhaps

Table 1.3 Distribution of employment by industries, 1960 and 1970

	1960		1970	
Sector	'000	%	'000	%
Agriculture	1581	61.5	1787	57.0
Mining	48	1.9	31	1.0
Manufacturing	250	9.7	380	12.1
Utilities	14	0.5	12	0.4
Construction	89	3.5	74	2.4
Commerce	369	14.3	427	13.6
Transport, Storage and Communication	68	2.6	84	2.7
Services	154	6.0	342	10.9
Total	2573	100.0	3137	100.0

Sources: CBS, *Population Census*, 1960 and 1970.

an explanation lies in the fact that both sectors are influenced to a large degree by government policies. For example, government's decision to expand educational, health, road and other infrastructural facilities immediately after independence might have caused an inflow of labour to that sector by 1960. After 1960 interest in the expansion of physical infrastructure had considerably waned and consequently the number employed in that sector fell. The same reason may explain the decline in the work force in the electricity and water supply sector.

Both manufacturing and services recorded appreciable employment growth rates. This is not surprising for manufacturing since the country had embarked on an accelerated industrialisation programme soon after independence. Government (both central and local), banking, educational, health and other services were also expanded in the same period.

In relation to occupational distribution by sex, in both 1960 and 1970 agriculture provided greater employment opportunities for both males and females than any other sector (see Table 1.4). Commerce is a traditionally female dominated occupation, and a quarter of the employed female labour force was engaged in this one occupation in both 1960 and 1970, compared to only 6 per cent of the male employed labour force engaged in commerce in 1960 and 3.7 per cent in 1970. Higher participation of the female labour force is also observed in manufacturing, mainly because of more female involvement in the informal food processing sectors such as kenkey and gari-making.

Table 1.4 Occupational distribution by sex, 1960 and 1970

	1960			1970		
Sector	Male	Female	Total	Male	Female	Total
Agriculture	63.7	58.0	61.5	59.1	54.1	57.0
Mining and Quarrying	2.9	0.2	1.9	1.7	0.1	1.0
Manufacturing	9.1	10.7	9.7	9.7	15.0	12.1
Utilities	0.9	Neg.	0.5	0.7	Neg.	0.4
Building and Construction	5.5	0.3	3.5	4.1	0.2	2.4
Commerce	6.0	27.5	14.3	3.7	25.7	13.6
Transport, Storage and Communication	4.3	0.1	2.6	4.8	0.1	2.7
Services	7.8	3.2	6.0	16.3	4.4	10.9
Total	100	100	100	100	100	100

Notes: The total may not always add to 100 because of rounding off.
Neg. = Negligible.
Sources: CBS, *Population Census*, 1960 and 1970.

Females understandably avoided mining and quarrying utilities, building and construction, and transport, storage and communication, contributing only a small percentage of the total labour force in these sectors. These sectors demand much strenuous work and have traditionally been avoided by women in the country. The sudden jump of male participation in services from 7.8 to 16.3 per cent of the male labour force between 1960 and 1970 is clearly due to the expansion of employment avenues in that sector, an expansion which both sexes took advantage of, though more so in the case of males than females.

2 Growth and Structure of the Economy

2.1 GROWTH OF GDP AND GDP PER CAPITA

The foundations of the present structure of the Ghanaian economy were laid between 1890 and 1910.[1] This twenty-year period witnessed an annual average growth of 1.8 per cent in GDP (Gross Domestic Product) per capita. Judged by the performance of the LDCs (less developed countries) at that time, such a growth rate was high and marked a significant improvement in living standards. As Omaboe observes:

> This was the period during which the export economy of the forest belt of the country was developed and transformed. Prior to this the country had a small export trade but this was based largely upon the collection of naturally-occurring forest produce such as palm fruits and kernels, kola nuts and wild rubber. These two decades saw the replacement of this export trade by the product of two major economic activities, gold-mining and cocoa-farming. They have dominated the economy of the country for more than half a century now and they have dictated the pace of economic growth and the present structure of the economy.[2]

The growth in GDP and the transformation of the economy continued after 1910, following the high rates of capital formation achieved in the gold-mining and the cocoa sectors and also in railways and construction. Although in subsequent decades the rates of growth in these activities slowed down, per capita real GDP doubled during the half century from 1911 to 1960, and this took place during a period of rising population.[3]

It was thus in the midst of highly favourable conditions that the new nation was born in 1957, and there was great optimism about the development of the economy as may be seen from the following observation:

> In terms of modern Western criteria, Ghana had a promising start as one of the richest, most successful and politically mature regions

of black Africa. Per capita income was reportedly the highest, real growth was satisfactory, sterling reserves substantial, and development plans were well formulated.[4]

The first few years of independence witnessed satisfactory rates of growth.[5] Real GDP grew at the rate of 4.6 per cent in 1957. It fell to 3.7 per cent in 1958, but this was followed by two years of substantial growth at 15.2 per cent in 1959 and 7.5 per cent in 1960. By the standard of developing countries this was an exceptionally good performance, and in 1960 Ghana was far ahead of many other developing countries. With a per capita income of £70 in 1960 Ghana fared better than, for example, Nigeria (£29), Egypt (£56) and India (£25).[6]

The growth of GDP continued during the 1960s. With the exception of 1966, which showed a fall in GDP of 4.3 per cent, GDP rose over the whole period; but the average annual real growth rate of 2.8 per cent during the 1960s was lower than the average growth rate for the period before 1960. Significant fluctuations in GDP growth occurred during the 1970s with four years showing negative growth rates. Overall growth in GDP at 1970 prices was only 3.7 per cent from 1970 to 1980, the average annual rate of growth being only 0.4 per cent. In 1981, 1982 and 1983, GDP fell by 4.2, 6.9 and 4.6 per cent respectively.

Table 2.1 summarises the growth in GDP and GDP per capita over the period, 1890-1983. Data on GDP growth for the period up to 1957 are not complete and exist mostly in the form of estimates. We have

Table 2.1 Growth in GDP and GDP per capita, 1890-1983 (percentages)

Period	GDP growth rate Total	GDP growth rate Annual average	GDP per capita growth rate Total	GDP per capita growth rate Annual average
1890-1910	—	1.8*	—	0.7
1910-1960	19.2†	6.0†	100*	3.5*
1960-1970	31.9	2.8	3.5	0.4
1970-1980	3.7	0.4	-19.7	-2.2
1980-1983	-14.9	-5.2	-21.2	-7.7

* Estimates.
† For 1957-60.
Sources: Birmingham (1966, p. 18); See Tables A.1 and A.2.

included the information as it enables us to see the development of the Ghanaian economy for a comparatively long period.

Ironically, the average per capita GDP growth rate was higher in the pre-independence than in the post-independence era. It is true that the average annual GDP growth of 2.8 per cent during the 1960s was not negligible, but the annual population growth of 2.4 per cent during this period meant an increase of only 0.4 per cent in real GDP per capita. With the low growth rate of GDP during the 1970s and the average population growth of 2.6 per cent per annum, there was a fall in real GDP per capita of 2.2 per cent. Significant falls in growth rates for both GDP and GDP per capita were observed during 1980–83.

With 1970 as the base year, the GDP index for 1957 was 63.6. It went up to 75.8 in 1960 and further to 89.0 in 1965. The highest recorded index was 112.2 for 1974, but this was followed by a fall of 12.4 per cent in GDP bringing the index down to 98.2 in 1975. The index rose to 103.7 in 1980, but fell to less than 90 in 1983. Table 2.2 shows GDP and GDP per capita at current and constant prices for the post-independence period.[7]

At the per capita level, because of the population growth rate at 2.4 to 2.6 per cent per annum and the failure of the GDP to rise faster, real GDP per capita has been lower since 1975 than it was at the time of independence in 1957. During the 1960s and the first half of the 1970s there were some signs of improvement in the level of GDP per capita. Thus for 1971, the best year, GDP per capita was ₡271 at 1970

Table 2.2 GDP and GDP per capita, 1957–84
(selected years)

	Gross domestic product (₡ million)		Gross domestic product per capita in ₡	
	Current prices	1970 prices	Current prices	1970 prices
1957	740	1437	118	230
1960	956	1713	142	255
1965	1 466	2011	193	265
1970	2 259	2259	264	264
1975	5 283	2219	544	228
1980	42 854	2342	3 814	212
1982	86 451	2090	7 443	180
1984	270 561	2167	22 139	177

Sources: Tables A.1 and A.2.

prices. However, the per capita index in 1971 was only 2.7 above the base year index. Between 1971 and 1983 GDP per capita declined considerably, at an average annual rate of about 4 per cent. In other words, GDP per capita declined by about 35 per cent from 1971 to 1984. Thus the failure of the economy to grow and the consequent deterioration of the living standards of the average Ghanaian formed a major feature of post-independence Ghana.

According to World Bank estimates, in 1984 the per capita income of Ghana was US $350.[8] Thus, a country which in 1960 was one of the leading developing countries has now joined the ranks of the low-income developing countries with neighbours such as Nigeria (US $730 in 1984), the Ivory Coast ($610) and Liberia ($470) showing much higher per capita income.[9]

The causes of the economic decline of Ghana are discussed at some length in the Overview section. It is perhaps sufficient to mention here that Ghana's economic decline is largely self-inflicted. For instance, the adoption of policy measures controlling prices of goods and the foreign exchange rate of the cedi gave rise to a seriously distorted price structure which, accompanied by administrative problems, resulted in an unsatisfactory economic performance.

2.2 GDP: DISTRIBUTION BY USES AND INDUSTRIAL ORIGIN

An analysis of the uses of GDP shows that at current prices private consumption in general increased from around 73 per cent of GDP during the late 1950s to over 80 per cent during the late 1970s. In 1982, it was about 90 per cent of GDP. On the other hand, government consumption as a percentage of GDP, which increased from about 9 per cent in 1957 to about 17 per cent in 1968, declined from 14 per cent in 1969 to 6.5 per cent in 1982. The effect of the fall in the share of government consumption is reflected in the deteriorating quality of public services and government administration, following the inability of the government to adjust wages and salaries of its employees in line with inflation.

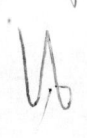

The general decline in the ratio of gross fixed capital formation (GFCF) to GDP is particularly disturbing. At current prices, the ratio of GFCF to GDP increased from 15 per cent in 1957 to 20 per cent in 1960. It was above 20 per cent in 1961, and varied between 17 per cent and 18 per cent during 1962 to 1965. The second half of the 1960s and the first half of the 1970s saw the proportion of GFCF to GDP varying between 8 and 12 per cent. From 10.7 per cent in 1972 the proportion declined sharply to 3.5 per cent in 1982.

Poor performances are also observed in the export and import sectors. The value of exports was as high as one-quarter of the GDP at current prices during the late 1950s. From 1960 on, the export proportion of GDP generally declined and during the late 1970s it was about 10 per cent. As for imports, the ratio of total imports to GDP also fell drastically, from about 31 per cent of GDP in 1960 to less than 10 per cent of GDP in 1980. Imports in 1982 represented only about 3 per cent of GDP (see Table 2.3).

Table 2.3 Percentage distribution by uses of GDP at current prices, 1960-82
(selected years)
(current prices: percentages)

Uses	1960	1965	1970	1975	1980	1982
Private consumption	72.59	77.26	73.68	73.31	83.50	89.78
Government consumption	10.04	14.45	12.85	12.03	11.87	6.48
Gross fixed capital formation	20.29	18.07	12.01	11.62	4.79	3.53
Changes in stocks	+2.30	−0.24	+2.14	+1.11	+0.60	−0.15
Exports	25.73	17.12	23.16	19.36	8.59	3.36
Imports	−30.96	−26.66	−23.84	−18.43	−9.36	−2.98
GDP	100	100	100	100	100	100

Note: The total may not always add to 100 because of rounding off.
Source: Table A.3.

Since there are differences in the impact of inflation on the various sectors of the economy, the distribution by uses of GDP at current prices needs to be interpreted with caution. At current prices the growth of, say, private consumption appears different from that of fixed capital formation (with high import content). This is because of higher inflation rate in the consumer than in the capital goods sector. Due to the highly overvalued exchange rate of the cedi which was particularly marked until 1983, the inflation rate implicit in imports stood significantly lower than the inflation rate applying to local food, an important item in private consumption. For local food the inflation rate is known to have risen sharply by 869 per cent from 1970 to 1977.[10]

Table 2.4 shows the percentage distribution by uses of GDP at 1970 prices.[11] The fall in recent years in exports and imports as percentages of GDP is not as sharp as it appears when current prices are used. The fall in gross fixed capital formation in recent years (at constant prices) is also less marked than when current prices are used. However, even after necessary corrections for the sharp price rises which have affected different sectors differently, the proportions of capital formation, exports and imports to GDP are significantly lower for recent years than for, say, 1961 or 1966.

Table 2.4 Percentage distribution by uses of GDP at 1970 prices, 1961–84 (selected years)
(1970 prices: percentages)

Uses	1961	1966	1971	1976	1982	1984
Private consumption	76.0	69.6	71.7	72.9	68.0	75.9
Government consumption	10.5	15.7	11.6	13.8	20.3	16.7
Gross fixed capital formation	20.2	15.3	13.1	11.3	7.1	7.6
Changes in stocks	−2.0	−0.1	+1.6	−0.9	−0.1	+0.1
Exports	27.9	24.2	21.4	21.1	13.8	7.9
Imports	−32.5	−24.7	−19.3	−18.2	−9.0	−8.2
Total	100	100	100	100	100	100

Note: The total may not always add to 100 because of rounding off.
Source: Table A.4.

Distribution of GDP at current prices by industrial origin is shown in Table 2.5. Agriculture had the highest share, accounting for about 41 per cent of GDP in 1965, rising to 61 per cent in 1980. Agriculture's contribution to GDP increased over the years, and in the early 1980s it had a share of around 60 per cent of GDP. The performance of the different sub-sectors within the broad agricultural sector was, however, not uniform. The share of the cocoa sub-sector increased from 8.4 per cent in 1965 to 14 per cent in 1970 and then fell sharply to 5.5 per cent in 1980 and was 1 per cent in 1982. The low share of cocoa is mainly due to the artificially overvalued cedi.[12]

The share of industry in GDP was 18.6 per cent in 1965. Falling slightly to 18.3 per cent in 1970, it went up to 21.0 per cent in 1975, but fell again, quite drastically, to 11.4 per cent in 1980 and further to 6.2 per cent in 1982. The GDP share of all the sub-sectors, except that of electricity and water, fell sharply from 1975 onwards. The increase in power supply from the hydro-electric source helped to maintain the overall share of the electricity and water sub-sectors in total GDP. The drastic fall in mining, an export sector, is attributable, as in the case of the cocoa sub-sector, mainly to the valuation of exports at the artificially overvalued foreign exchange rate of the cedi.

The share of the services sector declined from about 41 per cent of GDP in 1965 to about 28 per cent in 1980, but it jumped sharply to 36 per cent in 1982. The 1982 rise in the share of the services sector is due to the rise of the share of the wholesale and retail sub-sector from 14.2 per cent in 1980 to 26.5 per cent in 1982. This sharp rise

Table 2.5 Percentage distribution of GDP by industrial origin at current prices, 1965–82
(selected years)
(current prices: percentages)

Sector	1965	1970	1975	1980	1982
1. Agriculture	<u>40.8</u>	<u>46.5</u>	<u>47.7</u>	<u>61.0</u>	<u>57.3</u>
Agriculture and livestock	27.1	28.1	29.7	48.4	50.9
Cocoa production and marketing	8.4	14.0	10.9	5.5	0.7
Forestry, logging and fishing	5.3	4.4	7.1	7.1	5.7
2. Industry	<u>18.6</u>	<u>18.3</u>	<u>21.0</u>	<u>11.4</u>	<u>6.2</u>
Mining and quarrying	2.4	1.7	2.0	1.4	0.3
Manufacturing	9.7	11.4	13.9	7.2	3.6
Electricity and water	0.4	1.3	0.6	0.6	0.6
Construction	6.0	4.2	4.5	2.3	1.7
3. Services	<u>40.7</u>	<u>35.3</u>	<u>31.4</u>	<u>27.6</u>	<u>36.4</u>
Trade and transport	23.1	19.4	17.9	17.8	30.1
Finance and business	8.7	6.9	5.3	2.1	1.6
Government services	8.9	8.9	8.2	7.8	4.7
GDP	100	100	100	100	100

Note: Due to rounding off the total figures may show slight discrepancy.
Source: Table A.5.

was a result of high price rises during this period. The shares of other sub-sectors have in general declined over the years.

Distribution of GDP at constant prices by industrial origin is shown in Table 2.6 for the period from 1968 to 1984.[13] At both current and constant prices, the share of the agricultural sector was the largest and was on the increase, but unlike at current prices the rise at constant prices continued to 1982 mainly because of an increase in the share of the agriculture and livestock sub-sector. There was a slight fall of the agricultural sector in 1984 because of the fall in the share of the cocoa sub-sector. At constant prices, the cocoa sub-sector contributed about 11 per cent of GDP in 1968 and 15 per cent in 1975, but its share fell in later years, though not as drastically as at current prices.

At *constant* prices, the share of industry in GDP declined, though not so sharply as in the case of *current* prices. This is mainly because of a realistic valuation of manufactured goods which probably had

Table 2.6 Percentage distribution of GDP by industrial origin at 1970 prices, 1968–84
(selected years)
(percentages)

Sector	1968	1973	1978	1982	1984
1. Agriculture	41.8	51.2	49.4	54.8	53.9
Agriculture and livestock	26.1	29.8	32.9	38.5	39.5
Cocoa production and marketing	10.9	15.1	9.3	8.7	6.6
Forestry, logging and fishing	4.8	6.3	7.2	7.6	7.9
2. Industry	20.4	20.0	18.4	12.6	11.6
Mining and quarrying	2.5	2.3	1.5	1.2	1.1
Manufacturing	12.6	12.7	12.4	7.4	7.2
Electricity and water	1.0	0.7	0.6	0.9	0.7
Construction	4.3	4.3	3.9	3.0	2.6
3. Services	37.9	28.8	32.2	32.7	34.5
Trade and transport	17.7	18.1	15.2	14.1	15.1
Finance and business	8.4	4.8	4.4	6.0	6.5
Government services	11.8	5.9	12.6	13.9	12.9
GDP	100	100	100	100	100

Note: Due to rounding off the total figures may show slight discrepancy.
Source: Table A.6.

been recorded at low artificial (controlled) prices for measuring GDP at current prices. The correction for the effect of significant distortions in the foreign exchange rate also made the fall in mining output less drastic at constant prices. The construction sector showed a fall in its share of GDP mainly because of a decline in government development expenditure which affected this sub-sector. Reduced availability of building materials also contributed towards its fall.

The services sector had started to recover after recording a fall in its share of GDP from 37.9 per cent in 1968 to 32.2 per cent in 1978. In 1984 its share of GDP was 34.5 per cent. As measurement at constant prices removes the effect of inflation, instead of the sharp jump as recorded at current prices for 1982, wholesale and retail trade showed a decline.

2.3 SAVINGS

The part of the GDP which is not used for private or government consumption is gross domestic savings. Nurkse and Myint, among others, have argued that in the low income countries there is a vicious circle of low income leading to low savings, low investment and back to low income.[14] Failure to break the vicious circle by raising the saving proportion of income results in perpetual stagnation and poverty.

In any development pattern (capitalism or socialism), for a country to develop it must exert a major effort to improve its savings ratio. Most socialist countries have shown that by following development strategies based on central planning it is possible to achieve and maintain a very high ratio of savings to GDP.[15] In the non-socialist developed countries, the ability to generate enough savings to maintain the rate of growth has brought what has been termed self-sustained development.[16] Developing countries like South Korea, India and the Ivory Coast have also been able to raise the savings ratio significantly during the last ten years or so.[17]

Since independence, the average savings rate has fluctuated widely in Ghana. At constant prices it was about 14 per cent of GDP in 1957, a high figure by the standard of developing countries at that time. It increased to 22 per cent in 1964; a high record, but since then it has been generally declining – at times rapidly. For 1971 and 1972, however, the saving rates were unusually high at 17 and 20 per cent, respectively. This is not perhaps surprising considering the big jump in export earnings in 1970 and 1972 (see Table A.16) which enabled higher imports of investment goods.

Table 2.7 shows average and marginal rates of savings at 1970 prices for different time periods since independence. As can be seen from the table, incremental savings were positive during the first three periods and negative in the last two. The negative incremental savings (−₡95 million) in the fourth period, occurred in a situation of high negative incremental GDP (−₡158 million). On the other hand, high negative incremental savings in the fifth period (1976-80) occurred in a situation of positive incremental GDP (₡201 million). This accounts for the high negative marginal rate of savings in this period.

Table 2.7 Average and marginal gross domestic savings at 1970 prices, 1957-80

Period	1957-60 1	1961-65 2	1966-70 3	1971-75 4	1976-80 5
Total savings (₡ million)	943	1715	1210	1815	1120
Incremental savings $(S_p - S_0)$* in ₡ million	162	45	20	−95	−149
Average savings Rate (% of GDP)	18.4	18.0	11.8	15.4	9.8
Marginal savings Rate†	0.37	0.19	0.06	—	−0.74

* Where S_p is saving in the last or present year and S_0 is saving in the first or original year in each period.
† $(S_p - S_0)/(Y_p - Y_0)$.
Sources: Tables A.1 and A.4.

Table 2.8 shows savings and investment for selected developing countries, including Ghana, in 1984. As little as 5 per cent of GDP was saved by Ghana in that year. As external capital inflows remained low (only 1 per cent of GDP in 1984), gross investment was only 6 per cent of GDP, the lowest figure achieved in our selected group of countries. With the exception of Burkina Faso, with a negative 13 per cent domestic savings, but with a respectable investment rate of 14 per cent because of large external capital inflows, none of the other countries in our table showed a single digit gross domestic savings rate or investment rate. Investment data for over 100 countries provided by the World Bank for 1984 show that Ghana had the lowest investment rate of 6 per cent.[18]

Growth and Structure of the Economy 55

Table 2.8 Savings and investment in selected developing countries, 1984 (percentages of GDP)

	Gross domestic savings	Gross investment	Reserve balance
Burkina Faso	−13	14	−27
India	22	24	−2
China	30	30	Nil
Guinea	13	10	3
Ghana	5	6	−1
Kenya	20	22	−2
Liberia	14	20	−6
Zambia	15	14	1
Indonesia	20	21	−1
Ivory Coast	28	13	15
Philippines	18	18	Nil
Egypt	12	25	−13
Nigeria	15	12	2
Thailand	21	23	−2
Turkey	11	20	−9

Source: World Bank (1986), p. 188.

The present low rate of savings and investment can only reinforce the economic decline of Ghana, given the high growth rate of population (2.6 per cent) and high incremental capital-output ratios.[19] Any attempt to raise the per capita income will demand an increase in domestic savings, other factors remaining the same. Increased inflows of external capital can, of course, raise the investment to GDP ratio as in the case of Burkina Faso, but the availability of such funds cannot be guaranteed and even if they were forthcoming the amount would not in all probability be sufficient to satisfy the growth objective. In Part IV we shall deal in detail with Ghana's domestic savings, external capital and investment. It is sufficient to mention here that, in the immediate future, in order to raise the investment ratio to adequate levels, increased inflows of external capital will be required to overcome the present savings and investment gap arising from the very low domestic savings level and the minimum amount of investment required to stop the economic decline. External capital inflows will also be required to overcome the foreign exchange gap which will arise if the country fails to increase its export earnings rapidly because of slow growth in the volume of exports or a fall in the terms of trade, or both.[20]

2.4 INCOME DISTRIBUTION

In this section we shall discuss briefly inter-personal income distribution. The approach usually adopted to measure this is to define it in relative terms. Income inequality between different income groups is measured by comparing the population share with the proportion of income accruing to each group – for example, the share of income going to the top 20 or 30 per cent of the population. In socialist countries, in general, about 25 per cent of total income goes to the bottom 40 per cent of income earners. This compares with an average 16 per cent of total income accruing to the bottom 40 per cent of earners in non-socialist countries.[21] As far as under-developed countries are concerned the degree of inequality in the distribution of income is generally more marked than in the developed countries, but there is a wide range of variation among these countries.

Ahluwalia classified LDCs into three income inequality groups – low, moderate and high.[22] In the low inequality group, on average, 18 per cent of the income goes to the lowest 40 per cent income group, while in the high inequality category the income share of the lowest 40 per cent is as low as 9 per cent of the total income.

The above measure of relative inequality, however, fails to show the extent of absolute poverty existing in many less developed countries. Estimates available on population below the poverty line show that 30.5 per cent of the total population in Africa had an income per capita of less than $50 in 1969, while about 48 per cent had an income per capita below $75. The corresponding figures for Asia are 36.1 per cent and 56.4 per cent, respectively. As observed by Ahluwalia:

> Much of the poverty problem is a direct reflection of low levels of per capita income, but skewed distribution patterns are also important. Observed differences in the degree of inequality are such as to offset per capita incomes which are two or three times higher. It follows that development strategies which succeed in raising the level of per capita income may not have much impact on the poverty problem if they are accompanied by a deterioration in relative income shares.[23]

Data on income distribution in Ghana are not readily available. No systematic sample surveys have yet been undertaken to measure the degree of income inequality in the country. In the absence of such data the best one can do is to depend on partial evidence available

Growth and Structure of the Economy

from different sources. Killick took such an approach.[24] From the enormous amount of partial and incomplete data he was able to collect, he reached a conclusion that no generalisation was possible:

> Overall, the very incomplete evidence defies any neat generalisation about trends in the distribution of Ghana's national income. There is evidence in both directions and there are enormous measurement problems. All that can be said is that the pattern was complex and diffuse, and the changes seemed to be generally well tolerated by society, failing to give rise to the fierce political tensions generated by this issue in other countries.[25]

In Ghana, as indeed in many other developing countries, a study on income distribution will have to take into consideration the distributive role and welfare impact of the extended family system. Such extended family obligations, though on the wane in Ghana, 'remain a potent distributive force'.[26] Killick's findings are highly relevant for the 1960s and in one area – urban–rural income distribution – he was able to reach the conclusion that 'the allegedly large inequalities between the towns and the villages were largely mythical, and the view that these were becoming larger was almost certainly wrong'.[27]

However, the inflationary pressure of the 1970s seems to have increased the degree of income inequality. During the early 1980s there were talks of 80 per cent of income being concentrated in the hands of the top 10 per cent of the population. Such statements are difficult to substantiate and could even exaggerate the real case. But one cannot deny the worsening income distribution pattern that has been rapidly emerging. For example, once prosperous cocoa farmers have suffered significant losses in income, with the real value of the cocoa producer prices in 1981 estimated at 20 per cent of what the farmer received in 1960.[28] The income of the salaried and wage earners in the industrial sector increased only by an annual average of 23 per cent from 1970 to 1980, while the average annual inflation rate was 40 per cent during the same period (see Chapter 11). On the other hand, the income of the self-employed group was able to keep pace with inflation and much of their income, usually unrecorded, was not subject to income taxation, thus making the salaried and wage earners still worse off as they bore the brunt of income taxation.

Recent information on the pattern of income distribution in primary production, as available from a 1982 study conducted in the Biriwa district of the Central Region, reveals a wide divergence in income.[29]

Table 2.9 Income distribution by source and class in primary production in the Biriwa area, 1982 (percentages)

	Income source		
Income class	Farming	Fishing	Total
Lower (Up to ₡5000)	63.8 (256)	0.5 (2)	64.4 (258)
Middle (₡5000–50 000)	28.9 (116)	1.03 (4)	29.9 (120)
Upper (Over ₡50 000)	0.5 (2)	5.28 (21)	5.7 (23)
Total	93.2 (374)	6.73 (27)	100 (401)

Note: Figures in brackets show the numbers of persons engaged in the occupational activities.
Source: Centre for Development Studies (1983), p. 22.

According to this study, the income of the majority in the community – the farmers – was significantly lower compared to that of the minority fishing community. Annual median net farm income was ₡3650.00 (375 respondents) against the corresponding figure of ₡306 800.00 for the fishermen (27 respondents). Table 2.9 shows the income inequality in the area by income source and income class. The income of the farmers as shown in the table is probably not strictly comparable with that of the fishermen, as the former is likely to be under-stated because of its subsistence nature. But the difference between the two is so wide that the conclusion of high income inequality would stand even after the necessary corrections are made.

The above study also demonstrates the magnitude of the poverty problem existing in the area. Over 60 per cent of the population is in the low income bracket, which is indeed very low, and must be below the poverty line.[30]

2.5 REGIONAL IMBALANCES

As in the case of inter-personal income distribution, sufficient information is not available to determine differences, if any, in inter-regional income distribution in Ghana. It is important to note, however, that there are significant inter-regional imbalances in areas such as urbanisation, education and employment, implying inter-regional differences in income levels and living standards.

In 1970, about 29 per cent of the total population of Ghana lived in urban areas, but the level of urbanisation is uneven as may be seen from Table 2.10. Similar regional differences are observed in wage employment. Eighty-two per cent of the total industrial labour force is employed by three regions, Greater Accra, Ashanti and Western Regions, which have a population share of only 36.0 per cent. These three regions also have 77 per cent of non-industrial consumers of electricity.[31]

Table 2.10 Inter-regional differences in urbanisation, 1970

Region	Total population '000	Urban population '000	% of urban population
Western	770	207	26.9
Central	890	259	29.1
Greater Accra	852	727	85.3
Eastern	1262	310	24.6
Volta	947	151	16.0
Ashanti	1482	441	29.8
Brong-Ahafo	767	169	22.8
Northern	728	148	20.3
Upper	861	61	7.1
Ghana	8559	2474	28.9

Source: CBS, *Population Census 1970*.

Table 2.11, on the regional distribution of school enrolments at primary, middle and secondary levels in 1981, shows the existence of significant differences between regions. Indeed, judged by literacy rates in 1970, the Upper and Northern Regions fared very badly, with literacy rates of 12.1 and 11.1 per cent respectively, compared to 64.5 per cent in the Greater Accra and over 50 per cent in the Eastern and Ashanti Regions.[32]

Similar regional differences are observed in water supplies and health facilities. With respect to public water supplies, as much as

Table 2.11 Inter-regional distribution of student enrolments at primary, middle and secondary school levels, 1978-79 (percentages)

Region	Primary	Middle	Secondary
Western (9.0)	9.75	8.79	9.48
Central (10.4)	10.62	11.26	10.70
Greater Accra (10.0)	9.63	10.64	16.11
Eastern (14.7)	18.85	20.22	16.54
Volta (11.1)	12.55	14.04	13.00
Ashanti (17.3)	20.34	20.94	19.56
Brong-Ahafo (9.0)	9.94	9.15	6.59
Northern (8.5)	4.30	2.32	3.38
Upper (10.1)	4.03	2.45	4.66
Total	100	100	100

Note: Figures in brackets show percentage distribution of regional population in 1970.
Source: Government of Ghana, *Government's Economic Programme* (1981-82 to 1985-86), p. 10.

88.3 per cent of the population in the Greater Accra Region enjoy this facility while, on average, 39.4 per cent of the population in the other regions are served by water supplies.

3 Infrastructure

The development of infrastructural facilities in the form of transport and communications systems, energy, water supply, health and education is a necessary precondition for investment in directly productive activities such as agriculture and manufacturing. The importance of social and economic overhead capital was emphasised by Adam Smith, who argued that it was the duty of the state to erect and maintain

> those public institutions and those public works which, though they may be in the highest degree advantageous to a great society, are, however, of such a nature, that the profit could never repay the expense to any individual or small number of individuals, and which it therefore cannot be expected that any individual or small number of individuals should erect or maintain.[1]

As the arbitrary involvement of the state might lead to incorrect development of infrastructural facilities, Smith suggested some formal tax or toll to pay for their expenses. By doing so, he thought, it would be possible to stop a 'magnificent high road' being built through a desert 'where there is little or no commerce, or merely because it happens to lead to the country villa of the intendent of the province'.[2] What Smith wanted to see was the development of only those infrastructural facilities which were just necessary for encouraging the development of directly productive activities. Thus, according to him, 'their grandeur and magnificence, must be suited to what that commerce can afford to pay. They must be made consequently as it is proper to make them'.[3]

In a developing country like Ghana a proper balance between investment in infrastructural facilities and that in directly productive activities is particularly important because of the severe shortage of capital, which is keenly competed for by different activities. Thus, while recognising the importance of social and economic overheads, care should be taken that these are not over-emphasised. It is also necessary that pricing policies relating to the provision of a number of these facilities are such that there is no improper use of resources in these areas, and that the revenue collected can help in maintaining and expanding these facilities.

Below we make a brief survey of the development of infrastructure in Ghana under three broad headings: physical, social and institutional.

3.1 PHYSICAL INFRASTRUCTURE

Table 3.1 presents statistics on physical infrastructural facilities. Each type of infrastructure appearing in the table is discussed in some detail below. It should, however, be noted that most of these facilities have registered declines in their services. Even electricity generation, which had witnessed phenomenal growth, fell drastically in 1984 (not shown in the table) on account of the power cuts necessitated by the fall in the water level of the Volta Lake.

3.1.1 Transportation

An in-depth investigation into the transport infrastructure of Ghana was carried out by the Nathan Consortium, and its report submitted in 1970 provides useful insights into the sector.[4] According to this report, Ghana's transport systems were essentially adequate to service the foreseeable needs of the 1970s without major expansion. The central network of internal transport (rail and road) provided adequate lines through the southern section and the highway system connected the major urban centres of the relatively sparsely populated north. The country's ports had under-utilised capacity in 1970, and there were regular airline and shipping services. However, Ghana's 'transport infrastructure, once among the best in Africa, suffered from economic and political stresses of the sixties' and each of the modal systems was found to be functioning below its potential capacity in 1970.[5] Below we examine the transport network separately for each of the systems.

(a) Road

Road transport is the most widely used form of transportation in the country, accounting for the bulk of passenger and freight traffic. According to an official estimate, as much as two-thirds of the transportation and communication sector's gross value added derives from road transport services.[6]

After 1900, the British started converting the ancient routeways (which were nothing more than footpaths) into broad roads with

Table 3.1 Physical infrastructure, 1960–80
(selected years)

	1960	1965	1970	1975	1980
Power					
Generation of electricity (mil. kw)	374	636	2919	3958	5315
Kwh per person	56	84	341	407	482
Railway					
Length (km)	963	1 286	1 291	1327	
Locomotives (No.)	155	128	138	90	100
Passenger coaches (No.)	183	209	219	196	196
Goods vehicles (No.)	3 014	3 135	3 375	3 139	3 118
Million-passenger km	253	499.1	508.4	488.8	459.6
Million-cargo km		347.8	306.1	205.4	106.4
Road transport					
Total length of roads ('000 km)	7.1	8.9	10.0*	12.0	26.7†
Bitumen (km)	3.1	3.5	3.9	4.0	5.8
Non-Bitumen (km)	4.0	5.4	6.1	8.0	20.9
No. of vehicles newly registered	12 731	6 129	12 050	15 254	14 568
Air transport					
Aircraft movement (Accra)	10 997	14 640	22 124	7 460	4167
Aircraft movement (Ghana)	14 353	45 279	26 008	11 118	5 217
Ocean transport					
Ships cleared	1 550	1 779	1 880	1 879	911
Tonnes loaded ('000)	2 068	2 385	2 300	2 331	1 611
Tonnes unloaded ('000)	1 873	2 822	2 546	2 876	2 750
Posts and telecommunications					
Letters and postcards carried inland (mil.)	60.2	112.2	199.3	43.3	43.0
Registered Articles (mil.)	3.2	6.4	8.9	3.2	0.50
No. of telephones in use ('000)	23.8	35.0	48.2	59.9	68.8
No. of post offices	151	172	189	236	254
No. of postal agencies	584	640		713	717
Pipe-borne water supply					
Water works (No.)	75	107	236	318	353
Gallons supplied (mil.)		10 743	16 991	23 294	28 276

Table 3.1 continued on next page

Table 3.1 (continued)

	1960	1965	1970	1975	1980
Population served by pipe-borne water (mil.)	1.4‡	1.5	1.7	3.1	3.6

* Refers to 1967/68 data.
† Includes feeder roads.
‡ 1962 data.
Sources: CBS, *Statistical Year Book*, 1965–66; *Economic Surveys* 1967, 1968, 1969, 1972–74, 1975–76 and 1981; and Table A.2.
Personal communications with Ghana Railway Corporation; and Ghana Highway Authority.

laterite surfacing.[7] The routeways linked Kumasi, the capital of the Ashanti Empire, with other parts. Most of the modern roads still converge on Kumasi. The British also constructed new roads where these did not exist previously, and wherever possible they built bridges across streams and rivers. By the end of World War I, therefore, several hundred kilometres of motorable roads had been laid, mostly in southern Ghana (including Ashanti and Brong-Ahafo). However, the only functioning road in northern Ghana in that period was the Great North Road linking Tamale with Kumasi via Salaga.

More kilometres of roads were constructed in the inter-war years. Some of the winding roads were straightened. In addition many kilometres of trunk roads were improved with bitumen or gravel surfaces. Since World War II other major roads have been constructed. They include the coastal Accra-Cape Coast-Takoradi road, the Takoradi-Tarkwa and the Tema-Akosombo roads linking the port towns with the important gold mining and electricity generating towns.

By 1980, there were about 27 000 kilometres of roads including primary, secondary, feeder and town roads. Over half of the total road mileage was under the management of the Ghana Highway Authority, which in 1982 maintained 5.9 thousand kilometres of bitumen road and 7.9 thousand kilometres of non-bitumen roads.[8] Only about 22 per cent of the total mileage of motor roads is bitumen-surfaced and therefore usable throughout the year.[9] These are termed first class roads. In northern Ghana there are only three principal trunk roads. However southern Ghana has a dense network of roads, especially in the Central and Eastern Regions. Feeder roads connect the main roads with the farming communities in the countryside. Where rivers are encountered they are crossed either by bridges or by ferries. Two important bridges span the Volta at Sogakofe and Adomi.

Until World War I there were three main ways of conveying goods by land. These were by using pack animals (especially in the grasslands of northern Ghana where the animals could feed on the vegetation and were also free from the tsetse fly); by headloading; and by putting the goods into casks which were then rolled along the footpaths or roads. These methods were laborious and inefficient. Motor vehicles first appeared in the country in 1903.[10] All types of motor vehicles including articulated trucks currently operate on the roads in the country. Government expenditure on roads and waterways (recurrent and development expenditure)was 1.1 per cent of GDP in 1965, but it fell below 0.5 per cent by the early 1980s.[11]

According to the *Nathan Report*, the vehicle fleet of the country as at 1970 formed a major and costly element in the economy, because about 80 per cent of the running and standing expenses involved foreign exchange.[12] Moreover, the fleet was getting older with the passage of time. In 1967, for example, the age of vehicles averaged 5.7 years as against 3.5 years in 1960.[13] The trend still persists owing to the fact that there are only a few vehicle assembly plants (whose operations are limited by foreign exchange constraints), and the amount of foreign exchange for direct import of vehicles has been severely reduced.

The road transport network cannot be said to have functioned as expected, and this can be attributed principally to the lack of maintenance in the past. The *Nathan Report* recommended that a modest investment was required to compensate for the past neglect of the roads.[14] However, the maintenance of roads still continues to be a major bottleneck.

(b) Railway

The first railway lines were laid with a view to promoting the development of the gold mining industry in the Western Region. Railway construction was begun in 1898 from the surf port of Sekondi and reached the gold mining town of Tarkwa in 1901, Obuasi (another gold mining town) in 1902 and Kumasi in 1903.[15] A branch line from Tarkwa to the gold mining town of Prestea was completed in 1911. An eastern line from the surf port and capital town of Accra to Kumasi was started in 1909 and completed in 1923. A branch line on the western line was constructed from Huni Valley to Kade in the heart of the cocoa, kola-nut and timber producing areas of the Eastern and Central Regions. In 1943 another line was constructed from Dunkwa to the rich bauxite deposits at Awaso. In 1956 the Achiasi-Kotoku and Achimota-Tema lines were completed as part of the preparations for the Volta River Project. Another line was laid from Tema to the Shai Hills for the transportation of quarried rock for the construction of the Tema harbour.

For a long time the railway system was able to stimulate the Ghanaian economy. Apart from gold mining, manganese and bauxite mining also benefited tremendously from the railway network. The cocoa, timber and kola-nut industries were also major beneficiaries.

Rail transport, when efficiently operated, should normally provide the least-cost mode of transportation for bulk produce. However, in recent years the total tonnage of freight conveyed by the railways has been declining at an alarming rate. This decline is not due to competition from more efficient road transportation, since that sector too has its own peculiar problems. It is mainly the result of the railway system itself failing to function properly owing to a number of problems such as the age of engines, poor condition of tracks and shortage of spare parts. In 1968 the railways carried about 20 per cent less cargo than in the early 1960s, and had by that time been running deficits of about ₵2 million per year even without adequate provision for renewals.[16] The losses incurred by the Ghana Railway were often substantial, and had to be met by government subventions which increased from ₵10 million in 1976-77 to ₵31.3 million in 1978-79, and further to about ₵150 million in 1983.[17]

In the light of such inefficient operations, the Nathan Consortium recommended, *inter alia*, that an investment of about ₵3.8 million was needed for the installation of new signalling and communications equipment, the efficient utilisation of the rolling stock (which was found to have an unacceptably high 'out of service' rates) and improvements in administrative practices. It was recommended that the railway sector be separated from the responsibilities of the ports.[18] In 1981 the railway service was given statutory autonomous status (Railway Corporation) and separated from the Ports Authority. The Ghana Government also obtained a loan of US $42 million from the World Bank and the African Development Bank, and technicians from the Indian Railway to help improve the management of the railways.[19]

(c) *Water transport*

Ocean transport is a major form of transportation between Ghana and the rest of the world. Many foreign ships call regularly at the country's two modern artificial harbours at Takoradi and Tema. In addition, the state-owned Black Star Line has ships travelling to the Mediterranean, the Atlantic and Great Lakes ports of North America and to Britain and other countries of Europe. However, in recent years the Black Star Line's performance has declined significantly because of the age of its vessels and the frequent breakdown of machinery and management.

Takoradi, whose harbour was opened in 1928 and expanded in 1953, is the main exporting port, handling bulky exports like manganese, bauxite and timber, and cocoa from the Western, Ashanti and Brong-Ahafo Regions. Tema harbour, which was opened in 1962 as part of the Volta River Project, now receives over 80 per cent of the country's imports. For instance, most of the imported capital goods for the manufacturing firms in Accra and Tema and for the VALCO smelter at Tema are naturally unloaded at Tema.

On the whole, the two ports of Tema and Takoradi should have the capacity to handle all the assignments expected of them, but their performance is currently very poor. The port of Tema is, for example, said to be suffering from operating inefficiencies owing to the shallowness of harbour drafts in some areas. This limits the size of vessels that can enter the harbour. There are also cumbersome customs and cargo handling procedures, especially with the Tema port. In addition, in common with other modes of transport – particularly railways – the ports have management problems, a situation which reinforced the Nathan Consortium's demand for the separation of the Ports Authority from the railways system.[20]

As far as inland water transport is concerned, in the past the Ankobra and Tano Rivers were important for transporting logs to the coast. However, they have been superseded by road and rail transport, mainly because of convenience and costs. The Volta too used to serve as a transport artery along which canoes and steam launches could carry cotton, palm oil and rubber downstream and salt and dried or smoked fish upstream. Its usefulness was, however, reduced considerably by the rapids at Kete Krachi and Kpong. However, with the formation of the Volta Lake water transport may become significant again. The Volta Lake Company has been established to handle the development of transportation to the lake. It has established landing facilities at Akosombo, Kete Krachi, Yapei and other places. There have been moderate increases in the cargo traffic between the south and the north. This is in the interest of the economy as water transport is one of the most energy-saving means of transport. The economic development of northern Ghana could also be greatly enhanced by the development of water transport on the lake.

(d) Air transport

As in other modes of transport, development in aviation was rapid during the post-independence period. Internal air transport is handled by Ghana Airways. It carries mainly passengers and small lightweight goods. The only towns and cities served are Tamale, Sunyani, Kumasi, Takoradi and Accra, which also has an international airport. The Tamale airport is being upgraded to international level. There are direct air routes from Accra to many other African countries, Europe and Asia, served by various airlines such as Ghana Airways, Swissair, British Caledonian, Air Afrique and Nigeria Airways. As of 1970, the Ghana Airways Corporation carried about 52 per cent of all international passengers to and from Ghana.[21]

Owing to problems such as that of dealing with local currency some airlines, including Pan-American, have discontinued their services to Ghana. Ghana Airways itself has a number of problems, the principal one being lack of efficiency in management. This has resulted in recurring heavy annual losses which ran at ₵7.8 million in 1980.

3.1.2 Power

Electricity is so vital to modern living that its output per capita is often used as an index of the standard of living of a society. Growth in the supply of electricity is often directly related to the growth of manufacturing plants. In the case of developing countries, the relationship could be one of a vicious circle, in the sense that facilities for the production of electricity will not be created in the absence of effective demand from manufacturing plants, and the plants will not be established in the absence of an assured supply of electricity. Ghana boldly broke the circle in the 1960s with its Volta River Project, by securing an increased demand from Kaiser Aluminium and also by securing the necessary funds for a viable hydro-electric project.[22] Prior to the 1960s, oil-fired generators were used to provide electric power for the mines and for a few urban centres such as Accra, Kumasi and Sekondi-Takoradi.

The Volta River Authority (VRA), a statutory body, was established in 1961 and charged with the primary functions of constructing a dam at Akosombo on the Volta River for the generation of hydro-electric power and the distribution of the power so generated to meet the industrial and domestic requirements of the country. The Authority has now completed another generating facility at Kpong, lower down the river. It currently supplies electric power to the VALCO aluminium smelter at Tema, which takes as much as 60 per cent of the total power supply from Akosombo. The VRA also supplies electric power to the Electricity Corporation of Ghana, the mines and the Republics of Togo and Benin. Southern Ghana is linked by a national grid for the purpose of electricity distribution.

Measured in terms of crude oil equivalent (fuel), the average energy consumption in Ghana is estimated to have increased at 4.6 per cent per annum from 1970 (2.6 million tons) to 1979 (3.9 million tons).[23] In per capita terms, the average energy consumption in Ghana was 343 kg of oil equivalent in 1979 compared to the African average of 300 kg, and the world average of 1500 kg. According to an official estimate, the distribution of energy supply from different sources in Ghana is as follows:

Petroleum products	30 per cent;
Electricity	37 per cent; and
Wood and charcoal	33 per cent.[24]

The bulk of the electricity is supplied by the Volta River Hydro-Electric Project; of the total electricity generated in 1981 (5382 mil kwh) over 99 per cent was obtained from Akosombo hydro-electric power.

3.1.3 Water supply

The country's public water supply system came into being shortly before the outbreak of World War I, but was limited to the capital, Accra. During the period of the war public water supply services were extended to Cape Coast, Winneba and Kumasi. The extensions were exclusively to other urban centres until 1948, when the Department of Rural Water Development was established. Its main activities were bore-hole drilling and well-sinking for rural communities.

In 1959 the government of Ghana signed an agreement with the World Health Organisation (WHO) for a study to be conducted into the water supply and sewerage development of the country. A WHO team subsequently visited Ghana and recommended that on the grounds of health, industrial development and normal population growth, the existing facilities needed to be expanded and modernised for the current and future requirements of a developing economy. It also suggested the establishment of a corporation to handle the task.

In 1965, therefore, the Ghana Water and Sewerage Corporation (GWSC) was established and charged with the responsibility for the provision, distribution and conservation of water for public, domestic and industrial purposes. It was also entrusted with the operation and control of sewerage services.

There is a strong correlation between good health and productivity. Good health, however, may be difficult to ensure in countries where the population is vulnerable to water-borne diseases, and where human waste disposal is still a problem. Furthermore, industries such as manufacturers of soft drinks and breweries need a dependable supply of good quality water. The work of the GWSC is thus vital to Ghana's rapid economic development. Although the number of people served by pipe-borne water increased from 1.5 million in 1965 to 3.6 million in 1980, the services have deteriorated significantly in recent years, mainly as a result of the deterioration in pipe connections and pumping systems.

3.1.4 Posts and telecommunications

The development of a postal service in the country started in 1843. Since then it has been expanding *pari passu* with improvements in road, rail and air communications. In 1981 there were 252 post offices and 726 postal agencies throughout the country. However, there has been a sharp fall in the number of letters and postcards carried by the post offices. In 1981 only 44 million such items were carried, the corresponding figure for 1970 being 199 million; the quality of service has fallen so dramatically that people have lost confidence in the system.

Telecommunication service in the country began in 1881 between Cape Coast Castle and Elmina Castle.[25] It was extended to Christianborg Castle in Accra the same year. A public telegraph line of 4 kilometres was installed between Christianborg and Victoriaborg in 1882. By 1912 the country had 2387 kilometres of telegraph lines and 48 telegraph offices.

The first telephone exchange was opened in Accra in 1890 with 70 lines. Another exchange with 13 lines was opened in 1902 in Cape Coast. In subsequent years other exchanges were established in Sekondi, Tarkwa, Dodowa and other commercial centres. Automatic exchanges were installed in Accra in 1953, Kumasi in 1957, Sekondi-

Takoradi in 1961 and Tema in 1962. By 1972 16 automatic telephone exchanges had been established in the country.

The following telecommunication services are now provided between Ghana and the rest of the world: radio telephone, telegraph, photo telegraph, telex and maritime radio communication. London provides transit facilities for all services to almost all parts of the world. A satellite earth station has also been constructed near Accra.

As in postal service, the telecommunications system has deteriorated to a serious degree, reaching a near non-functional level mainly because of failures in maintaining the efficiency of the system.

3.2 SOCIAL INFRASTRUCTURE

Education and health facilities form the two main components of social infrastructure. Both are essential to economic development: one in enhancing the skill of the labour force and the other maintaining higher productivity. The education aspect has been extensively studied, and people like Schultz are convinced that the social rate of return accruing from education is high, although it is not very easy to quantify some of the benefits from such an investment.[26]

Table 3.2 contains data on Ghana's education and health. It shows that there were significant improvements in the educational and health services in the 1960s and 1970s. However, as will be found below, the quality of these services seems to have been deteriorating with economic decline, partly as a result of a fall in government expenditure in this sector.[27]

At the time of independence Ghana's investment in education as a percentage of GDP was quite considerable, far higher than that of any other tropical African country. In 1965 over 5 per cent of GDP was spent by government on education.[28]

Expansion in education was accelerated in the 1960s and 1970s when many new secondary schools and training colleges were established; and the University College of Ghana at Legon and the College of Technology in Kumasi were upgraded into full-fledged universities in 1961. Another University College was established in Cape Coast in 1962 and later upgraded to a University in 1972. The average annual rate of growth in enrolment between 1965 and 1980 was 2.3 per cent for elementary schools, 8.3 per cent for secondary schools and 6.1 per cent for higher educational institutions.

As the country's educational facilities expanded, more trained teachers were produced to man the schools. Between 1965 and 1975

Table 3.2 Statistics on education and health, 1965-80 (selected years)

	1965	1970	1975	1980
Education				
Enrolment in elementary schools ('000)	1 313	1 400	1 491	1 858
Enrolment in secondary schools ('000)	33.1	49.2	74.8	107.1
Enrolment in technical schools ('000)	4.8	7.6	10.4	—
Enrolment in higher educational Institutions ('000)	3.4	4.8	6.0	8.3
No. of secondary school leavers	—	7 771	11 144	—
No. of trained teachers in primary and middle schools	14 973	30 350	44 005	41 669*
Health (Average '000)				
Population per hospital	61.3	—	—	106.3
Population per trained nurse	2.9	1.2	—	1.2
Population per public health nurse	66.1	58.6	—	—
Population per midwife	4.7	3.0	—	6.3
Population per dentist	217.2	208.8	—	115.3
Population per doctor	13.4	12.8	—	7.1

— Data not available.
* For 1981.
Sources: CBS, *Statistical Year Book* 1969-70 and *Quarterly Digest of Statistics* 1983; Ghana Education Service, *Digest of Education Statistics*; Ministry of Education, *Ghana Education Statistics* 1970-71; and Table A.2.

the average annual rate of increase in the output of trained teachers was 11.3 per cent, as against 0.8 per cent between 1975 and 1980.

However, there has been a massive exodus of trained teachers in recent years, particularly to Nigeria. The problem is very serious in the secondary schools which are normally manned by university graduates. There are many vacancies in the secondary schools which are being taken up by retired teachers on contract and by part-time teachers. For the primary and middle schools untrained teachers have generally filled the gap. The annual growth rate in the recruitment of untrained teachers was 17.5 per cent from 1977 to 1981.

The number of trained teachers in primary and middle schools steadily increased from 1965 to 1976, after which it started declining. The average annual rate of decline in the supply of trained teachers was 2.75 per cent from 1977 to 1981. The rate of decline was faster in the case of male teachers (4.26 per cent) than in the case of female teachers.

As far as health facilities are concerned, the aim is to maintain and improve the health of the labour force and of the population in general. Ill-health leads to reduced working capacity which in turn causes low productivity. As in most less developed countries, the state of health facilities in Ghana leaves a lot to be desired. There is a serious shortage of hospitals, doctors, nurses and drugs.

For the training of doctors, both the University of Ghana (Legon) and the University of Science and Technology (Kumasi) run medical schools. But after completing their courses, some of the graduates go to work abroad. A substantial number do however remain in the country, mainly working in the more lucrative private practices. This accounts for the improvement in the population/doctor ratio as shown in Table 3.2 above.

Total government expenditure on health (both recurrent and development expenditure) as a percentage of GDP was 1.6 per cent in 1965. The corresponding figure in 1980 was around 1.0 per cent.[29]

3.3 INSTITUTIONAL INFRASTRUCTURE

Institutional infrastructural facilities (in the form of a developed banking and insurance system, basic research, etc.) though not conventionally viewed as social overhead capital, are as important as physical and social infrastructure because in the absence of such facilities investment in directly productive sectors will not be fully effective.

Ghana's institutional infrastructure was rapidly developed during the post-independence period. The Council for Scientific and Industrial Research (CSIR), for instance, was set up in 1968 to reorganise into an effective group the various research institutes that had been established earlier by the Nkrumah Government. The Council is charged with the responsibility of advising the government on scientific and technological matters of importance to the utilisation and conservation of the nation's natural resources; encouragement of scientific research to aid industry, agriculture and medicine; co-ordination of research in the country; and collation, publication and dissemination of the results of research and other useful technical information. It has separate research institutes on Animal, Food, Building, Forest Products, Soil and Aquatic Biology. There are also other independent research centres, for instance the Cocoa Research Institute under the Cocoa Marketing Board.

The country's three Universities also undertake research activities as well as teaching. Their various faculties undertake scientific and social research. In addition, the Technology Consultancy Centre (TCC) at the University of Science and Technology, Kumasi; the Institute of Statistical, Social and Economic Research (ISSER) of the University of Ghana; and the Centre for Development Studies (CDS) of the University of Cape Coast are devoted almost exclusively to research and the dissemination of research findings, the TCC mainly at the practical engineering level while the other two operate at the academic level dealing with current socio-economic problems. There is also a Regional Institute of Population Studies at the University of Ghana providing training in demography for students from West African countries.

The Ministry of Finance and Economic Planning is responsible for the preparation of the national annual budgets and development plans. The Central Bureau of Statistics (recently renamed as Statistical Service), and the Research Department of the Bank of Ghana are two other important bodies involved in data collection and analysis.

Another important institution is the Ghana Investments Centre (formerly the Capital Investments Board) which was originally established in 1963. Its major function is investment promotion through identification of investment opportunities and the granting of tax and other concessions to newly established foreign-owned enterprises and the promotion of such ventures. The Ghana Enterprises Development Commission is another agency that encourages small-scale Ghanaian investors.

Central and commercial banks are discussed at some length in Chapter 8. One of the success stories in the financial structure of the country has been the experience of rural banks, the number of which has been rapidly increasing in recent years with rural people taking an active interest in banking. Being development-oriented, these rural banks are discussed together with other development banks, namely the Agricultural Development Bank and the National Investment Bank (under Development Banking in Chapter 9).

The non-bank financial institutions include the First Ghana Building Society, the Social Security and National Insurance Trust, the National Trust Holding Company and the insurance companies. Like its counterparts all over the world, the Social Security and National Insurance Trust performs the functions of a social security system.

The National Trust Holding Company was formed in 1976 by the government with the main purpose of encouraging domestic investment. Another function of the company has been to act as an embryonic stock exchange. It provides facilities for buying and selling of shares in local enterprises. However the low level of economic activity, the small volume of shares on the market and the lack of the share holding habit are obstacles to the development of an active stock exchange market. By far the largest insurance company is the State Insurance Corporation, which was established in 1962 and is state-owned. Other privately owned Insurance Companies co-exist with the State Insurance Corporation.

Different financial and non-financial institutions, though comparatively well-developed by the standard of many developing countries, are currently passing through a difficult period because of severe constraints resulting from the economic decline of the country.

3.4 CONCLUSIONS

Ghana's development of physical, social and institutional infrastructure in the 1950s and early 1960s was commendable and, as mentioned earlier, the *Nathan Report* expressed with confidence that Ghana's transportation system was adequate for the needs of the 1970s. There are, of course, others who argue that in the 1960s Ghanaian infrastructure was over-extended and, according to Seidman, 'the infrastructure created was not always the most appropriate ... [It failed to] attract adequate private investment to set off a chain of economic growth, [and] it could not have paid for itself for decades'.[30] The Volta hydro-electric project 'was expected to facilitate extensive national economic growth',[31] but Dickinson has argued that while the benefits to the Kaiser Corporation (which uses cheap electricity for its aluminium smelter at Tema) are clear, from the Ghanaian point of view the project

> has not been a success. It did not generate new industry, provide cheap electricity, or give Ghana any stake or influence in the world aluminium industry. It continues to burden the Ghanaian economy and has tied up capital which might well have been better used in modest schemes more specifically directed to overcoming obstacles to industrialization.[32]

However, the economic decline of the 1970s and early 1980s has caused such a deterioration in infrastructural facilities that development of other sectors has been adversely affected. The road and rail networks could no longer carry all the foodstuffs from the farm gates to the consuming urban centres and processing factories; cocoa was being left in the village where it grew while the output of the mines piled up (especially at Nsuta and Awaso) simply because the railway system was either incapable of conveying it to Takoradi harbour or the mining companies could not pay the railway freight charges. Timber too could not reach the ports and the local milling factories.

Perhaps the low prices charged by the government for the use of these public facilities partly explain the deterioration. (See the opening Overview section for an analysis of this argument in a broader context.) For example, the road tax imposed on private cars (the roadworthiness certificate) has failed to increase with inflation. It was ₵10.00 ($3.64 at the official exchange rate) in 1982 and had risen to only ₵20.00 ($0.67 at the then official exchange rate) by mid-1984. Tolls on the few bridges and roads on which they are collected have been very low until recently. Postal and telephone rates have remained unbelievably low. For instance, in mid-1984, even at the overvalued exchange rate of the cedi, it cost 7 pence to mail an air letter to Britain, while it cost 26 pence in the other direction (that is, Britain to Ghana)! As early as 1970 Ghana Railway was advised to raise its rates and charges as, for example, charges on the transport of cocoa and manganese were lower than the costs of providing the service.[33] No wonder then that enough funds have not been generated from these sources to finance extension and maintenance of the physical infrastructure.

The proposal to raise charges for government-supplied services, in a situation where the real income of fixed wage earners has fallen drastically (see Chapter 12), is an unwelcome one. There is also the question of linking price increases with the quality of services offered and, given the fall in the quality of services in transport, post and telecommunications, and it becomes difficult to justify price rises in line with inflation. Moreover, the comparison of different charges (of government supplied services) prevailing in *developed* countries with those in *developing* countries needs to be approached with caution, because of significant differences in respective income levels. One should not however be too critical of a realistic price structure, particularly in view of the need to develop the infrastructure rapidly. Realistic pricing will enable both the suppliers and users of different services to be conscious of the opportunity costs involved and it will make the services 'suited to what the commerce can afford to pay', as suggested by Adam Smith 200 years ago.[34]

There is also the serious issue of the foreign exchange constraint posed by the shrinking volume of exports, resulting mainly from the overvaluation of the cedi over the years. Even if more funds were collected in local currency from the users of infrastructure, it is doubtful whether the government could come up with the foreign exchange equivalent for importing the necessary inputs for their maintenance and expansion.

In relation to the social infrastructure there has been a massive migration of the trained and experienced labour force to Nigeria and other countries. Low salaries in Ghana have also resulted in low morale for those who stay – and this exerts a downward pull on productivity.

The institutional infrastructural facilities face similar problems. The research institutes, like the universities, have lost high calibre personnel largely to neighbouring countries. The Central Bureau of Statistics, for instance, which used to publish *Economic Surveys* annually, has failed to do so regularly since the early 1970s mainly because it lacks adequate personnel and other inputs.

In conclusion it may be noted that the decline in infrastructural facilities, which was mainly caused by the general decline of the economy, has in turn become a causative factor in the country's economic crisis.[35]

Part II
Directly Productive Sectors

4 Agriculture

Agriculture is by far the largest sector of the Ghanaian economy. At constant prices, its contribution to GDP was 46.5 per cent in 1970; by the early 1980s its share stood at around 55 per cent (see Chapter 2). In 1960 the sector engaged 61.5 per cent of the total labour force. By 1970 the share had declined slightly to 57.0 per cent. About 23 million hectares of land are suitable for agriculture, although currently only about 3 million hectares are under cultivation.[1]

Throughout the country, rainfall is the main determinant of cultivation. The southern equatorial rain forest is thus able to support the cultivation of tree crops such as cocoa, coffee, kola nut and rubber; fibres such as jute and sugar cane; starchy staples such as several types of yam, cassava, plantain and banana; also oil palm and coconuts and many types of vegetables. The northern savanna grasslands, on the other hand, are well suited to the cultivation of cereals such as maize, millet, guinea corn and rice; root crops like yam, and groundnuts and tobacco. The Accra Plains in the southeast, with its scanty rainfall, is able to support crops that can withstand long periods of drought, such as cassava, or crops that require only a short growing period, such as maize and vegetables like tomatoes and okra that can be cultivated quickly during the heavy rains.

This and the next chapter are devoted to agriculture. The latter concentrates on cocoa, which is the main cash crop of the country and has accounted for about one-half to two-thirds of the total export earnings. Both chapters draw heavily on sample survey data. Such data on Ghanaian agriculture, which is characterised by country-wide variations, cannot be fully relied upon unless the coverage of the samples is very large. Considering that some of the samples were not as large as one would have liked, their findings should be taken with caution.

4.1 PRODUCTION

Total production of the important food crops over the last three decades is shown in Table 4.1. The table reveals that until the mid-1970s production of cereals increased significantly. In 1974, for instance,

Table 4.1 Production of important food crops, 1950–83
(selected years)
(thousand metric tonnes)

Crops	1950	1961	1966	1970	1974	1978	1980	1983
Maize	169	153	359	442	486	218	382	172
Rice	23	—	30	65	73	108	78	40
Millet	99*	132*	67	93	154	93	82	40
Guinea Corn	—	—	109	86	177	121	132	56
Cassava	512	1778	1170	1644	3606	1895	2322	1728
Cocoyam	518	508	441	1099	1510	726	643	720
Yam	482	1107	748	1459	850	544	650	866
Plantain	1276	437	—	935	2024	940	734	342

* Millets and guinea corn.
Sources: Birmingham et al, (1966); CBS, Economic Surveys, 1969, 1972–74 and 1981; and CBS/Statistical Service, Quarterly Digest of Statistics, September 1983 and September 1986.

total cereal production (maize, rice, millet and guinea corn) was 890 000 tonnes compared to 686 000 tonnes in 1970.

One of the most important cereals produced in the country is maize. It grows very well in all vegetational zones of the country, and two crops a year can be raised in the southern forest zone. It is the main ingredient for a variety of foods. Increases in maize production, like those of all the other important food items, featured prominently during the government's 'Operation Feed Yourself' and 'Industries' campaign of the early 1970s.[2] The total output of maize was as much as 486 000 tonnes in 1974, compared to less than 200 000 tonnes in the early 1960s. The country was thus almost self-sufficient in that important cereal.

Interest in rice cultivation in northern Ghana heightened in the 1960s and 1970s. Big commercial farmers entered that occupation, using modern capital equipment such as tractors, combine harvesters and rice mills. With the introduction of improved planting methods, rice production recorded significant increases, reaching a total output of 108 000 tonnes in 1978.

The starchy staples such as cassava, yam, cocoyam and plantain all showed rapid expansion in output in the first half of the 1970s. The second half of the 1970s and the early years of the 1980s, however, witnessed a significant decline in the output of most of the important crops. Currently, however, soaring food prices resulting from excess demand are stimulating production once again.

Apart from the above important crops, minor crops and fruits such as sweet potatoes and groundnuts, oranges, mangoes, avocadoes and pineapples are cultivated; so are vegetables such as tomatoes, garden eggs, shallots and hot pepper.[3] An important group of agricultural products that is currently being developed or cultivated commercially for the agro-based industries includes oil palm, coconut, rubber, tobacco and cotton.

4.2 LAND USE

Table 4.2 shows the land area devoted to the cultivation of the various important food crops. It shows that the areas for most of the crops increased consistently in the 1970s, reaching a peak in about 1980 for rice and maize and then declining for a number of other crops. The massive migration of the young to the urban centres and to neighbouring countries (especially to Nigeria) and the lack of other inputs were considered responsible for this sharp decline in the area devoted to foodcrops.

Land productivity appears to have fluctuated over the years. Output per hectare of maize, rice and cassava increased from 1963 to 1969, but has generally fallen since then. The most important factor in higher output in Ghana is rain. Inadequate and irregular rainfall will cause a decline in agricultural output and thus a fall in land productivity.

Food production in the country is undertaken largely by small-scale farmers with holdings of less than 2.5 hectares per farmer. In the *Ghana Sample Census of Agriculture* (GSCA) 1970, conducted by the Ministry of Agriculture, it was estimated that 30.6 per cent of farm holdings were less than 0.8 hectares, 55 per cent were less than 1.6 hectares and 82 per cent were less than 4.0 hectares per farmer (see Table 4.3).[4]

The percentage distribution of size of holdings by regions is shown in Table 4.4. The table reveals that in the Western and Brong–Ahafo Regions around two-fifths of the holdings fell below 1.6 hectares in size and in the Volta Region nearly three-quarters of holdings were in this size category. Again in the Volta and Northern Regions very few holdings were above 8 hectares in size. In the Central, Eastern and Upper Regions less than 5 per cent of holdings were more than 8 hectares in size while in the Ashanti Region the figure was as high

Table 4.2 Land area under important food crops and land productivity, 1963-83 (selected years)

Crops	1963 '000 ha	1963 Output per ha (mt)	1969 '000 ha	1969 Output per ha (mt)	1975 '000 ha	1975 Output per ha (mt)	1980 '000 ha	1980 Output per ha (mt)	1983 '000 ha	1983 Output per ha (mt)
Maize	202.5	0.9	275.4	1.1	319.7	1.1	440	0.9	400	0.4
Rice	32.4	1.0	40.5	1.5	78.5	0.9	99	0.8	40	1.0
Millet	101.3	0.7*	162	0.5	198.7	0.6	139	0.6	175	0.2
Guinea Corn	162	—	145.4	0.6	208	0.7	261	0.5	220	0.3
Cassava	160	7.5	162	9.3	284.5	8.4	315	7.4	339	5.1
Cocoyam	68	5.0	214.7†	10.8†	205.2	5.4	137	4.7	113	6.4
Yam	162.4	6.8	—	—	177.4	4.0	113	5.8	143	6.1
Plantain	—	—	121.5	6.7	229.9	5.4	123	6.5	143	2.4

Notes: Data up to 1969 shown in acres in the source have been converted to hectares.
ha = hectare; mt = metric tonne.
* For millet and guinea corn.
† For cocoyam and yam.
Sources: CBS, *Economic Surveys* 1968, 1969 and 1981; and CBS/Statistical Service, *Quarterly Digest of Statistics*, September 1983 and 1986.

Table 4.3   Distribution of farm holdings by size, 1970

Size of holdings (hectares)	Number of Holdings	As % of total
0–0.8	246 100	30.6
Over 0.8–1.6	194 200	24.1
Over 1.6–2.4	105 200	13.1
Over 2.4–3.2	71 800	8.9
Over 3.2–4.0	42 100	5.2
Over 4.0–6.0	55 000	6.8
Over 6.0–3.1	31 600	3.9
Over 8.1–12.1	27 200	3.4
Over 12.1–20.2	17 900	2.2
Over 20.2	14 100	1.8
Total	805 200	100

Note: Holdings shown in acres have been converted to hectares.
Source: Ministry of Agriculture, *Report on Ghana Sample Census of Agriculture 1970*, vol. 1.

Table 4.4   Percentage distribution of holding size by regions, 1970 (sample census)

Region	0–1.6	Over 1.6 to 4.0	Over 4.0 to 8.1	Over 8.1	Total
Western	40.9	28.7	21.0	9.4	100
Central	68.4	19.9	7.3	4.4	100
Eastern	62.6	25.2	7.8	4.4	100
Volta	71.9	21.9	5.0	1.2	100
Ashanti	48.4	21.8	12.4	17.4	100
Brong-Ahafo	40.4	32.5	15.2	11.9	100
Northern	47.5	42.7	7.1	2.7	100
Upper	47.6	34.4	13.3	4.7	100
Total	54.7	27.2	10.7	7.4	100

Note: Greater Accra Region is not separately shown in the source, and is most probably included in the Eastern Region.
Source: As Table 4.3.

as 17 per cent. However, in the cultivation of oil palm in the south and rice in the north, the average farm size is large. The production of these two commodities is dominated by the state (oil palm) and private commercial farmers (rice).

The GCSA 1970 also provides useful data relating to the number of farms per farmer (holder) and the ownership pattern in different regions of the country, as shown in Tables 4.5 and 4.6 respectively. The average number of farms per holder for the whole country was

Table 4.5 Average number of farms per holder by regions, 1970

Region	No. of holder	No. of farms	Average no. of farms per holder
Western	68 100	175 000	2.6
Central	81 100	212 000	2.6
Eastern	148 200	429 000	2.9
Volta	108 600	317 000	2.9
Ashanti	147 700	462 000	3.1
Brong-Ahafo	71 600	185 000	2.6
Northern	61 200	114 000	1.9
Upper	118 700	261 000	2.2
Ghana	805 200	2 155 000	2.7

Note and Source: As Table 4.3.

Table 4.6 Percentage of farms by land tenure by regions, 1970 (percentage of farms)

Region	Owned or ownerlike possession	Rented	Held under some other system	Total
Western	94.3	5.5	0.2	100
Central	85.6	13.8	0.6	100
Eastern	87.4	11.4	1.2	100
Volta	92.2	7.1	0.7	100
Ashanti	97.3	2.5	0.2	100
Brong-Ahafo	99.0	0.5	0.5	100
Northern	99.5	0.4	0.1	100
Upper	98.3	1.1	0.6	100
Ghana	93.7	5.7	0.6	100

Note and Source: As Table 4.3.

2.7, although the range of regional variation was from 1.9 to 3.1.[5] The Northern Region had the lowest average number of farms (1.9) per farmer. The Ashanti Region and the Eastern Region each had an average of 2.9, while it was 2.6 for Western, Central and Brong-Ahafo Regions, and 2.2 for Upper Region.

The pattern of land tenure in the regions does not show a significant variation in terms of ownership of land - 93.7 per cent of farms are possessed in the form of outright ownership or ownership of similar nature. Such a system accounts for between 97 and almost 100 per cent of the total farms in four regions (Northern, Brong-Ahafo, Upper and Ashanti). For the Volta and the Western Regions it is 92.2 and 94.3 per cent respectively, while for the Central and Eastern Regions the respective figures are 85.6 and 87.4 per cent, these two regions having the highest percentage of rented farms (11.4 and 13.8 per cent). The rented farms comprised only 5.5 to 7.1 per cent of the total in the Western and the Volta Regions respectively, 2.5 per cent in the Ashanti Region and around 1 per cent or less in the other regions.

The proportion of farms held under some other tenure system (that is, other than owned or rented) has been negligible in Ghana - only 0.6 per cent for the whole country. The highest figure was 1.2 per cent for the Eastern Region and the lowest was 0.1 per cent for the Northern Region.

The research on land use undertaken by the Centre for Development Studies, University of Cape Coast, in 1982 revealed that yam and maize, the two major crops, were cultivated on 544 out of 815 farms in three selected districts, thus representing 67 per cent of the sample size. Other crops in descending order of importance were rice (10 per cent), cassava (9 per cent), groundnuts (6 per cent), maize-cassava combination, millet, guinea corn and plantain (together 8 per cent). (See Table 4.7.)

The pattern of mixed or inter-cropping, based on the sample survey data (see Table 4.8), reveals inter-regional variations - Swedru (Central Region) and Tamale (Northern Region) showing a high concentration of farms with over one crop to two crops, while the Atebubu District (Brong-Ahafo Region) has a high concentration of farms with only one crop. Inter-cropping is common in the subsistence farms, while single cropping is commonly practised by commercial farmers (Atebubu falling in the latter category). However, in all three districts of the sample survey inter-cropping intensity above two crops is on the low side - 16 per cent in Tamale and 9 and 7 per cent, respectively, in Atebubu and Swedru.

Table 4.7 Number of farms in the cultivation of major crops in three selected districts, 1982

Crops	Swedru No. of farms	%	Atebubu No. of farms	%	Temale No. of farms	%
Yam	—	—	168	56	116	38
Maize	106	50	51	17	103	34
Rice	—	—	63	21	15	5
Cassava	61	28	9	3	5	2
Groundnuts	14	6	4	1	33	11
Maize-Cassava	24	11	—	—	—	—
Millet	—	—	4	1	15	5
Guinea Corn	—	—	1	—	14	5
Plantain	9	5	—	—	—	—
Total	214	100	300	100	301	100

Source: Honny (1982).

Table 4.8 Inter-cropping: Number of crops per farm in three selected districts, 1982

No. of crops	Swedru No. of farms	%	Atebubu No. of farms	%	Tamale No. of farms	%
1	40	28	136	68	34	28
Over 1 to 2	105	65	45	23	68	56
Over 2 to 3	9	5	17	8	16	13
Over 3	3	2	1	1	4	3
Total	157	100	199	100	122	100

Source: As Table 4.7.

4.3 TECHNOLOGY

As much as 70 per cent of the farm holdings are engaged in subsistence cultivation – small farms producing food crops using traditional methods[6] or 'low level of technology' as described by Amonoo.[7] The equipment of small-scale farmers consists of hand tools, mainly hoes, cutlasses and mattocks and a few harvesting containers for holding or head-carrying farm produce. In the Twifu Prasu area of the Central Region, it was found in 1983 that on average a household owned 8 hand tools and 7 harvesting containers. According to this report:

> As there may be only one or two of a given type of tool in the number owned by a household, and, considering that the average size of household is approximately 10, it would appear that 8 tools per household is rather meagre. Scarcity of farm tools is therefore a major problem to be confronted, and farmers never stopped saying so.[8]

The small-scale farmers do use fertilisers and improved seeds and seedlings when these are made available to them, but they do not yearn for these modern inputs. The big commercial farmers, the State Farms Corporation and the Food Production Corporation, on the other hand, are eager to get hold of pesticides, fertilisers, tractors, harvesters, improved seeds and seedlings and other modern inputs. However, the use of these is not well developed in the country owing to erratic supplies and transportation difficulties, not only in terms of reaching the farms but of reaching them at the appropriate period of the production cycle. The output from these big farms, however, comprises only a small proportion of total output. For example, the State Farms Corporation produced only 0.5 per cent of palm fruits and 0.02 per cent of cassava produced in the country in 1979 and 1980. The Food Production Corporation produced only 1600 tonnes of maize in 1979, compared to the total production of 380 000 tonnes in the whole country (that is, only 0.4 per cent).[9]

There is scope for choice of technology at different stages of agriculture (for example, in land preparation and tillage). For land preparation and tillage, the choice is mainly between the traditional hoe and cutlass, especially in the forest zones, and mechanical land preparation machines, tractors and harvesters. Animal traction is not much used and is confined to the interior savanna, which is free of the tsetse fly.[10] In the open savanna areas of Northern Ghana and the Accra Plains the trend now is to rely on tractors for initial clearing.

4.4. LABOUR

Labour services, as one would expect, are an important input in agricultural production in Ghana. Farmers rely both on family and hired labour. In a study conducted in 1983 (referred to earlier) on the

Table 4.9 Hired and family labour in selected districts, 1982

District	Total man hours	Family Man hours	%	Hired Man hours	%
Swedru	234	187	80	47	20
Atebubu	401	237	59	164	41
Tamale	433	403	93	30	7

Acreages: Swedru 490; Atebubu 1422; and Tamale 1285.
Source: As Table 4.7.

agricultural sector of three districts (Swedru in the south, Atebubu in the middle belt and Tamale in the north) a wide regional variation was observed in the use of hired labour – between 7 and 41 per cent of the total labour used. The rest was supplied by the farmer and his family. This is shown in Table 4.9.

With rapid urbanisation in the post-independence period, massive migration to Nigeria and rapid inflation in the 1970s and early 1980s, the traditional small-scale farmers have been confronted with soaring costs of hired labour. Given their universal conservatism and their view of farming as a way of life rather than as a commercial undertaking involving revenue and costs, they have been hesitant about employing hired labour at the prevailing high wages. Currently, however, with the prevailing high prices of farm produce, many small-scale farmers have joined the commercial farmers in hiring labour at the going price. This is particularly true of the Atebubu district, which is one of the country's main food crop producing and marketing areas.

A strong feeling prevailing in Ghana is that there is a shortage of agricultural labour. It is difficult to understand how such a situation could arise when there is open unemployment and under-employment in the country and many people, including unskilled labourers, have been emigrating in large numbers. The low use of hired labour relative to family labour in small farms probably provides the argument supporting the labour shortage thesis. It is obvious that total farm labour has two components. That is,

$$L_t = L_f + L_w \ldots \qquad (1)$$

where L_t is total farm labour, L_f is household (family) labour and L_w is wage (hired) labour. Household labour depends on the household size, age-group and sex-distribution in the household. The average household size for one area (Twifu Prasu) was found to be 9.7.[11] The hired labour (L_w) component in L_t is dependent not only on the availability of wage labour but also on the cost of it. Labour shortages in agriculture may be interpreted by some as the non-availability of wage labour to the farmers. But this is not the case as 'most farmers feel [that] they would, with money, obtain the labour they need' as is evident from one survey conducted in 1983.[12] It is the high labour cost that prevents them from doing so.

The relative cost of hired labour, as reflected in the prevailing farm wages, is reported to have been as high as 3 to 4 times the official minimum wages in early 1980s. On the other hand, the cost of capital had been significantly lower as agricultural machinery, equipment and inputs such as fertilisers and water had been provided at artificially low prices. Thus until the 1983 budget, which introduced heavy surcharges on imports (see Chapter 10), it had been very cheap to use imported equipment such as tractors and harvesters because of their low relative prices, even though if in some cases these and other imported inputs were not readily available. One can thus infer that high wages keep the L_w component in equation (1) low, giving the impression of low L_t.[13] (The policy of keeping the price structure highly distorted and its harmful effects on resource allocation are discussed in some detail in the opening Overview chapter.)

For small subsistence farmers the situation is different from that of the large (commercial) farmers, in that family labour is readily available to the former. Their dependence on family labour arises for a number of reasons. First, the opportunity cost of family labour is considered low because of various imperfections in the labour market. Second, the family labour which includes even the young children can be used during hours which are not conventionally used in cases of hired labour. Thus one finds a negative correlation between the size of holding and the use of family labour. As found in the GSCA 1970, with an increase in the holding size wage employment increased – being 81 per cent for 10 acres or over, while it was only 52 per cent for farm size of 2 acres or less.[14]

4.5 OTHER INPUTS

There is no doubt that water is the most important input in agricultural production in Ghana, apart from labour. Almost all agricultural production depends on rainfall. Crop yields are invariably poor when the rains fail, or come too early or too late. In Southern Ghana, with two rainy seasons in the year (May-July and September-October), there are two farming seasons - the major period of activity being from April to August and the minor one from September to November. Northern Ghana, with only one rainy season, has only one period of intense agricultural activity from May to August. In the second half of the 1970s agricultural production declined significantly in the country, mainly because of poor rains.

National recognition of the importance of water in agricultural production has led to efforts to reduce the high dependence on climatic factors. Thus an Irrigation Authority has been established and irrigation projects such as Vea, Tono, Dawhenya, Aveyime, Ashiaman, Weija, Okyereko, Komenda and Asutsuare, among others, are being undertaken.[15]

Table 4.10 shows the number of continuing irrigation projects and the total area under the irrigation schemes. Area developed under irrigation in these projects covered about 4000 hectares by 1981, out of the total projected irrigated area of over 12 000 hectares in 23 irrigation projects in different regions of the country. The total land area under irrigation is still less than 0.5 per cent of the total cultivated land. The main items produced under irrigation are rice and vegetables. As observed by Atsu, irrigation has 'yet to make any impact on both total crop acreages planted and outputs'.[16] Between 1980 and 1982, maize production under irrigation covered only about 0.4 per cent of

Table 4.10 Continuing irrigation projects and total projected irrigated area by regions, 1981

Region	No. of projects	Already developed	Yet to be developed	Total
Greater Accra	3	502	1698	2 200
Eastern	2	980	620	1 600
Volta	5	320	1360	1 680
Brong-Ahafo	2	260	1200	1 460
Northern	4	102	704	806
Upper	2	2500	1000	3 500
Ashanti	2	80	420	500
Central	3	15	585	600
Total	23	4759	7587	12 346

Source: CBS, Economic Survey 1981.

the total maize output. The respective figures for groundnuts, vegetables and rice were 2, 5 and 15 per cent. Irrigation facilities are localised and capital investment is considered high. For example at the post-1983 budget level, the cost of extending irrigation facilities per hectare is estimated at ₡350 000.[17]

Other inputs of technological advancement include fertilisers, pesticides and other chemicals, improved seeds and seedlings and extension services. The use of fertilisers, when available, has been well-established in the country, the main types imported over the years being 15-15-15, 20-20-20 and sulphate of ammonia. However, as Table 4.11 shows, imports of fertilisers have been erratic. For instance, of the total import of fertilisers, 15-15-15 alone comprised 48 to 80 per cent from 1971 to 1975, yet in 1977 the import of 15-15-15 comprised only 9 per cent of total fertiliser imports. This increased to 82 per cent by 1979 (not shown in the table). Again in 1977 approximately 36 per cent of the fertiliser imports was the single super phosphate, but after that year none of that type was imported. The quantity and type of fertiliser imported depends critically on government policy in relation to the country's import programme. Farmers have therefore learnt over the years not to put too much reliance on the ready availability of fertiliser in their planning programmes. As noted by Amonoo, 'the level of fertiliser input in agriculture is low because fertilisers are not easily made available to them (farmers)'.[18]

In a study conducted in 1982 on food production and resources on the traditional food farms, Atsu and Owusu found that Ghana's small-scale traditional farmers were well aware of the numerous limitations of the traditional methods of agricultural production.[19] They realised the value of applying chemical fertilisers to the land, now that population pressure has reduced considerably the fallow period through which farm land traditionally regained its fertility. These farmers are also aware of the value of improved farming techniques such as high yielding seeds and seedlings and better seed-bed preparation through ploughing of the land. But the main constraint on the adoption of these modern techniques is finance.[20] The accessibility of institutionalised credit, such as that provided by the Agricultural Development Bank and the other financial institutions, is highly limited and concentrated on the commercial farmers, and the rural banks which are being established right in the farming communities have yet to make their impact (see Section 4.6 of this chapter).

The improved seed-fertiliser technology has been found to raise crop yields and net farm incomes substantially in situations where it

Table 4.11 Fertiliser imports 1971–81 (selected years)

Type of fertiliser	1971 Million tonnes	1971 As % of total	1973 Million tonnes	1973 As % of total	1975 Million tonnes	1975 As % of total	1977 Million tonnes	1977 As % of total	1980 Million tonnes	1980 As % of total	1981 Million tonnes	1981 As % of total
15-15-15	4922.0	55.85	14 287.0	47.96	16 075.0	79.43	4 200.0	9.02	8 000.0	66.67	39 600.0	65.50
20-20-20	544.20	6.81	3 418.0	11.47	1 475.0	7.29	14 000.0	30.08	—	—	—	—
Sulphate of Ammonia	1942.50	32.04	8 100.0	27.19	256.5	1.27	2 900.0	6.23	4 000.0	33.33	17 980.0	29.74
Calcium Ammonium Sulphate	—	—	—	—	—	—	4 300.0	9.24	—	—	1 480.0	2.45
Single Super Phosphate	459.75	5.22	1 500.0	5.04	1 025.0	5.06	16 700.0	35.88	—	—	—	—
Triple Super Phosphate	210.45	2.39	300.0	1.01	900.0	4.45	780.0	1.68	—	—	400.0	0.66
Nutrient of Potash	178.75	2.03	200.0	0.67	5.0	0.02	70.0	0.15	—	—	500.0	0.83
15-20-15	159.2	1.81	930.0	3.12	500.0	2.47	3 000.0	6.44	—	—	500.0	0.83
Others	98.0	1.11	1 056.0	3.54	2.5	0.01	250.0	0.54	—	—	—	—
Kieserite	298.0	3.38	—	—	—	—	350.0	0.75	—	—	—	—
Total	8812.85	100	29 791.0	100	20 239.0	100	46 550.0	100	12 000.0	100	60 460.0	100

Source: Ministry of Agriculture (unpublished data).

has been properly adopted and rainfall and cultural practices (such as planting densities with control) have been satisfactory.[21] This is supported by recent nationwide studies. By using improved seed and fertiliser the average crop yield, in kg per acre, increased in recent years from 450 to 1050 for maize, 327 to 655 for sorghum, 109 to 436 for cow-peas and 450 to 818 for groundnuts.[22] Net income per acre increased from ₵1480 to ₵4184 for maize, ₵1127 to ₵2555 for sorghum and ₵5272 to ₵7797 for groundnuts, while for cow-peas the increase was spectacular – from ₵112 to ₵3612. However, it was found that

> most farmers do not make use of improved varieties [of seeds] because the seed multiplication unit of the Ministry of Agriculture is not able to produce enough seeds to meet farmers' demand. Instances were also recorded where farmers refused to adopt new varieties [of seeds] because consumers did not show any preference for improved varieties [of the product]. This was certainly the case of improved maize crops. So the variable in the model relates to indigenous seeds.[23]

Agricultural extension services in the country started around 1900 with travelling instructors who were trained at Aburi.[24] These extension officers advised on cultivation of cash crops and also supplied inputs to farmers. However, in 1958 a change of approach was effected with the adoption by the Ministry of Agriculture of the western concept of extension, which put much emphasis on the educational approach with no involvement in the supply of inputs. By 1966 the government had found this approach unacceptable and started searching for a more realistic approach. The help of USAID was therefore sought. Ray Johnson of USAID subsequently advised that a policy of 'Focus and Concentrate' be followed whereby the government would focus its agricultural advisory service on the areas with the greatest development potential in terms of both human and natural resources.[25] According to Brown, the rationale underlying the Focus and Concentrate Programme is simply this:

> it is designed to concentrate advisory services, improved cultural and soil fertility practices along with sufficient credit for the procurement of inputs on a limited geographical area with considerable potential for increased agricultural production. The goal of the project ... is to try this new approach to agricultural development for a period of three years and to apply any useful lessons learnt from the pilot area to the rest of Ghana.[26]

This new approach was accordingly extended to all the regions of the country in 1970 after the completion of the three-year pilot scheme in selected areas of a few regions. Nevertheless, at present there is inadequate government support for farm extension services owing to lack of staff and inputs and the poor transportation network to the farming communities.

4.6 AGRICULTURAL CREDIT

Shortage of capital has been a serious constraint on the development of agriculture in Ghana. The constraint is reflected in both fixed and working capital, limiting the use of better agricultural implements, improved seeds and fertilisers. As found by Amonoo, the limited use of improved inputs is essentially due to the low and irregular cash incomes of the majority of small holder food crop farmers in Ghana.[27] His findings based on the Upper and Northern Regions show that, because of the low level of cash income, part of the seed stock earmarked for planting is very often used for household consumption during the crucial production period of clearing and planting. Lack of ready cash and the absence of adequate credit facilities are the two main factors compounding the problem of capital scarcity in the agriculture sector.

In Ghana, non-institutionalised agricultural credit is normally provided by relatives and friends, private money-lenders, traders, distributors of farm inputs and processors of agricultural products. It is customary, for instance, for a man who is about to begin a cocoa farm in a new location perhaps far from his home town or even outside his region to be given a substantial grant by the extended family, as a symbol of their full support and approval of the venture. Friends will usually advance interest-free loans when the need arises. Initially non-institutional credit was mainly provided for consumption loans. Later on it was extended to cater for the needs of the cocoa farmers, a system legalised by an ordinance in 1940.

For years money-lending by both licensed and unlicensed money-lenders has been a flourishing business in the country. A Government Committee on Agricultural Indebtedness stated in 1958 that 'socially, the money-lender is an asset to the village farming community and he is held in high regard. Besides, he is approachable and he is ready to lend at short notice'.[28] The committee also pointed out that the money-lenders of the time were more successful in their businesses than the institutional credit agencies because the money-lenders had a more effective system of collection of loans. Money-lenders who provide agrarian credit live in the farming communities, are in close touch with their debtors and frequently call on them, especially at the time of harvest and sale of agricultural produce. A 1973 study conducted by the Agricultural Development Bank confirmed that moneylenders 'are very highly respected in their respective communities and wield a lot of influence'.[29]

Loans advanced by money-lenders are of two types: production and consumption loans. They disburse the loans mainly in cash. However, a loan may be disbursed in kind, especially in the case where the lender is at the same time a shop-owner and has available in his store the items the borrower intends to purchase with the loan.

Traders dealing in commodities such as maize, rice, yam, cassava and fish; processors of commodities like cassava, pineapple and tomatoes; and distributors of agricultural inputs such as farm machinery, poultry feed and day-old chicks also grant loans to their creditworthy customers. Traders usually advance loans to farmers and fishermen on the expressed understanding that the debtors will sell their produce to the creditors at harvest time and at prices determined when the loans were made. The lenders' prices are usually lower than the ruling market prices. The recovery rates for such loans, as was found in the ADB research, were very high.[30]

The main criticism levelled against these non-institutional agricultural credit agencies is that their rate of interest is very high (varying from 25 to 100 per cent per annum) and that they are, therefore, exploiting farmers and fishermen. However, since their activities are shrouded in secrecy and the agreements between them and their customers are confidential, attempts at the quantification of their operations have to date proved futile. Perhaps, the view expressed by Shepherd in 1936 on the rate of interest and terms of repayment of money-lenders as being exorbitant and harsh is still the case.[31] The government has therefore over the years been encouraging the development of institutionalised agrarian credit.

The first attempt at institutionalising agricultural credit was the co-operative credit programmes that were started in the early 1930s. For instance, in 1931-32 the Cocoa Co-operative Societies made their first loans to members for both production and consumption purposes. In 1938 a co-operative bank, known as the Kpeve Bank, was established in the Volta Region. However, this bank could not survive the problems created by the Second World War. In 1946 the Gold Coast

Co-operative Bank was established by the Cocoa Co-operative Societies in the country. The bank operated smoothly and on sound commercial lines until 1960, when its registration was cancelled by the Nkrumah Government on political grounds. It was liquidated in 1961 and its assets and liabilities taken over by the Ghana Commercial Bank. Its loan-granting activities too were transferred to the United Ghana Farmers Co-operative Council (UGFCC), which was established in 1961 as the sole cocoa buying agent in the country.

At the same time the cocoa co-operative societies owned by the Co-operative Bank were abolished. The UGFCC gave three types of loans: consumption loans (to finance children's education, funerals and clothing for the family, for example); loans in kind through the provision of tools and other inputs like machetes, insecticides (gammalin) and fertilisers; and loans in the form of hiring services (for instance for hiring tractors at specific rates per hectare for land clearing and land preparation).

Another institutional agency that had engaged in agricultural credit was the Cocoa Purchasing Company (CPC). It was established in 1952 as a subsidiary of the Ghana Cocoa Marketing Board primarily to purchase cocoa in competition with the expatriate cocoa purchasing firms then operating in the country. It decided to add credit granting as an auxiliary function. In principle, genuine cocoa farmers who chose to sell their produce to the CPC instead of the expatriate firms were to be given both consumption and production loans. Loans were advanced against the next season's deliveries of cocoa. The CPC went into liquidation in 1957. At the time it had made production loans totalling about £3.3 million and only £7127 of the total interest charges of £294 000 had been recovered.[32]

The Jibowu Commission, appointed in 1956 to enquire into the activities of the CPC, was able to establish that a large proportion of the loans had been made to 'ghost' farmers. The CMB (Cocoa Marketing Board) took over the task of loan recovery, but failed. For instance, during 1962-64 it cost the CMB £234 125 to collect only £47 820. No wonder then that the agricultural credit institutions that have since been established in the country have been extremely cautious and painstakingly thorough in adhering to procedures laid down for loan disbursement and loan collection.

Currently the two major institutions providing agricultural credit in Ghana are the Agricultural Development Bank (ADB) and the National Investment Bank (NIB). The NIB was to provide loanable funds for investment, particularly in industry, agro-industry and large-scale farming. The loans offered by the NIB are for medium and large-scale investment extending from 3 to 25 years. In agriculture it caters especially for the credit needs of agro-based industries and large-scale farming. The ADB was established with a view to expand institutional credit to small and medium-scale farmers whose credit needs were not satisfied by the NIB. In order to qualify for an ADB loan the farmer was required to cultivate a minimum of 10 acres of land and in addition had to provide an acceptable security. According to the GSCA 1970, only 30 per cent of farm holdings were over 10 acres, so by definition 70 per cent of farm holders were disqualified from ADB loans because of the first requirement. To overcome this problem, in 1969 the ADB introduced the 'commodity credit scheme' or group lending scheme under which small farmers are grouped on the basis of single commodities, thus waiving minimum farm size and security requirements.[33]

Since 1976 the Bank of Ghana has been establishing unit banks called rural banks within farming and fishing districts of the country, with a view to enhancing local mobilisation and local utilisation of investible financial resources. Up to 1983, 75 rural banks were established. Rural banking is discussed in some detail in Chapter 9.

The commercial banks in the country also advanced credit to the agricultural sector, although virtually all the loans from this source were confined to large-scale farmers. However, in 1976 the Ghana Commercial Bank (GCB) introduced a scheme called the GCB Farmers Association, similar to the ADB Commodity Credit Scheme.[34]

Table 4.12 Loans and advances to the agricultural sector by commercial banks, NIB and ADB, 1966–81
(selected years)
(₡ million at current prices)

	1966	1970	1974	1979	1981
Commercial Banks	14	9	38	308*	582†
NIB	0.4	0.7	9.6	0.6	3.4
ADB	1	5	15	43	NA
Total	15.4	14.7	62.6	351.6	—

* December figure.
† June figure.
Sources: Bank of Ghana, *Annual Report* (various issues); NIB, *Annual Report* (various issues); and ADB, *Annual Report* (various issues).

Table 4.12 shows the growth of institutional credit to the agricultural sector over the years.

The rate of interest charged by the financial institutions is controlled by the government. In comparison with other sectors, the interest rate on agricultural credit for farming has been deliberately kept low by government policy. The objective is to stimulate investment in agriculture. Both the ADB and the NIB have charged lower rates of interest on farm loans as against, for example, agricultural loans having foreign exchange components to agro-industries.

It was however found that only 2 per cent of the farmers' estimated credit needs are satisfied by institutional credit.[35] Non-institutional sources of credit are therefore likely to play a very important role in satisfying farmers' credit needs. Rural non-banking institutions in the past financed as much as 90 per cent of agricultural investment.[36]

4.7 LIVESTOCK

Ghana's varied vegetational zones can support a variety of livestock such as cattle, sheep, goats, pigs and poultry. Table 4.13 shows the growth of different types of livestock population over the years, while Table 4.14 shows livestock population and its percentage distribution by regions in 1982.

The country produces about one-third of its beef requirement. The remaining two-thirds are imported from traditional beef import sources like Mali and Burkina Faso and also from distant countries like Argentina and Australia. Ghana also imports almost all of its requirement of milk and dairy products.

Table 4.13 Domestic livestock population by type, 1960–82
(selected years)
(number: thousand)

Type of livestock	1960	1965	1970	1975	1979	1982
Cattle	480	511	926	898	780	—
Sheep	500	355	1 315	1317	1 880	1691
Goats	500	380	1 356	1362	1 896	1210
Pigs	44	51	280	270	370	265
Poultry	—	348	10 200	8784	12 384	4863

Sources: CBS, *Economic Surveys* 1961, 1969, 1972–74, 1981 and 1982.

The northern savanna grasslands and the Accra Plains are well suited to cattle rearing. The tsetse fly, which causes the trypanosomiasis animal disease, inhibits extensive cattle rearing in the forest zone. However, even in the naturally suitable areas problems confront the industry. One such problem is the low nutitrional value of the local grasses. Another is the lack of adequate water supplies, especially during the long dry season in the Northern and Upper Regions where about 80 per cent of the cattle are raised. Construction of dams and dugouts to store water has eased this problem somewhat. Other constraints to rapid expansion in cattle production include difficulties in controlling pests, parasites and other diseases, and inadequate supply of credit facilities. Finally, the indigenous West African short-horn cattle has low genetic potential in growth rate and ultimate size, a problem which is caused partly by the close inbreeding which results from the existing open range conditions.

Tables 4.13 and 4.14 also show the sheep and goat population. These browse on nearby bush and also feed on food materials that would otherwise go to waste. Goats, particularly, have the ability to browse on a wide variety of plants and eat any edible material. Compared to cattle, sheep and goats have a shorter gestation period and hence the capacity to multiply more rapidly. Although nutritional deficiencies and management problems contribute to losses of sheep and goats, by far the most important cause of death of these animals is disease.

The rearing of pigs, like the rearing of sheep, goats and poultry, exhibits the characteristics of technological dualism. Well-run pig farms where the animals are reared in confinement exist side-by-side with open range farms where the animals range freely and feed on anything edible. This is obviously due to the rapid increase in demand for livestock products with increasing population and urbanisation. The supply of maize constitutes the main bottleneck to increased production of pigs. Again the supply of iron is essential to piglets reared in confinement. Piglets without iron supply develop anaemia which makes them weak and susceptible to most types of pig diseases.

Considerable progress was achieved in poultry production, which increased from 348 000 in 1965 to over 12 million in 1979, although there has been a sharp drop to about 5 million in 1982 (see Table 4.13.) While formerly the country relied heavily on imports from Holland for eggs and day-old chicks, Ghana is now almost self-sufficient in these poultry items. Apart from diseases, the main bottleneck to the further expansion and efficient running of the poultry

Table 4.14 Livestock population by regions, 1982

Region	Cattle* Nos. '000	Cattle* As % of total	Sheep Nos. '000	Sheep As % of total	Goats Nos. '000	Goats As % of total	Pigs Nos. '000	Pigs As % of total	Poultry Nos. '000	Poultry As % of total
Upper	303	38.85	408	24.13	316	26.12	83	31.32	1010	20.77
Northern	310	39.74	254	15.02	244	20.17	61	23.02	697	14.33
Brong-Ahafo	36	4.62	258	15.26	198	16.36	43	16.23	453	9.32
Ashanti	2	0.26	252	14.90	114	9.42	15	5.66	682	14.02
Eastern	75	9.62	159	9.40	91	7.52	14	5.28	331	6.81
Greater Accra	12	1.54	84	4.97	24	1.98	15	5.66	923	18.98
Volta	40	5.13	27	1.60	120	9.92	20	7.55	351	7.22
Central	1	0.13	122	7.21	36	2.98	5	1.89	216	4.44
Western	1	0.13	127	7.51	67	5.54	9	3.40	200	4.11
Total	780	100	1691	100	1210	100	265	100	4863	100

* Figures for 1979.
Source: CBS, *Economic Surveys*, 1981 and 1982.

industry is the shortage of maize which constitutes over 50 per cent of the ingredients of poultry feed. But there has been competing demand for maize, arising from its consumption by both humans and livestock. In the event, the country's failure to meet the excess demand for maize will continue to be a constraint on the expansion of poultry production. As to diseases, the Newcastle disease still attacks even those birds that have been vaccinated against it. Salmonella is still carried by some chickens, especially the local domestic chickens.

Several of the facilities essential to the modern poultry industry have already been established in Ghana. A number of small, medium and large-scale producers, both in the state and private sectors such as Pomadze and Darko Farms respectively, have improved upon the quality and quantity of their poultry and eggs. The major difficulty now is the inadequate supply of maize and also of antibiotics and other drugs.

Ghana has an impressive array of institutions that provide the requisite support services for the livestock and poultry sub-sector. For instance, the Animal Husbandry Division of the Ministry of Agriculture provides extension services to livestock farmers. The Animal Research Institute is engaged in improving grasses and legumes for multiplication. Livestock are also being improved, especially through the importation of high-yielding varieties from other countries and by cross-breeding. The Veterinary Services Division, also part of the Ministry of Agriculture, fights livestock diseases. The National Pig Multiplication Centre at Adidome supplies breeding stock and offers advice on general management and feeding of pigs. The universities undertake research into problems associated with livestock. In relation to credit the Agricultural Development Bank and the other financial institutions advance loans to the sub-sector. The central bank, the Bank of Ghana, has even established a ranch to demonstrate to livestock farmers the usefulness of the application of modern technology in livestock production. However, as in many other areas of the economy, most of the above institutions are presently passing through a difficult period caused mainly by the inadequacy of trained staff and the lack of chemicals and other inputs.

4.8 FISHING

Fishing is a well-established occupation in Ghana. Fishing takes place in the rivers, Lakes Bosomtwe and Volta, the lagoons and the sea.

Table 4.15 Catch of fish from sea and Volta Lake, 1965–82 (selected years) ('000 tonnes)

	1965	1970	1975	1980	1982
Marine Catch					
Metric tonnes	59.7	157.0	212.7	184.1	199.1
% of total catch	(74.5)	(79.7)	(83.5)	(82.2)	(85.0)
Volta Lake catch					
Metric tonnes	20.4	39.9	41.9	40.0	35.0
% of total catch	(25.5)	(20.3)	(16.5)	(17.8)	(15.0)
Total catch	80.1	196.9	254.6	224.1	234.1

Sources: CBS, *Economic Surveys* 1968, 1972–74, 1975–76, 1981 and 1982.

However, the most important sources are marine fishing and fishing in Lake Volta. Table 4.15 gives the figures for the catch from these two important sources.

The fishing industry is characterised by technological dualism, with large companies such as the State Fishing Corporation and Mankoadze Fisheries, using trawlers, co-existing with the traditional fishermen who depend heavily on canoes. The traditional fishermen hardly venture beyond the continental shelf and yet the share of canoes in the domestic output of marine fish increased from 60 per cent in 1976 to 77 per cent in 1980.[37]

The Volta Lake catch increased in metric tonnes, from 20 000 in 1965 to about 40 000 in 1970. (The highest recorded was 61 000 tonnes in 1969.) The volume of catch has declined in recent years. In 1982 it was 35 000 tonnes. The marine catch increased from 20 000 tonnes in 1965 to 157 000 tonnes in 1970 and further to 213 000 in 1975; it fell to 184 000 tonnes in 1980 but increased again to 199 000 tonnes in 1982. The fall in output in recent years is a result of the country's inability to solve the problems confronting the industry. Currently the problems include the shortage of imported inputs such as outboard motors and their spare-parts, fishing nets and ropes. For instance in 1980, out of the total number of 435 outboard motors only 260 were operational.[38] There is also a lack of chemicals used as refrigerants for cold storage facilities and fish cannot be transported to the interior owing to the lack of insulated vans.

4.9 FORESTRY

Tropical rain forest covers about 8.2 million hectares or 34 per cent of the total land area of the country and forest products have contributed about 5.5 per cent of GDP per annum in recent years.[39] Many enterprises are engaged in this sector. As many as 150 known commercial species of timber are found in the forest zone. Of these only about 25 to 30 are currently being extracted. Indeed, 5 species account for over 75 per cent of the total earnings from timber. These are the well-known red-woods (Mahogany, Odum, Utile, Sapele and Wawa). These are all included in the group known as the primary species. Wawa accounts for the greater percentage of timber exports.

Next to cocoa, timber used to be Ghana's most important export. Between 1970 and 1976 its share of total exports was 9.6 per cent. It dropped to 4.1 per cent by 1979.[40]

For a long time most of the timber companies did not establish processing units, and even in the ones that did, the processing was done only to the lumber stage. Only a few companies manufactured veneer and plywood. Thus most of the country's timber was exported as raw material for factories outside Ghana.

As the figures in Table 4.16 show, the timber industry, like almost all other sectors of the Ghanaian economy, has been stagnating in recent years. A major problem confronting the sector is the lack of

Table 4.16   Number of firms, installed capacity and capacity utilisation in different sectors of the timber industry, 1982

	No. of firms	Installed capacity ('000 m^3)	Production ('000 m^3)	Capacity utilisation (%)
Logging	600	2500	600	24
Sawmilling	164	1100	275	25
Plywoodmilling	9	100	63	63
Veneermilling	12	45	20	44
Chipboardmilling	1	15	6	40
Furniture manufacturing (medium and large-scale)	65			
Flooring	4	150	50	33
Doors	6			

Source: E. B. Karkari, *et al* (1983).

vital inputs such as glue and v-belts and other imported items, and a scarcity of labour. The over-valuation of the cedi was also a major problem, a problem that could not be remedied by the meagre export bonus and tax concessions granted to the industry. Another problem has been the inability of Ghana Railway to transport timber to the processing factories and the ports.

In 1972 therefore a move was initiated to ban the export in log form of fourteen major primary species. However it was not until January 1979 that the ban came effectively into force, and it has remained in force since then. The main aims of the ban are the conservation of the primary species which are not being replaced by replanting, and the re-routing of the primary species to the local mills, which have been working under-capacity, to enable them to export products with higher value-added. Processed timber products command higher prices on the world market. For example, in 1979 the price of logs stood at $93.00 per cubic metre, while sawn timber fetched $264.2 and plywood as much as $536.4 per cubic metre. The country thus stands to earn more foreign exchange by exporting timber products instead of logs. Table 4.17 shows the production and export of logs and sawn timber from 1960 to 1982. A fall in the production and export of logs is apparent, the decline being particularly marked after 1970; but for

Table 4.17 Production and export of log and sawn timber, 1960–82
(selected years)
(million cubic metres)

Item	1960	1965	1970	1975	1980	1982
Logs						
Production	1.62	1.57	1.55	1.31	0.71	0.42
Export	0.81	0.55	0.59	0.43	0.08	0.06
Export as % of Production	50.00	35.03	38.06	32.82	11.27	14.29
Sawn timber						
Production	0.37	0.45	0.36	0.39	0.21	0.15
Export	0.23	0.23	0.24	0.16	0.06	0.04
Export as % of Production	62.16	51.11	70.59	41.03	28.57	26.67

Note: Data shown in cubic feet in the source for the early years have been converted into cubic metres.
Sources: CBS, *Economic Surveys* 1965, 1969, 1972–74, 1975–76, 1981 and 1982.

sawn timber, in both production and export, there were ups and downs until 1975, and a sharp and steady fall after that.

The forestry sector directly employed a labour force of 70 000 during the early 1980s.[41] However, there are only a few well-established furniture and joinery companies in the country. The local demand for timber mainly comes from small-scale carpentry workshops which invariably use labour-intensive techniques.

5 Cocoa

For well over half a century cocoa has been the backbone of the Ghanaian economy and will probably remain so for many years to come. According to the *Ghana Sample Census of Agriculture* 1970, 3.6 million acres (1.45 million hectares), or about 50 per cent of the total cultivated land area, was under cocoa production.[1] As may be seen from Table 5.1, cocoa is grown in six regions of the country. Of the total land under cocoa cultivation in 1970, the Ashanti Region had the highest share with 40 per cent of the total, followed by Brong-Ahafo, the Eastern and the Western Regions. The Central and Volta Regions contributed 7 and 3.9 per cent, respectively, to total land under cocoa cultivation in 1970. The cocoa sector accounts for the employment of about 17 per cent of the total labour force of the country.[2]

5.1 PRODUCTION

The production of cocoa in Ghana, which was 322 000 tonnes in 1959/60, increased to about 429 000 tonnes in 1962/63 and further to 566 000 tonnes in 1964/65. After suffering a decline in subsequent years it increased to 470 000 tonnes in 1971/72, but has generally

Table 5.1 Percentage distribution of cocoa by region according to farming areas, 1970

Region	Total area (hectares)	As % of total
Western	202 500	13.9
Central	101 250	7.0
Eastern	202 500	13.9
Volta	56 700	3.9
Ashanti	587 250	40.4
Brong-Ahafo	302 535	20.8
Total	1 452 735	100.0

Source: Ministry of Agriculture, *Report on Ghana Sample Census of Agriculture* 1970, vol. I, p. 61.

Table 5.2 Output, nominal and real producer prices of cocoa, 1959/60–81/82
(selected years)
(producer price in cedis)

	'000 tonnes	% of world output	Per load of 30 kg Nominal (current) price	Real 1963 price
1959/60	322.1	30.5	—	—
1962/63	428.8	36.5	5.40	4.99
1965/66	416.6	34.0	4.40	2.33
1968/69	339.3	27.4	7.00	3.85
1971/72	470.0	29.6	8.00	3.53
1974/75	382.0	24.6	15.00	3.67
1977/78	271.0	18.0	36.29	1.51
1979/80	296.0	18.1	120.00	2.17
1980/81	258.0	15.5	120.00	1.00
1981/82	225.0	13.0	360.00	2.45

Notes: The discrepancy between nominal and real prices in 1962/63 (1963 = 100) is due to nominal price data being based on financial year and the price index on calendar year.
Sources: Government of Ghana, *Five-Year Development Plan* (1975/76–1979/80) Part II; CBS, *Economic Survey* 1969–71, 1972–74, 1981 and 1982; *Quarterly Digest of Statistics* September 1983; and *Statistical News Letter* No. 18/83, 23 November 1983; and Nyanteng (1980).

declined since then. In 1981/82, production of cocoa was only 225 000 tonnes (see Table 5.2).

While all the cocoa producing regions in Ghana have registered declines in output, the falling trend in the case of the Volta Region is particularly striking, as revealed in Table 5.3. Output from the region fell from 14 000 tonnes in the mid-1970s to around 1500 tonnes in the early 1980s. Brong-Ahafo has also suffered a significant fall in cocoa output. It is believed that the declining trends in cocoa output in the Volta and Brong-Ahafo Regions would probably not have been so drastic but for smuggling to Togo and the Ivory Coast, which are contiguous to the respective cocoa regions in Ghana.[3]

Cocoa is exported both in the form of raw cocoa beans and in processed forms of cocoa butter and cocoa paste. Following the fall in cocoa output in the country and the success achieved by other nations in expanding production, Ghana's percentage share of the world output of cocoa has been declining, as shown in Table 5.2. Ghana produced as much as one-third of total world cocoa output in

Table 5.3 Cocoa production in Ghana by regions, 1975/76–81/82 (thousand tonnes)

Region	1975/76	1976/77	1977/78	1978/79	1979/80	1980/81	1981/82
Ashanti	124.3	105.8	89.6	86.9	100.4	91.5	70.8
Brong-Ah.	88.6	79.5	69.5	50.4	74.9	47.6	49.7
Eastern	68.6	54.2	41.3	50.2	45.0	46.6	36.9
Central	49.7	39.1	21.5	25.7	19.0	25.6	22.1
Western	55.6	41.0	42.0	45.9	52.3	45.1	43.7
Volta	14.0	9.4	7.4	6.0	4.8	1.5	1.7
Total	401.0	329.0	271.3	265.1	296.4	258.0	224.9

Source: Ghana Cocoa Marketing Board as quoted in CBS, *Economic Survey 1982*, p. 32.

the early 1960s. In the early 1980s the corresponding figure was less than one-seventh of world output. Currently the major importers of Ghanaian cocoa, in descending order, are the USSR, the Netherlands, the USA, West Germany and the United Kingdom.

5.2 DEMAND ELASTICITY

The world market price of cocoa has been characterised by sharp fluctuations. For instance, in July 1977 the spot price for Ghana cocoa was as high as £3427 per metric tonne, and the average for 1978 was £2005. But there was a continuous and precipitous fall of cocoa prices from £1607 in January to £987 in December 1980.[4]

In the immediate post-World War II period, demand for the commodity exceeded supply with the result that the world market price rose very rapidly to the then all-time record of £400 in 1954.[5] With new plantings in the producing countries supply finally caught up with and overtook demand in the late 1950s (given a time-lag of 5 to 6 years between planting and first harvesting).

Long-term trends in international trade show that export prices of manufactured goods (which usually come from the industrialised advanced nations) have been rising faster than those of primary commodities (from the less developed countries). One explanation for this is that demand for primary commodities has been expanding relatively slowly.

The demand for Ghana's cocoa, a derived demand from final products such as chocolate and chocolate products, has a low income elasticity. In other words, an increase in real income in the cocoa importing nations leads to a less than proportionate increase in their demand for cocoa. Empirical studies have revealed that income elasticity of demand for cocoa is positive but less than unity, thus conforming to 'Engel's Law'.[6] It has been suggested that the demand for cocoa tends to reach a saturation point beyond which the demand elasticity becomes zero at higher income levels.[7] In the United States, for instance, the post-war increase in real income per capita has not led to any increase in the per capita consumption of chocolate products. Increases in imports of cocoa are wholly accounted for by increases in population, not in real incomes.

Technological progress in the advanced industrialised countries has led to the development of alternative products that have been replacing some of the natural or original products from the less developed countries. For instance, since World War II, cocoa has been facing stiff competition from substitutes such as groundnuts. This was especially the case in the 1950s when the world market price of cocoa shot up. As expected manufacturers started economising on their use of cocoa.

The industrialised (advanced) nations also have a high degree of effective tariff protection on imports of semi-processed products from the less developed countries. For instance, while the tariff on cocoa beans have been low (4 per cent in the European Common Market), cocoa products such as cocoa butter and cocoa paste have faced higher tariffs – 12 and 15 per cent respectively.[8] These factors have obviously contributed to the low demand for cocoa in the importing countries.

The long-run price elasticity of world demand for cocoa has been estimated to be 0.4, suggesting that a reduction in unit price of cocoa leads to a less than proportionate increase in demand.[9]

5.3 GOVERNMENT POLICY AND PRODUCER PRICES

In 1948 the British colonial government established the Cocoa Marketing Board (CMB) as the sole buyer, grader, seller and exporter of cocoa produced in Ghana. In addition to Ghana, marketing boards for various export commodities were also established in other British colonies in West and East Africa and Malaysia.

For over a decade up to 1982, government support for the cocoa sector had not been in proportion to its importance to the economy.

This is in sharp contrast to the vigorous efforts made by the government in the 1950s and early 1960s to stimulate the industry. In those years farmers were paid for the loss of diseased trees that were cut down to contain the swollen-shoot disease. Insecticides, spraying machines and other inputs were supplied at subsidised prices, and prices paid to the producer were relatively attractive. It is true that successive governments have revised upwards the producer price of cocoa. For instance, whereas in 1963 the nominal producer price was ₵5.40 per load of 30 Kg, by 1980 it was ₵120. The price was raised to ₵544 in December 1982 and ₵900 in mid-1984, but in real terms prices paid to cocoa farmers have generally declined, at times drastically until 1982 (see Table 5.2).

Cocoa is a commodity which in raw and processed forms contributed in the 1960s 12 per cent of GDP,[10] 60–70 per cent of the total export earnings, and through duties and taxes 30 per cent of the government revenue.[11] Using shadow prices (that is, social opportunity costs), net returns to the economy from a hectare planted with cocoa are estimated to be 100 times as high as for maize and 15 times as high as for rice. In other words, given the exceptionally high return from cocoa, Ghana would aim to expand its production. But a study conducted in 1983 by the Centre for Development Studies (University of Cape Coast) has shown that farmers' preferences were definitely shifting from cocoa cultivation to food crops such as rice and maize.[12] There are many reasons for the shift in farmers' interest, but the increasing relative profitability at market prices must be an important one, as is apparent from Table 5.4.

The government was effective, albeit to the disadvantage of the economy at large, in regulating the producer price of cocoa through

Table 5.4 Comparative return to farmers on cocoa, rice and maize, 1980
(amount in cedis)

	Cocoa	Rice	Maize
Price to farmer (per kg)	4	2.4	1.3
Yield per hectare	500	1300	2250
Farmer's gross revenue per hectare	2000	3120	2925
Production cost per ha.	999	890	716
Net returns to farmer per ha.	1001	2230	2209

Source: Council of State, *Reviving Ghana's Economy*, p. 42.

the CMB at a level which in 1980-81 was only 20 per cent of what the farmer received in 1962-63 (see Table 5.2).[13] But government controls over the prices of food crops were not effective, so that farmers were able to charge high open market prices for these crops. Moreover, a shift of land and other resources from cocoa to other forest-belt crops, especially maize, cassava and plantain, is likely to be favoured by the farmers as the returns on investment are quicker than with the latter crops. Hence the policy of neglect of Ghana's comparative advantage in cocoa production. Hence too the implication for the economy in terms of opportunities foregone. Ghana, indeed, almost killed the goose that laid the golden eggs.

The low producer price of cocoa encouraged the smuggling of the commodity into the neighbouring countries, where the price is substantially higher when converted at the unofficial exchange rate. In 1978-79 one load of cocoa (30 Kg) was estimated to fetch ₡523.71 in Togo and ₡647.98 in the Ivory Coast, compared to ₡72.00 in Ghana.[14] Not surprisingly, there developed a large smuggling trade in cocoa which was estimated by the Bank of Ghana at 80 000 tonnes in 1980-81.[15] A new type of smuggling was even reported to be emerging in which all risks were borne by the foreigners who came to the doorsteps of farmers offering them over three times the price paid by the CMB.[16]

5.4 INPUTS

In a survey conducted in 1977, the input requirements of cocoa such as insecticides, spraying machines, seed and seedlings were systematically estimated by Nyanteng.[17] It was found that, in total, 4.5 million litres of insecticides per annum were needed whereas only 875 000 litres (that is, 19 per cent of total requirement) were supplied to farmers in 1977-78. The supply of Gammalin 20, the main insecticide distributed to farmers during the 1970s, varied around 20 per cent of the estimated demand, the only exception being 1976-77 when 56 per cent of the estimated demand was satisfied. The demand for spraying machines was estimated at 98 000 units in 1977-78 as against 37 000 machines found in operating condition in the possession of farmers. The quantity of cocoa seed, pods and seedlings distributed to farmers fell far below their demand. Of the farmers interviewed, 60 per cent reported that they produced their own seed and seedlings although these were mainly the unimproved Amelondo variety.

Data relating to essential inputs for the cocoa industry for recent years are not available. Given the decline of the industry one might,

Table 5.5 Farmers' reaction to input price subsidy versus output price increase, 1977

Condition	Percentage of farmers who favour
Output price increase	74.9
Input price subsidy	25.1
Total	100.0
N = 227	

Source: Nyanteng (1980), p. 96.

however, assume a decline in the demand for inputs as well. And given the loss of cocoa farms because of the swollen shoot disease, and the need to replace the old cocoa farms, the demand for seed and seedlings is likely to be higher today than it was in the late 1970s. For definite conclusions we must await survey findings from the research department of the Cocoa Marketing Board.

An important issue in this area is the continuation or withdrawal of the input price subsidy. The farmers would of course like to have an input price subsidy as well as an output price increase. However, the 1977 survey showed that the farmers' preferences for output price increases were stronger than for input price subsidy, as may be seen from Table 5.5. The implication of this preference is noted succinctly by Nyanteng as follows:

> The output price increase may not only encourage production but may also help to reduce the level of cocoa smuggled out annually to the neighbouring countries. On the other hand, removing the subsidies on inputs and allowing the factor market to establish the price levels would also reduce smuggling of the inputs to the neighbouring countries, making more of the inputs available for local use.[18]

5.5 PROBLEMS AND CONSTRAINTS

The problem created by sharp fluctuations in the world market price of cocoa and the problem of the long-term prospects of demand for cocoa are of much relevance to Ghana. The heavy dependence on this single commodity for foreign exchange earnings renders the Ghanaian economy, including its prospects for development, vulnerable to external influence. It is foreign exchange largely derived from cocoa exports which is used to finance imports for economic development. But while the economy's prospect of development still depends critically on this one crop, Ghana has already lost her position as the world's leading producer, and indeed her share of the world total output has been shrinking. Other nations such as Brazil, Malaysia and the Ivory Coast are rapidly expanding their output.

The internal transport of cocoa from the producing areas to the ports for export remains a major problem. Hence large quantities are left behind in the bush and do not become commercially available to the outside world.

International cocoa agreements between producers and consumers for stability in output and price levels have not been functioning as efficiently as expected. Nor is the Cocoa Producers' Alliance an effective instrument. It was formed in 1962 by Brazil, Cameroon, Gabon, Ghana, Ivory Coast, Nigeria and Togo. As a group, the members produce about 80 per cent of the world's cocoa output. The objectives of the Alliance include the exchange of technical and scientific information, discussion of problems common to producing countries and the promotion and expansion of cocoa consumption. Under the auspices of UNCTAD (United Nations Conference on Trade and Development) an agreement was concluded between producers and consumers in 1973.

As long as the internal producer price remains low relative to the prices of other commodities which compete with cocoa for resources, prospects for substantial increases in cocoa output will remain dim indeed. This would establish the case for a realistic pricing policy and more effective economic management than hitherto, as discussed earlier in the opening Overview chapter. Until the recent increases in producer price (the impact of which is yet to be fully felt), the price offered little incentive in relation to the value of food crops or the cost of many household items. This led to a shift of production to other food crops. Factors that influence the world market for cocoa generally lie beyond the control of any single producer nation. However, Ghana has established a long tradition of top grade cocoa production. Perhaps she can take advantage of this tradition to regain her lost share of the world market.

5.6 COCOA MARKETING BOARD[19]

The Cocoa Marketing Board (CMB), like its counterparts in the British Empire, was empowered to set producer prices below the world market price, with the aim of keeping the difference in a stabilisation fund. Such a fund was to be used to maintain the producer prices and ostensibly the producers' income, in case the world market price fell to disastrous levels. The fund could provide savings to finance the country's development. It would also drain off excess purchasing power and thus prevent inflation in the country.

The reserve fund thus accumulated by the CMB was, prior to the exchange control regulations introduced in 1961, simply invested in long-term British government securities. In other words, Ghana was lending money to Britain. The interest rate paid on these securities was extremely low; 0.5 per cent before 1950 and 2 to 4 per cent after 1952.[20] Again, the dollars earned by Ghana in exporting to the Dollar area, especially to the USA, were used to meet Britain's high demand for dollars to finance imports from the USA to develop her war-devastated economy. Thus dollar earnings which could have been used to import capital goods for Ghana's own development were diverted to Britain. This diversion in the form of investments in British government securities at low rates of interest and in the form of dollar reserves for the Sterling Area slowed down Ghana's own infrastructural and industrial development. According to Fitch and Oppenheimer,

> The CMB... continued to levy what were in effect huge export taxes, send cocoa profits to Great Britain, and thus help Britain maintain the pound while renouncing, or at best postponing, attempts to start Ghana in the direction of economic independence and development.[21]

The Ghana government had relied heavily on the CMB as a tax agency. For instance, in normal years, as much as 50 per cent of the total revenue realised by the CMB was taken by the government as export duty.[22] The heavy tax burden on cocoa farmers over the years has been one of the causes of the decline in output. For while producers of other food crops hardly paid any taxes the cocoa farmers, through the deliberately low producer price policy of the CMB, have been taxed heavily. No wonder then that farmers' interest in the cocoa industry declined considerably.

In addition to cocoa beans, the CMB also exports cocoa butter and cocoa paste from its three processing plants (two located at Takoradi and one at Tema). The lower the world market price for beans, the larger the quantity that it processes prior to export. The objective is to export higher value-added products and thus earn more foreign exchange for the country. The CMB also handles the export of coffee, kola nuts and sheanuts.

The performance of the CMB regarding the distribution of inputs and collection of output is far from satisfactory. It is true that the CMB is not able to control factors such as poor port facilities and internal transport bottlenecks. But given an efficient management the CMB could have achieved much better distribution of inputs than has so far been the case.[23] The inability of the CMB to transport cocoa quickly from farming villages to railway sidings and the ports has not only caused much inconvenience to the cocoa producers, but also contributed to the smuggling of the product to the neighbouring countries.

The monopoly enjoyed by the CMB has, of course, not helped in improving the efficiency of cocoa marketing in Ghana. However, any attempt to introduce competition in this area will require serious studies taking into account scale economies, national and farmers' interests and the ease and effectiveness of tax administration. The experience of other major cocoa producing countries, such as the Ivory Coast, Brazil and Malaysia, can be of great help in guiding policy decisions and improving the efficiency of cocoa marketing in Ghana.

6 Manufacturing

At current prices, the manufacturing sector accounted for 9.7 per cent of GDP in 1965; the share increased to 12.6 per cent in 1968 and after some decline during the early 1970s it went up to 13.9 per cent of GDP in 1975, the highest recorded at current prices. But since then the share of manufacturing in GDP has been falling consistently and it was only 3.5 and 3.8 per cent in 1982 and 1983 respectively. Within the manufacturing sector, according to the *Five Year Development Plan* (1875/76-1979/80):

> Large-scale manufacturing industries have played a major role in the growth of the industrial sector. Their contribution to Gross Domestic Product has generally been slightly more than thrice that of medium- and small-scale industries. In other words, large-scale manufacturing industries have provided more than 75 per cent of the sector's contribution to Gross Domestic Product.[1]

6.1 GROWTH OF MANUFACTURING

The manufacturing sector expanded rapidly during the 1960s and continued to increase up to the mid-1970s. At current prices, gross output increased from ₵93.1 million in 1963 to ₵10 105.5 million in 1983 (see Table 6.1). At constant (1970) prices, however, there were less impressive increases in gross output – from ₵261.3 million in 1968 to ₵535.6 million in 1975, after which there has been a marked decline. By 1983 the gross output was only ₵196 million. In terms of value added at current prices, the sector showed a phenomenal increase from ₵52.5 million in 1963 to ₵4563 million in 1983. However, at 1970 prices, value added increased from ₵134.9 million in 1968 to ₵180.6 million in 1978, after which it fell sharply, reaching ₵56.8 million in 1983.

The manufacturing sector includes enterprises producing agricultural inputs like pesticides, machetes and other farm implements. Others are engaged in food processing and canning, beverages and tobacco manufacturing. In addition, machine repairs, vehicle assembly and repair of electrical appliances are undertaken. Other manufacturing activities include petroleum refining, alumina smelting and

Table 6.1 Gross output, value added, employment, wages and salaries in medium and large-scale manufacturing, 1963–83 (selected years)

	1963	1966	1968	1970	1972	1975	1978	1980	1982	1983
1. Gross output (mil ₡)										
Current prices	93.1	142.0	222.9	441.0	572.4	1274.6	2649.5	5204.7	5767.4	10 105.5
1970 prices			261.3	441.0	455.0	535.6	429.1	373.7	286.8	196.0
Index (1970 = 100)			59.3	100	103.2	121.5	97.3	84.6	65.0	44.4
2. Value-added (mil ₡)										
Current prices	52.5	85.1	115.1	155.7	204.1	428.7	1115.2	2342.4	2588.9	4 563.0
1970 prices			134.9	155.7	162.2	180.1	180.6	168.0	128.7	88.5
Index (1970 = 100)			86.6	100.0	104.2	115.7	116.0	107.9	82.7	56.8
3. Employment ('000)	39.5		47.0	57.6	60.7	77.0	85.2	80.3	67.7	58.3
4. Value-added per employee (₡'000) 1970 prices			2.87	2.70	2.67	2.34	2.12	2.09	1.90	1.52
5. Wages and salaries (mil ₡)										
Current prices		20.8	27.8	45.5	57.5	114.1	287.8	473.6	716.2	1 286.9
1970 prices		22.9	30.9	45.5	47.8	52.6	22.6	16.1	9.2	7.5
6. Wages and salaries per employee										
₡ 1970 prices		579.7	657.4	789.9	787.4	683.2	265.3	200.5	136.1	128.0

Note: Manufacturing deflator (Table A.10) was used to calculate 1970 prices, except for wages and salaries for which Consumer Price Index was used (1970 = 100).
Sources: CBS Statistical Service, *Industrial Statistics* (various issues); *Economic Survey* 1967, and Table A.10.

aluminium wares, pharmaceuticals, soap, detergents and cosmetics production. Locally produced building and construction materials include concrete products, sawmilling products, bricks and tiles, ceramics, cement, paint and iron and steel rods. Tyres and tubes, batteries, leather, plastic products and textiles are also locally produced.

As output expanded in the manufacturing sector, employment also increased. The total labour force in the medium and large scale manufacturing increased from 39 500 in 1966 to 85 200 in 1978, after which it gradually declined, reaching 58 300 in 1983. This decline in the level of employment was due to a fall in investment. However, the decline in output has been much faster than that of employment, thus leading to a marked deterioration in labour productivity. At 1970 prices, value added per employee declined from ₵2870 in 1968 to ₵1520 in 1983.

The whole of the medium and large-scale manufacturing sector has been divided into seven broad categories as shown in Table 6.2 for two selected recent years, 1981 and 1984. In terms of value added, the three principal sub-groups are (a) food, beverages and tobacco; (b) chemical, oil and rubber products; and (c) metal, machinery and transport, which in 1984 respectively contributed 63.7, 13.0 and 3.4 per cent of value added. In terms of gross output, the leading sub-sector in 1984 was the food, beverages and tobacco group with 41.5 per cent, followed by the chemicals, oil and rubber products group with 33.4 per cent of total gross output. The wood and wood products sub-sector was the highest employer in 1984, providing 32.4 per cent of total employment, followed by food, beverages and tobacco (23.3 per cent) and textiles and footwear (20.1 per cent).

A comparison between fixed assets and value added for 1981 shows that the sub-groups having high shares of fixed assets, such as metal, machinery and transport (33.1 per cent) and wood and paper products (26.3 per cent) have lower shares of value added, 9.7 and 7.2 per cent respectively, while the sub-groups with high value added shares, for instance food, beverages and tobacco (54.4 per cent) and chemical, oil and rubber products (18.4 per cent) have low shares in fixed assets, 16.7 and 5.3 per cent respectively. This type of relationship is not, however, borne out by the data for 1984 as all the sub-groups having the highest shares of fixed assets (for instance, food, beverage and tobacco) do not have low shares of value added.

Table 6.3 shows the leading medium and large-scale industries in terms of value added in 1984. It also shows ranking by employment,

Table 6.2 Sectoral distribution of employment, value added, gross output, and fixed assets in medium and large-scale manufacturing, 1981 and 1984
(value in current prices)

Sector	Year	Employment ('000)	%	Value added ₵ mil.	%	Gross output ₵ mil.	%	Fixed assets ₵ mil.	%
Food, beverages and tobacco	1981	17.2	22.3	1 646.3	54.4	2 450.5	38.4	220.0	16.7
	1984	13.0	23.3	6 538.1	63.7	10 264.8	41.5	1346.2	29.2
Textiles and footwear	1981	18.3	23.7	235.3	7.8	564.4	8.8	179.5	13.6
	1984	11.2	20.1	755.4	7.4	2 137.1	8.7	467.5	10.0
Wood and paper products	1981	21.3	27.6	218.5	7.2	517.7	8.1	347.0	26.3
	1984	18.1	32.4	911.4	8.9	2 071.9	8.4	1512.4	32.8
Chemicals, oils, rubber products	1981	7.0	9.1	556.1	18.4	1 774.7	27.8	70.0	5.3
	1984	5.6	10.0	1 338.0	13.0	8 247.8	33.4	847.0	18.4
Building materials	1981	2.6	3.4	72.8	2.4	240.7	3.8	63.9	4.8
	1984	2.6	4.7	334.9	3.3	1 076.9	4.4	230.0	5.0
Metal, machinery and transport	1981	10.6	13.7	292.9	9.7	835.6	13.1	436.8	33.1
	1984	4.7	8.4	351.2	3.4	814.5	3.3	201.7	4.4
Others	1981	0.2	0.3	2.7	0.1	5.8	0.1	0.6	0.04
	1984	0.6	1.1	40.5	0.4	91.0	0.4	10.0	0.2
Total	1981	77.1	100	3 024.6	100	6 389.4	100	1318.0	100
	1984	55.8	100	10 269.5	100	24 704.0	100	4614.8	100

Sources: CBS/Statistical Service, Industrial Statistics, 1979–81 and 1982–84.

gross output, productivity and capital intensity in these industries. In terms of gross output, the leading industry in 1984 was petroleum, followed by tobacco (malt liquors and malt came third, followed by textiles and sawmilling). For inter-temporal comparison, 1984 data, shown in Table 6.3, have been compared with those of four other selected years, 1964, 1975, 1981 and 1983.[2] In terms of employment, sawing and sawmilling has been a leading sub-sector since the early 1960s. In 1964 it was the largest employer in manufacturing, employing a labour force of 13 250. The sawmilling sub-sector maintained its leading status by employing 17 103 workers in 1975, 15 085 in 1981, 13 749 in 1983 and 13 120 in 1984. In terms of value added, it ranked second in 1964, sixth in 1975, fifth in 1981 and first in 1983 and 1984.

Tobacco processing has featured prominently in medium and large-scale manufacturing. This sector enjoyed the first position in terms of value added in all the selected years except 1975, when it ranked second to chocolate and other allied cocoa processing. The tobacco industry was also the second largest contributor to gross output for four of the selected years (1964, 1981, 1983 and 1984) and third in the other (1975). Another industry that has been contributing significantly to national output, in terms of value added, is petroleum products processing. It moved from the fourth position in 1964 to the third position in 1975 and 1981, and to the first position in 1983 and 1984.

The most capital intensive sector in 1981 was non-ferrous metal manufacturing, followed by cement, malt liquors and malt and grain mill products in that order.[3] In 1984 the most capital intensive sector was carpet and rug manufacturing (not shown in Table 6.3), followed by soaps and perfumes, dairy products and drugs and medicines, in that order. The wood and wood products group which has traditionally been the largest employer among the medium and large-scale manufacturing industries had a relatively low capital intensity in the mid-1970s. For example, in 1975 it ranked 17th in terms of capital intensity for the selected industries, but in 1981 it ranked sixth (not shown in Table 6.3), probably the result of installation of more machinery in the sector or the adoption of more capital intensive techniques of production. The ranking for the industry group in 1984 was eleventh.

For the industry groups shown in Table 6.3, the ranking of value added has been correlated with that of gross output, employment and capital intensity. The Spearman coefficient of rank correlation is very high (0.9218) and significant at 1 per cent level in the case of the ranking of value added and gross output. This result appears to suggest

Table 6.3 Output, employment and capital intensity in 20 major medium and large-scale manufacturing industries, 1984

ISIC	Industry	Value added ₡ mil.	Ranking	Value added per employee ₡'000	Ranking	Gross output ₡ mil.	Ranking	Employment No.	Ranking	Fixed asset/employee ₡'000	Ranking
3112	Dairy products	261.6	7	360.3	5 (5)	493.4	10 (10)	726	15 (17)	280.7	2 (3)
3115	Vegetable and animal oils/fats	102.5	16	47.8	19 (42)	387.7	12 (12)	2 147	5 (5)	68.2	10 (24)
3116	Grain mill products	258.7	8	476.4	4 (4)	629.6	9 (9)	543	20 (26)	103.2	8 (17)
3117	Bakery products	75.7	19	118.9	11 (15)	124.6	19 (23)	637	18 (21)	18.1	16 (46)
3119	Chocolate etc.	228.7	10	112.2	13 (18)	1155.7	6 (6)	2 039	6 (6)	41.6	11 (29)
3131	Spirits	105.2	15	87.8	14 (22)	253.7	14 (15)	1 197	10 (10)	23.4	15 (41)
3133	Malt liquors	2148.1	2	894.3	2 (2)	2752.0	3 (3)	2 402	4 (4)	226.0	3 (4)
3134	Soft drinks	65.7	20	79.8	15 (28)	111.8	20 (25)	823	13 (15)	11.9	17 (51)
3140	Tobacco	3149.7	1	2341.8	1 (1)	3840.0	2 (2)	1 345	7 (7)	150.8	5 (8)
3211	Textiles	557.3	4	71.3	17 (31)	1751.0	4 (4)	7 821	2 (2)	32.0	12 (32)
3215	Cordage and rope	98.9	17	73.7	16 (29)	144.7	18 (21)	1 342	8 (8)	6.0	19 (54)
3311	Sawmills, planing mills etc.	685.5	3	52.2	18 (39)	1502.0	5 (5)	13 120	1 (1)	103.6	7 (16)
3420	Printing and publishing	150.4	13	45.4	20 (43)	354.8	13 (13)	3 309	3 (3)	11.4	18 (52)
3522	Drugs and medicine	231.9	9	178.5	9 (12)	450.2	11 (11)	1 299	9 (9)	171.0	4 (5)
3523	Soaps and perfumes	261.9	6	236.5	7 (8)	680.6	8 (8)	1 107	11 (11)	476.7	1 (2)
3530	Petroleum	519.0	5	807.9	3 (3)	6475.1	1 (1)	641	17 (20)	3.4	20 (56)
3551	Tyres and tubes	159.4	12	197.5	8 (10)	239.2	15 (16)	807	14 (16)	27.6	14 (35)
3692	Cement	221.2	11	347.3	6 (6)	802.2	7 (7)	637	18 (21)	112.7	6 (14)
3811	Cutlery and handtools	105.8	14	124.8	10 (14)	226.6	16 (17)	848	12 (14)	28.9	13 (34)
3819	Fabricated metal products except machinery and equipment	82.5	18	114.6	12 (17)	152.8	17 (20)	720	16 (19)	82.0	9 (20)

Note: The rankings shown in brackets and value added are based on the 59 industry groups given in the source.
Source: Statistical Service, Industrial Statistics, 1982-84.

that the industry groups with the highest values of gross output are those employing more of capital and labour and less of inputs that do not arise from the current factor services. However, the low rank correlations between value added and employment (0.312) indicate that the factor-use implication of increases in value added mainly relates to factors other than labour – mainly, in this case, capital. The rank correlation between value added and capital intensity (0.5549) suggests that the factor contributing more to increases in value added is capital. This evidence is duly confirmed by the results of the following regression analyses.

$$\log V = -0.4735 \log L + 0.54915 \log K \quad (1)$$
$$(0.07945) \quad (0.11773)$$

$$R^2 = 0.9682$$
$$DW = 1.83516$$
$$F = 289.83$$

where

V = value added,
L = labour input, and
K = services of capital stock.

The regression shows the existence of an inverse relationship between labour and value added in the production function, so that a 1 per cent reduction in labour input in an apparently overmanned industrial sector will be expected to give rise to an increase in value added by one half of 1 per cent. On the other hand, it can be read from the production function that an increase in capital stock through a flow of investment would give rise to an increase in value added, the elasticity of output with respect to capital being 0.5492. The significance of the contribution of capital to value added is further shown when labour productivity (V/L) is regressed on capital intensity (K/L).

$$\log (V/L) = 1.22135 \log (K/L) \quad (2)$$
$$(0.09056)$$

$$R^2 = 0.9054$$
$$DW = 2.1292$$
$$F = 181.89$$

The evidence appears to suggest that, *ceteris paribus*, a certain proportionate increase in capital intensity almost invariably translates into more than a proportionate increase in labour productivity, hence reinforcing the case for more investment in the Ghanaian manufacturing industry. The argument for increased investment in manufacturing can be sustained on grounds that the expansion of the sector accounting for a growing share of GDP would provide greater employment opportunities than hitherto, provided that investment decisions showed concern for the adoption of appropriate technologies.

It is of interest to single out the engineering sector of the economy for further, more detailed discussion. At the beginning of the 1960s the industry was mainly carrying out repair work and the manufacture of simple metal products. Over the last two decades the industry has become capable of manufacturing simple machinery and equipment, although because of acute shortages of basic inputs, production has suffered severely in recent years. Thus, in the context of the four-stage development of the engineering industry in LDCs described by Huq and Prendergast, Ghana may be considered to have passed the first stage of repair work and the manufacture of simple metal products and has reached the second stage of producing simple machinery and equipment.[4] (At the third stage, the production of industrial machinery and equipment expands and diversifies and the manufacture of machine tools forms an important part of the industry; while at the fourth stage, the engineering industry will be fully developed.)

Information collected for 1981 showed that there were 29 medium and large-scale metal engineering industries located in Accra, Tema and Kumasi. Of these, four were under state ownership, 13 were private Ghanaian including five owned by Ghanaians of foreign origin, and the rest were joint ventures with foreign and domestic partners (private/state).[5]

Two companies established in recent years are the GIHOC Foundry in Tema and Western Castings in Takoradi. The latter commenced production in 1984. The Railway Workshop at Sekondi has a well-equipped foundry, but its services are mainly used to maintain Ghana Railway. The level of capacity utilisation in the medium and large-scale engineering industries was on the low side in 1981, averaging 25 to 30 per cent. Their range of products included iron sheets, mild steel rods, aluminium sheets, hand tools, simple agricultural implements such as disc ploughs and harrows, simple machines, block making machines and cassava graters. Fuel tanker and tipper and vehicle bodies are also made.

The metal engineering industry also comprises hundreds of small engineering workshops scattered all over the country, the major locations being Suame Magazine (Kumasi), Kokompey (Accra), another Kokompey (Takoradi) and Siwudu (Cape Coast) (that is, in the cities and large towns). These workshops mainly perform servicing and repairing jobs for motor vehicles. Employment in the small-scale engineering sector is considered to be very high. Suame Magazine has the highest concentration of small engineering workshops (estimated at 8000) and is thought to employ 40 000 to 50 000 people.[6] The engineering industry is heavily dependent on imported raw materials; all the basic inputs including steel plates, steel bars and electric motors are imported. Local raw material components are confined to simple items such as wood and plywood, paint, scrap iron, electrodes, gas, welding rods, aluminium ingots, rubber and foam.

The number of units in the manufacturing sector employing 30 to 99 workers increased to a peak of 305 units in 1977. However, the number had fallen to 231 in 1981. For those enterprises employing more than 99 workers, the peak in numbers was also reached in 1977 with 177 enterprises, falling to 156 units by 1981.[7] The main reason for the declining numbers of medium and large-scale manufacturing enterprises is the acute shortage of imported raw materials and other imported inputs due to lack of foreign exchange.

Table 6.4 End-use of manufacturing output, 1968

	₵ million (current prices)	Percentage distribution
A. Total intermediate products	236.2	33.15
(a) Input into manufacturing	104.7	14.69
(b) Input into other areas	131.5	18.45
B. Total other uses (final demand)	476.4	66.85
(a) Current expenditure	274.4	38.51
(i) Consumers	(238.5)	(33.47)
(ii) Government	(35.9)	(5.04)
(b) Gross capital formation	96.3	13.51
(i) Fixed	(67.1)	(9.42)
(ii) Stocks	(29.2)	(4.10)
(c) Exports	105.7	14.83
Total of A + B	712.6	100.00

Source: CBS, Input-Output Table of Ghana 1968.

In the absence of more recent data we present information on the end-use of manufacturing output in 1968, provided by the CBS (Table 6.4). It shows that in that year the manufacturing sector absorbed only 14.7 per cent of its own output, while as much as 33.5 per cent went to consumption, 13.5 per cent to gross capital formation and 14.8 per cent to exports. Thus, inter-industry linkages were not very well developed by 1968. There is no reason to suppose that the picture has changed much since that period.

6.2 INDUSTRIALISATION STRATEGY

In Ghana industrialisation has for a long time been an official policy of the government. As far back as 1947, a statutory body – the Industrial Development Corporation (IDC) – was established to 'foster industrial growth'.[8] Among its functions was 'securing the investigation, formulation and carrying out of projects for developing industries in the Gold Coast'.[9] In 1952 the government requested Arthur Lewis, a West Indian economist, to inquire into the problems and prospects of industrialisation in Ghana. His report, which was published in 1953, had a great influence on subsequent government industrialisation policy.[10]

Following the attainment of independence in 1957 the Ghana Government's interest in industrialisation increased tremendously. In the *Second Development Plan* 1959–64 (later suspended) the government declared that 'specially high priority will be given within the next five years to promoting the establishment of not less than 600 factories of varying sizes producing a range of over 100 different products'.[11]

The *Seven Year Development Plan* (1963/64–1969/70) was even more emphatic on industrialisation. It envisaged an increase of 83 per cent in industrial output by 1970. Import substitution was declared a policy objective within the Plan. Secondly, industries were to be established to process the agricultural and mining products that were being exported in unprocessed form. Thirdly, the Plan called for the establishment of such basic industries as chemicals and metals. Finally, the Plan envisaged the commencement, on a small scale, of electronics and machine tool industries.

With Nkrumah's socialist objectives, the government favoured the establishment of directly productive enterprises in the public sector. The government undertook some manufacturing ventures on its own, participated with the private sector in others, or encouraged the inflow of foreign private capital by granting tax holidays and other liberal concessions, and also shifted the import licensing programme in favour of domestic manufacturing firms.

Beginning from 1964, the Ministry of Industries routinely handed over to the State Enterprises Secretariat completed government-owned manufacturing plants. This agency, which succeeded the IDC, controlled all state-owned factories and all ventures jointly owned by the government and the private sector.

Following the 1966 military coup, the main interest of the new government (the National Liberation Council) was the stabilisation of the economy. It sold off to the private sector some of the state enterprises, and reorganised others under the Ghana Industrial Holding Corporation (GIHOC) which was established in 1968. It did not undertake any new investment in manufacturing. Instead, the utilisation of manufacturing capacity was improved by providing more import licences for raw materials and spare parts.

In 1961 the country had its first serious balance of payments deficit. In subsequent years the problems encountered by the nascent manufacturing industries in acquiring raw materials and spare parts assumed alarming proportions. The problems posed by the balance of payments constraint had led to less emphasis being put on industrialisation as a development strategy by the government.

We will now consider Ghana's industrialisation strategy under the following headings: (a) import substitution; (b) incentives; (c) protection; (d) export expansion; and (e) location.

(a) Import substitution

Over the years the country's policy of industrialisation has been import substitution. In his book *Africa Must Unite*, first published in 1963, Nkrumah declared: 'Every time we import goods that we could manufacture... we are continuing our economic dependence and delaying our industrial growth'.[12]

In the *Seven Year Development Plan* (1963/64–1969/70) too, as mentioned earlier, import substitution was a declared policy objective, with the government hoping to produce domestically 'the staples of consumer demand' that the country imported in large quantities.[13]

Imported items with high consumer demand were easily identifiable. At the time of the launching of the *Seven Year Development Plan* such imported items included processed food, drinks and tobacco; textiles, clothing and footwear; petroleum and chemicals; cement and other building materials; soap and detergents; face powder and other cosmetics; and leather and plastic bags. Though import substitution of many of these products had already started, production was on a minor scale and the process was in much need of a big push from the government. The big push was provided by the Nkrumah Government in the first half of the 1960s.

There are several ways of measuring the degree of import substitution. One common measure is the percentage of total supplies of manufactured products, both domestically produced and imported, which is supplied by domestic production. But an obvious problem with such a measure is that the import content of some domestically produced manufactured goods may be extremely high. For example, in Ghana, television sets, transistor radios, refrigerators and deep freezers are domestically produced. But the process simply involves the assembling of imported components.

Despite the statistical (and conceptual) difficulties of measuring import substitution, a noticeable trend in the 1960s was that more and more manufactured products with the label 'Made in Ghana' kept appearing on the Ghanaian market. Value added in manufacturing was also increasing rapidly (see Table 6.1).

But this rising trend of domestically manufactured products contained its Achilles heel. As will be shown in Section 6.5 below, the high degree of dependence on imported inputs is thought to be an important factor in the low capacity utilisation of recent years.

(b) Incentives

To accelerate the industrialisation process, the Nkrumah Government in the 1960s introduced a package of incentives for the manufacturing sector. For instance, in 1963 the Capital Investments Act was passed. That Act offered a wide range of fiscal and other concessions to potential investors including tax holidays of up to five years, accelerated depreciation rates for buildings, plant and equipment; exemption from custom duties on machinery, raw materials, spare parts and fuel; deferment of company registration fees and stamp duty for up to five years on capital invested in approved projects; guaranteed remittance of capital and profits and employment tax credit for a period of up to ten years.[14]

The Capital Investments Board that was established to carry out the provisions of the Act also organised activities such as conferences and seminars for the stimulation of investments, and obtained exemptions and other facilities necessary for the establishment or expansion of approved projects.[15] The Board also established an Investment Centre in Frankfurt, West Germany, from where direct contact could be initiated between potential foreign investors and Ghana.

Judged by the substantial inflow of foreign and domestic capital into the manufacturing sector in the 1960s, these incentives had some impact in expanding the manufacturing sector. However, the Capital Investments Decree of 1973, which replaced the Capital Investments Act of 1963, did not have much impact because of the serious price distortions which became acute, particularly after the mid-1970s. It was in turn superseded by the Investments Code Act of 1981.

In 1963 the National Investment Bank Act was passed establishing the National Investment Bank (NIB) as a joint state-private institution to promote, finance and assist enterprises in all sectors of the Ghanaian economy. The NIB has been assisting manufacturing industries over the years by providing assistance in the establishment, expansion and modernisation of such enterprises; encouraging and facilitating the participation of domestic and foreign capital in such enterprises; providing advice on sound business management; identifying investment opportunities and bringing together competent management. The NIB also provides long-term credit to manufacturing enterprises.

In spite of the array of incentives for industry the profits to be obtained from importation rather than domestic production became ever greater as the over-valuation of the cedi and domestic inflation worsened during the late 1970s and early 1980s. With the 1983 Budget, however, the government took steps to rectify the situation by granting export bonuses to exporters and imposing heavy surcharges on imports. This move was followed in October of the same year by a devaluation of the cedi by about 1000 per cent in an attempt to discourage imports and encourage exports. Another round of devaluation took place in March 1984, when the cedi was devalued from ₵30 to ₵35 per US dollar. These moves, by raising import costs and export earnings, are directed towards establishing a realistic value of exports and imports. However, judged by the foreign exchange rates prevailing in the parallel market, estimated at four times the official rate in mid-1984, there was still a significant over-valuation of the cedi at the official exchange rate.

Another disincentive to local manufacturers was the insistence of the government on controlling prices in the face of rising production costs and raging inflation (see Chapter 11). Indeed, in some cases, the government controlled prices, which only took into consideration official prices of inputs and not the higher open market prices at which manufacturers sometimes purchased their inputs, were lower than the average cost.

(c) Protection

In pursuit of an import substitution strategy the Ghana Government has instituted over the years an elaborate system of protection for local industry. This involves the imposition of tariffs, quotas and out-right bans to restrict the inflow of manufactured products from abroad. The import licensing system, established in December 1961, became an important determinant of investment and production in manufacturing.[16]

Tariffs as applied before 1961 were on selected 'luxury' consumer goods such as alcoholic beverages, tobacco, jewellery, textiles and apparel. *Ad valorem* rates of 25 per cent or less were generally imposed. From 1956 to 1961 a number of new industries (nails, furniture, roofing sheets) were taken up for protection and the tariff rate on textiles and apparel was raised from 20 to 25 per cent. In 1961 and 1962 protection was extended to more goods and tariff rates were raised to 66.67 per cent on textiles and apparel and 100 per cent on cosmetics. The resultant tariff structure aimed to reduce non-essential imports. In 1966 a pyramidal pattern emerged 'with low duties on capital goods (2 per cent) and on inputs into production of capital and intermediate goods (9 per cent), and progressively higher rates on inputs into consumer goods industries (16 per cent), consumer durables (25 per cent), ordinary consumer non-durables (55 per cent) and luxury items (128 per cent)'.[17]

Leith has made a detailed study of effective protection rates for various products in Ghanaian manufacturing in 1968 and 1970.[18] His findings showed that while some industries enjoyed rather high effective rates of protection in 1968 and 1970, others did not. For instance, the export commodities such as kente weaving and fruit squash had negative or near zero rates of protection. The highest rates of protection were enjoyed by records, radio and television assembly, apparel, shoes and cosmetics. No wonder then that these industries flourished in the country. For example, by 1970 Ghana was almost self-sufficient in textiles. By the early 1980s, however, as shown in Section 6.5 below, the manufacturing sector was operating far below capacity. While value added, at 1970 prices, in the large and medium-scale manufacturing sector was about ₡156 million in 1970, rising to ₡168 million in 1980, by 1981 it had fallen to only ₡133 million (see Table 6.1). The over-valuation of the cedi also created a situation in which imports became significantly cheaper than locally produced items. According to an official estimate, the landed import price of a bag of sugar in 1982 was only 6.1 per cent of the wholesale farm-gate price of locally produced sugar, while for palm oil it was 8 per cent.[19]

(d) **Export expansion**

The industrialisation drive also involved the objective of increasing export earnings. For example, some amount of cocoa beans are now processed into cocoa butter and cocoa paste before export. The amount so processed varies inversely with the world market price for cocoa beans – the higher the price of raw beans the lower the proportion processed prior to export. Again the export in log form of some species of timber has now been banned. Instead they must first be processed into sawn timber and veneer. Other manufactured items with high export potential, especially to the neighbouring ECOWAS countries, include wood furniture components and other wood-based products, aluminium products, footwear, plastic goods, textile products, tyres and some preserved and canned food items. Indeed, some of these are currently being exported either legally or illegally to neighbouring countries. However, the over-valuation of the cedi at the official exchange rate in the 1970s and early 1980s was such that it discouraged the export of manufactures through official channels, and smuggling was resorted to on a large scale for products such as cocoa (see Chapter 5).

(e) **Location**

Whether planned consciously or unconsciously, the industrialisation strategy has resulted in the concentration of manufacturing industries at the tips of the Golden Triangle – Accra-Tema (Greater Accra Region), Kumasi (Ashanti Region) and Sekondi-Takoradi (Western Region). The result is that the three regions in which these metropolitan centres are located dominate manufacturing, with over 90 per cent of total value-added in manufacturing (see Table 6.5). These localities possess large, sophisticated urban markets and enjoy comparatively

Table 6.5 Percentage distribution of value added in medium and large-scale manufacturing by regions, 1970–84 (selected years) (percentages: current prices)

Region	1970	1975	1980	1982	1984
Western	17.9	18.9	34.0	23.4	31.4
Greater Accra	63.6	60.5	50.2	60.2	48.8
Ashanti	10.2	11.9	10.8	9.9	13.5
Other regions	8.2	8.6	5.0	9.9	6.3
Total	100	100	100	100	100

Sources: CBS/Statistical Service, *Industrial Statistics* 1970–72, 1975–77, 1979–81 and 1982–84.

efficient transportation and communication systems. They are connected to the national electricity grid and have fairly dependable pipe-borne water supplies and other social overheads. For those manufacturing industries that depend on imported raw materials, Accra-Tema and Sekondi-Takoradi with nearby port facilities are the logical places to locate manufacturing plants (other things being equal).

6.3 SIZE OF FIRMS

Ghana's manufacturing industry comprises firms ranging from small-scale through medium-sized to large-scale enterprises. According to the *Five-Year Development Plan* (1975/76–1979/80), the share of medium and small-scale enterprises in total industrial production ranged between 22 and 24 per cent during 1968–74. The Central Bureau of Statistics (CBS), on the other hand, lumps medium and large-scale enterprises together, making a direct comparison on the basis of scale rather difficult when using official data.

The size of a manufacturing firm may be measured by the number of workers employed, by capital stock, production or value added. The measure which is most often used is the number of workers employed. Using the employment criterion, the CBS defines small-scale firms as those employing 1 to 29 workers and medium and large-scales as those employing 30 or more workers.

Besides the medium and large-scale manufacturing firms, there are many small-scale firms for which data are not readily available. Those engaged in this sector are not generally covered in the *Industrial Statistics* published by the CBS. In the small-scale sector, there are enterprises which do not have wage employment (that is, paid labour). These along with the establishments employing 5 workers or less can probably be considered as belonging to the *informal* sector.[20]

Table 6.6 shows some relevant features of the small-scale sector in comparison with the medium and large-scale sector. As may be seen from the table, output and value-added per employee is significantly higher in the establishments with more than 99 employees than those with 99 or less employees. However, in the medium-scale sector, firms employing 10 to 29 persons show higher efficiency in terms of output and value-added per employee compared to those employing 30 to 99 persons. As against the high capital intensity in the large-scale sector (with over 99 employees), the capital intensity of establishments employing 10 to 29 persons is not significantly different from those employing 30 to 99 persons. One thus finds a case for encouraging the firms employing 10 to 29 persons as against those employing 30 to 99 persons. In the small-scale sector, the establishments without any wage employment show higher output and value-added per

Table 6.6 Employment, output, value added and capital per employee in small, medium and large-scale sectors, 1973

No. of employees	Small scale Without hired labour	1–9	Medium and large scale 10–29	30–99	>99
Average no. of total workers	2.6	6.7	21.0	54.1	436.0
Average full-time equivalent workers	1.3	4.9	19.1	54.0	435.9
Gross output (000 Cedi)	4.4	11.8	118.2	323.6	6 326.2
O/L (000 Cedi)	3.4	2.4	6.2	6.0	14.5
Value added (000 Cedi)	2.0	6.9	52.7	103.0	2 922.6
VA/L (000 Cedi)	1.5	1.4	2.8	1.9	6.7
Capital per employee* (in Cedi)	572	1388	6609	6595	14 504

O = Output. L = Labour. VA = Value Added.
* Original cost of fixed assets, adjusted to 1973 prices, is used as the measure of capital.
Source: Steel (1977), p. 90.

employee and also significantly lower capital intensity compared to the establishments with 1 to 9 employees.

Major research into small-scale industries was undertaken by the Centre for Development Studies (CDS) at University of Cape Coast in 1981, when extensive data were collected for three Regions of the country (Ashanti, Central and Volta).[21] These data had not been analysed at the time of writing this book. However, in an attempt to gain some insights into small-scale industries based on the 1981 survey, we took random samples of 20 for each of the regions surveyed. Our investigation enabled us to find out a number of important characteristics of small-scale industries in Ghana (see Table 6.7). There are significant similarities with Steel's 1973 study.[22] The important findings of the 1981 CDS survey are listed below.

1. Capital intensity per employee was very low, estimated at ₵278–335 on average. However, considering that the firms included were established at different dates over a long period an attempt was made to assess the capital intensity for those firms established in one year (1971–72), so as to overcome the vintage problem. It was found that the figure was very low for the Ashanti and the Central Regions, ₵375 and ₵769 respectively, compared to ₵1300 in the Volta Region. The figure for the Volta Region, based on only one observation, was still low compared to the ₵6595 to ₵14 504 range found for medium and large-scale firms by Steel (Table 6.6).

2. Establishments employed on average 4 to 5 persons, though it is likely that because of the requirements to pay 12.5 per cent Social Security contributions many establishments employing more than 5 workers probably under-reported the labour force. However, it can be safely concluded that the average employment would be on the low side.

3. Products (not shown in Table 6.7) comprised mostly consumer goods, but there were firms which also produced intermediate and capital goods. The common consumer products included palm oil, gari, soap, comb, kenkey, mat, pitto, furniture and akpeteshie. Intermediate and capital goods included trowels and chisels, metal trunks, beer crates, canoes and grating and crushing equipment (for gari and sugar-cane).

4. The firms depended heavily on the agricultural sector for raw materials. As much as 70 per cent of those surveyed in the Ashanti Region relied on agricultural inputs. The respective figures for the

Table 6.7 Selected characteristics of small-scale industries, 1981
(based on sample observations)

Characteristics	Ashanti Region	Central Region	Volta Region
Average labour per establishment	5	5	4
Average capital per employee (₡)	335	278	298
Raw materials (%)			
(i) *Type*	100	100	100
Agro-based	70	65	60
Non Agro-based	30	35	40
(ii) *Source*	100	100	100
Local	45*	85	50*
Urban	55†	15	50‡
Technology			
(i) *Type*	Simple	Simple	Simple
(ii) *Source*	Local (100%)	Local (85%) Urban (15%)	Local (100%)
Energy			
(Source by %)	100	100	100
Electricity	20	10	Nil
Manpower	80	90	100
Market for goods			
(% Distribution)	100	100	100
Local	70	80	65
Urban	30	20	35
Investment finance			
(Source by %)			
Own savings	100	70	100
Relatives	Nil	15	Nil
Banks	Nil	15	Nil

* Includes goods originating abroad and from local source (about 5% of the total in each case).
† Includes goods from local source (5%).
‡ Includes goods originating abroad (10%).
Source: Unpublished data from the 'Small Scale Industry Survey' conducted in 1981 by the Centre for Development Studies, University of Cape Coast.

Central and Volta Regions were 65 and 60 per cent. In the Ashanti Region, 45 per cent of the raw materials was obtained from the area in which the enterprises were located, while it was as much as 85 per cent in the Central Region and 50 per cent in the Volta Region. The greater dependence of small-scale manufacturing firms in the Ashanti Region on raw materials from the urban centres may be explained by the central location of Kumasi in that region, which enables the firms to acquire much more easily their raw material inputs from that city.

5. The firms depended mostly on domestic technology. As much as 100 per cent of the equipment required was obtained from the local areas in both the Ashanti and Volta Regions, while it was 85 per cent in the Central Region which obtained the remaining 15 per cent from urban sources. The source of energy was mainly manpower, ranging from 80 per cent in the Ashanti Region to 100 per cent in the Volta Region. In other words, their use of electricity ranged between zero (in the Volta Region) and 20 per cent (Ashanti Region).

6. Seventy per cent of the finance of the firms in the Central Region and 100 per cent in the Volta and Ashanti Regions came from owners' savings. Direct finance from relatives comprised 15 per cent of the total investment in the Central Region. Thus the development banks were not important sources of funds for the small-scale industries surveyed in the three regions.

7. Forty per cent or more of the firms had been established between 1976 and 1981, with as many as 55 per cent in the Volta Region, 85 per cent in the Central Region and 75 per cent in the Ashanti Region being established between 1971 and 1981 (not shown in Table 6.7).

8. The firms produced mainly for the local market, with 65 per cent of their products destined for the locality in the Volta Region, 70 per cent in the Ashanti Region and 80 per cent in the Central Region.

6.4 OWNERSHIP

The CPP Government which was in power from 1952 to 1966 became, in the early 1960s, more interested in the state assuming a dominant role in industrialisation by directly owning industrial enterprises.

Outside the state sector the CPP, led by Nkrumah, did not want to create indigenous capitalists who would be difficult to control politically.[23] It instead tended to encourage the development of foreign entrepreneurs through the granting of tax concessions. To further this end, the Capital Investments Board was established in 1963. By that time the expatriate group in the country already possessed considerable wealth which it had accumulated from commerce, mining and the then nascent manufacturing ventures. This group was also better able to attract foreign capital. Furthermore, it was politically impotent or neo-conformist.[24]

The Nkrumah Government tended to favour, in descending order of preference, the following categories of ownership.

1. State-owned manufacturing and other enterprises;
2. joint state-foreign and wholly-owned technically sophisticated, capital intensive enterprises which also possessed expert management;
3. intermediate firms which were more labour intensive such as light industries – for example, garment, furniture, plastic wares, electronic equipment (assembled), bottled perfumes and pharmaceutical products. These were dominated by Levantines, Indians and Ghanaians; and
4. small-scale (with less than 10 employees) enterprises such as food processing, tailoring and carpentry. These were to be reserved for Ghanaians.

Thus, in the government's view the Ghanaian as a private capitalist was to play a minor role in the industrialisation process. Nkrumah was worried that they might develop into a powerful capitalist group that could threaten his socialist régime. And from 1960 when the country became a republic, Nkrumah tended to draw closer and closer to the socialist camp both in word and deed.

Nkrumah and the CPP viewed state participation in manufacturing as 'a historical necessity, because there was no private indigenous accumulation of capital and therefore only the state was able to industrialize the country'.[25]

The NLC Government which took over in 1966 established, in 1968, the Ghana Industrial Holding Corporation (GIHOC) as a holding corporation to embrace originally 20 statutory corporations which were renamed divisions. At the time of GIHOC's establishment the accumulated losses of the 20 divisions amounted to approximately ₵15 million.[26] The decree establishing GIHOC charged it to embark upon the 'establishment and operation of manufacturing commercial enterprises... in an efficient and profitable manner'.[27] The Corporation quickly reorganised the divisions and in 1970 the government received its first dividend from GIHOC.

Over the years five divisions have left the umbrella of GIHOC. These include the Sheet Metal Division which was sold in 1969; the Cocoa Products Division which was transferred to the Ghana Cocoa Marketing Board in 1969; and the Sugar Products Division which left to constitute the Ghana Sugar Estates Limited in 1975. The Pharmaceutical Division joined the Corporation in 1968. In addition to this division, other important divisions currently include boatyards, brick and tile, cannery, distilleries, electronics, fibre bag, paper conversion, glass manufacturing, vegetable oil, paints, marble works, metal, steel works and meat products. GIHOC has also entered into joint ventures with private and other public enterprises. These joint ventures are in textile manufacturing, the assembly of bicycles, mopeds and outboard motors, glue, bottling of oxygen and gas and the manufacture of soft drinks.

Apart from its sole ownership in GIHOC and through GIHOC in other ventures, the Ghana Government is also a shareholder in partnership with foreign investors in certain manufacturing firms in the country. These firms include Lever Brothers (soaps and detergents); Ghana Cement Works (cement); Juapong Textiles (grey cloth); Ghana Sanyo Electrical Manufacturing Company (television and wireless sets, electric fans, fridges and cookers); Ghana Pioneer Aluminium Factory (aluminium cooking pots, pans, cutlery); New Match Factory (safety matches); Crystal Oil Mills (refined cooking oils, groundnut and copra oil); Dorman Long (petrol tanks); Tema Textiles Limited and Ghana Textiles Printing (printed textile cloths); Kumasi Brewery (beer and malt extracts); Ghana Aluminium Products Limited (aluminium roofing sheets, silos, boats); and Willowbrook Limited and Neoplan Limited (vehicle assembly).[28]

In terms of nationality of ownership, measured in terms of value-added in medium and large-scale manufacturing, the share of non-Ghanaian ownership declined from as high as 67 per cent in 1962 to less than 1 per cent in 1984 (see Table 6.8). The share of Ghanaian ownership was 26.8 per cent in 1962, declined to 18.4 per cent in 1970 and by 1984 it stood at 37.8 per cent. Mixed ownership has been rising consistently from 6.2 per cent in 1962 to 61.4 per cent in 1984.

Table 6.8 Percentage distribution of value added in medium and large-scale manufacturing by nationality and type of ownership, 1962-84 (selected years)

	1962	1966	1970	1975	1980	1982	1984
By nationality of ownership (%)	100	100	100	100	100	100	100
Ghanaian	26.8	26.9	18.4	26.8	43.9	43.1	37.8
Non Ghanaian	67.0	59.2	60.2	40.3	12.0	6.9	0.8
Mixed*	6.2	13.3	21.5	33.0	44.1	50.0	61.4
By type of ownership (%)	100	100	100	100	100	100	100
State	22.9	22.1	12.2	19.5	25.6	23.4	23.5
Joint State-Private	4.2	8.1	12.5	17.8	19.8	18.2	25.5
Co-operative	0.1	0.1	—	—	0.5	—	—
Private	72.9	69.7	75.4	62.7	54.1	58.4	51.0

Note: Due to rounding, the total may not always add up to 100.
* With Ghanaian father or mother.
Sources: For 1962 and 1966 see Leith (1974), p. 54. For other years see CBS/Statistical Service, *Industrial Statistics* 1970-72, 1975-77, 1979-81 and 1982-84.

In terms of type of ownership, again measured by value added, the state's share was 22.9 per cent in 1962, declined to 12.2 per cent in 1970 but increased to 23.5 per cent in 1984. The joint state-private type of ownership which was only 4.2 per cent in 1962, had risen to 25.5 per cent by 1984. Private ownership was 72.9 per cent in 1962, rose to 75.4 per cent by 1970 but had fallen to 51 per cent by 1984. Currently the state's interest in sole ownership of industrial enterprises seems to be waning. Between 1978 and 1984 only two public sector industrial firms were approved out of 330 total approvals.[29]

6.5 CAPACITY UTILISATION

There are serious conceptual problems in estimating capacity utilisation in manufacturing.[30] *Installed* capacity of machinery and equipment is normally stated on an hourly basis, but in the manufacturing sector the *stated* capacity is measured conventionally per day, per week or per year. However, such a measurement may involve production based on *one, two* or *three* shifts per day. It is rarely that manufacturing enterprises in developing countries operate three shifts even at full capacity production. Moreover, an exact measurement of installed capacity at plant level can create problems if one wants to examine installed capacities of individual machinery and equipment. *Stated* installed capacities at times can differ from effective *actual* installed capacities.

In the absence of in-depth studies on capacity utilisation in manufacturing in Ghana it is difficult to consider this issue realistically. However, the Central Bureau of Statistics has published figures of capacity utilisation in manufacturing. For the large and medium-scale manufacturing enterprises, in 1978 the average capacity utilisation was found to be as low as 40.4 per cent. It declined further to 33.1 per cent in 1979, still further to 25.5 per cent in 1980 and down again to 21.0 per cent in 1982.[31]

The almost insurmountable difficulties in acquiring imported inputs led to a situation where practically no new medium and large-scale industries were being established. Industries which use local raw materials such as palm fruits, tomatoes, sugar cane and oranges found it difficult to obtain them at the ridiculously low official prices. The low official prices which made smuggling a lucrative venture further restricted the availability of local raw materials. If manufacturers managed to obtain inputs at the high parallel market prices they ran the risk of making losses since the Prices and Incomes Board used official input prices in fixing output prices. Thus the insistence of the government on applying official prices for output also affected capacity utilisation adversely.

Thus it is clear that the general decline in the Ghanaian economy has hit the manufacturing sector too. A number of manufacturing firms have had to close down owing to an inadequate supply of foreign exchange to import raw materials and other inputs. For example, the number of establishments in the medium and large-scale industries declined from 482 in 1977 to 399 in 1979 and further to 387 in 1981.[32] For the same reason, those still in operation have been producing at very low capacity levels. As may be seen from Table 6.9, capacity utilisation in the textile factories plunged from 40 per cent in 1978 to

Table 6.9 Manufacturing industries: estimated rate of capacity utilisation, 1978–82 (large and medium-scale factories)
(percentages)

ISIC	Sub-sector	1978	1980	1982
3211	Textiles	40.0	20.1	10.0
3212	Garments	38.1	29.9	20.2
381	Metals	28.2	28.4	42.5
3833	Electricals	32.1	17.8	31.5
3560	Plastics	10.6	19.1	20.0
3843	Vehicle assembly	18.4	N.A.	15.0
313 & 314	Tobacco and beverages	50.0 }	30.0	N.A.
311	Food processing	40.8 }		N.A.
3231	Leather	31.3	20.9	18.0
3522	Pharmaceuticals	25.0	16.8	20.0
3525	Cosmetics	33.4	8.0	15.0
3519 & 3520	Paper and printing	31.0	28.4	25.0
3699	Non-metallic mineral manufacturing	47.0	29.7	15.0
35	Chemical	42.0	28.0	15.0
355	Rubber	21.6	16.4	27.0
3311	Wood processing	36.0	27.3	20.0
390	Miscellaneous	55.9	44.9	N.A.
3	All manufacturing industries	40.4	25.5	21.0

Note: Data for individual industries are obtained from Ministry of Industries; the estimate of capacity utilisation for all manufacturing industries is a weighted arithmetic average, using weights proportional to value of gross output in 1975.
Source: CBS, *Economic Survey*, 1982, p. 69.

Table 6.10 Percentage distribution of sources of inputs for medium and large-scale manufacturing, 1970–84 (selected years) (percentages: current prices)

Sector	1970 Domestic	1970 Foreign	1975 Domestic	1975 Foreign	1980 Domestic	1980 Foreign	1984 Domestic	1984 Foreign
Food, beverage and tobacco	71	29	50	50	64	36	57	43
Textiles and footwear	27	73	55	65	35	65	26	74
Wood and paper products	48	52	59	41	62	38	51	49
Chemical, oil and rubber products	12	88	8	92	4	96	6	94
Building materials	8	92	22	78	21	79	18	82
Metal, machinery and transport	5	95	8	92	6	94	27	73
Others	—	100	24	76	18	82	11	89

Sources: CBS/Statistical Service, *Industrial Statistics* 1970–72, 1975–77, 1979–81 and 1982–84.

10 per cent in 1982. Other important sub-sectors such as wood processing, pharmaceuticals, cosmetics and chemicals have had similar unsatisfactory experiences. It is only in plastics, metals and rubber that capacity utilisation has increased and it has remained unchanged in paper and printing.

As Table 6.10 shows, Ghanaian manufacturing depends heavily on imported inputs. In 1984 for instance, 73 per cent of the inputs of the metal, machinery and transport sub-sector was imported. For chemicals, oil and rubber products the figure was 94 per cent; and for building materials the figure was 82 per cent. It was only in food, beverages and tobacco, and wood and paper products that imported inputs fell below 50 per cent of total inputs in 1984.

Given the high import content of locally manufactured goods, the low rate of value-added to gross output in a sector indicates that the foreign exchange cost of the sector relative to its output is quite substantial. The country has also failed to develop strong links between the manufacturing and other sectors. For instance, the manufacturing sector has failed to provide the agricultural sector with adequate supplies of inputs such as cutlasses, insecticides and other chemicals. It has also failed to provide a ready market for the output of the agricultural sector. Thus, Ghanaian manufacturing is characterised by low capacity utilisation, low value-added and heavy dependence on imported raw materials.

The reported performance of the small-scale industries in 1981 was poorer than that of medium and large-scale establishments.[33] Their capacity utilisation varied from 10 per cent in electronics to 36 per cent in paper and printing; the overall reported capacity utilisation was 20 per cent. Unlike the medium and large-scale enterprises, small-scale manufacturing firms hardly ever obtain import licences to enable them to acquire cheaper imported inputs. However, with much lower overhead costs and higher ability to adapt to changing economic fortunes, small-scale industries are still being established to take advantage of the acute shortage of manufactured goods and the consequent high prices prevailing on the Ghanaian market.

6.6 TECHNOLOGY

The question of the choice of technology, mainly in the context of large-scale manufacturing, is discussed in some detail in Chapter 14.

Suffice it to mention here that no technology policy exists (in explicit terms) in Ghana. Although industrialisation has been pursued with vigour during the post-independence period, no government has attached much importance to the question of the type of technology transfer helpful for efficient industrial development.

The failure to evaluate technologies properly, by *types* and *sources*, has remained a major weakness in the investment decision-making process.[34] Lack of adequate information on alternative technologies and shortage of adequate personnel to carry out evaluation are some of the important reasons for the failure in this area.

In the medium and large-scale manufacturing sectors, sophisticated technologies were, in general, adopted. A large percentage of these enterprises are highly capital-intensive in nature. There are many reasons for the adoption of capital-intensive techniques, such as liberal concessions/incentives given by the government to the manufacturing firms encouraging the adoption of capital-intensive technology; overvaluation of the cedi making it less expensive to import machinery/equipment from abroad; presence of multi-national firms which bring capital-intensive techniques from the advanced countries; and the high cost of labour relative to capital.

The last point should, however, be viewed not in the context of the minimum government wage, but from the point of view of the prevailing market wage as against the prevailing interest rate and foreign exchange rate. If shadow prices for capital and labour were properly used no serious problem would have arisen, but such prices were not used, or only partially used, in project appraisal.[35] For example, the ADB (Agricultural Development Bank) uses the high market wage rate as it finds it difficult to get labour at the low official wage rate, while using the official foreign exchange rate in its project appraisals.[36] (See the opening Overview section which discusses, in the context of Ghana's economic decline, the consequences of such unrealistic pricing policies and argues for the adoption of a realistic pricing structure.)

In Ghana, as noted earlier, small-scale and the informal manufacturing enterprises employ labour-intensive techniques. As in the case of agriculture and mining, one can make reference to technological dualism. For example, in the production of soap, while the large-scale sector uses capital-intensive techniques, the small-scale and informal sector soap producers employ more labour-using methods. Product quality, of course, varies, though in some cases such as palm oil the product from less sophisticated technology was initially more accep-

Manufacturing

table to the consumers (as was found in the case of production from one oil plant in the Ashanti Region).[37]

6.7 EFFICIENCY OF INDUSTRIES

In the literature of development economics there has been much discussion regarding the efficiency of the industrial development pushed through by the governments of most less developed countries.[38] An important argument is that such development – which has taken place through high and non-uniform tariffs and through quantitative restrictions on imports – has often led to the development of industries in which the country concerned does not have a comparative advantage. Market rates of return, based on a distorted price structure, fail to reflect the social rates of return which, if properly calculated, would have shown the low or negative social contributions this type of industry is making to the economy.

Two criteria frequently used to examine whether there has been serious misallocation of resources are (a) Effective Rate of Protection (ERP) and (b) Domestic Resource Cost (DRC). The ERP shows the amount by which value added at protected domestic prices exceeds value added at world prices, that is $(v-w)/w$, where v is value added at domestic prices and w is value added at world prices.[39] A measurement of the DRC takes into account social opportunity costs of domestic resources used for the manufacture of a particular commodity. It can thus measure the number of units of local currency required to earn or save a unit of foreign exchange. The DRC is capable of considering the effect of price distortions on both inputs and output. Thus the DRC, like the ERP, takes into account the effect of the trade régime, but unlike the ERP it also takes into account the social opportunity costs, or shadow prices, of the primary factors of production, capital and labour.

Both the above criteria have been applied to examine the efficiency of industries in Ghana, Leith used the ERP measure with data from 1968 to 1970,[40] while Steel applied the DRC measure to 1967–68 data.[41]

Their findings for a number of selected industry groups are shown in Table 6.11. A general conclusion that emerges from both the estimates is the 'wide variation in the economic returns to Ghanaian import substitution industries'.[42] In both the estimates the range is wide indeed. Three things emerge from Steel's observations. First, the DRCs range widely, from −N₵25.00 per US dollar to over N₵30.00

Table 6.11 Domestic resource costs (DRC) and effective rates of protection (ERP) in manufacturing, 1967-70

ISIC	Industrial group		DRC 1967-68 (N₵ per US $)[1]	ERP 1968-70 (per cent)
2302	Gold mining			−12
2303	Bauxite mining			−3
2304	Manganese mining			−3
3113	Fruit squash			−13
3113	Beverage		0.71	
3116	Rice milling		1.15	
3116	Food manufacturing		4.45	
3140	Tobacco		1.15	1803
3210	Textiles[2]	a)	0.44	927 (knitting);
		b)	1.69	749 (wearing
		c)	1.87	apparel) and
		d)	−17.19*	−3 (kinte)
		e)	−12.27*	
3240	Footwear and clothing[2]	a)	1.40	1633 (shoes)
		b)	1.96	
		c)	3.08	
		d)	−25.55*	
		e)	−7.15*	
		f)	−1.65*	
3311	Wood and sawmills[2]	a)	0.67	−10 (sawn timber and timber)
		b)	1.53	
		c)	2.69	
3351	Chemicals[2]	a)	0.29	4 (industrial chemicals) and 19 (misc. chemicals)
		b)	0.44	
		c)	1.69	
		d)	6.97	
		e)	−1.93*	
		f)	3.39	
33	Non-metallic[2]	a)	1.00	
		b)	4.18	
		c)	−2.93*	
		d)	30.55	
3419	Paper		−7.82*	11
3559	Rubber products		4.05	16
3560	Plastics, misc.		23.96	70
381	Metal products[2]	a)	2.24	19 (nails) and 30 (aluminium ware)
		b)	2.26	
		c)	2.57	
		d)	3.59	

[1] N₵1.2 per US $; 50 per cent above the official exchange rate was assumed to be the likely over-valuation of the currency.
[2] More than one firm from the same industry was examined for estimating DRC, each firm identified as a, b, c, etc.
* Negative value added at world prices.
See text for definitions of DRC and ERP.
Sources: Steel (1972) and Leith (1974).

per US dollar. Second, in as many as eight cases, the DRCs value-added at world prices is negative. Third, significant differences in the DRCs were observed between firms. Leith's study shows that the ERP also varies widely (from −653 per cent to over 78 000 per cent; not shown in Table 6.11), and gives negative value-added at world prices in a number of instances.

The similarity of the conclusions from the two measures casts grave doubts on the efficiency of import-substitution industrialisation in Ghana as pursued up to 1970. Looking at the table it is apparent, particularly from the ERP measure, that export sectors like gold, bauxite and manganese have received unfavourable treatment with negative ERP in each. The conclusions reached by Steel from the DRC comparisons are equally damaging.

If tariffs and licensing had been removed at that time, only 15.4 per cent of the firms surveyed would have been competitive with imports at the official exchange rate, and devaluation by 50 per cent would have raised that share only up to 25.6 per cent. These results indicate that the existing structure and utilisation of manufacturing capacity represent a very costly and inefficient method of gaining foreign exchange and raising national income. Even worse, 24.0 per cent of output was produced at a net loss in foreign exchange, taking into account all foreign exchange costs of capital and domestically produced inputs. These firms represent a waste of investment funds and a failure of import substitution, if they are continually operated at the level and cost structure observed in 1967-8. As of that year, Ghana's industrialisation and import substitution policies were extremely unsuccessful in establishing a structure and level of manufacturing output which could efficiently reduce foreign exchange requirements and stimulate growth of GNP.[43]

No studies like the above have been made using recent data. However, considering the high inflation since the early 1970s and the high over-valuation of the cedi at the official exchange rate until April 1983, one can safely conclude that in most cases, unless they have good access to foreign exchange at official rates, both exporting and import-substituting sectors have been faced with negative rates of protection, the extent of which increased with inflation and over-valuation of the cedi as there were no corresponding measures in the form of subsidies to cushion the impact. As will be shown in Chapter 10, by the end of 1982 it became significantly cheaper to import rather than to produce at home, the landed cost of imported goods being as low as 8 or 10 per cent of locally made goods.

However, it was extremely difficult to get import licences, and even after these had been obtained there still remained the problem of establishing letters of credit and getting the imported goods onto the Ghanaian market. Thus, given the administrative barriers encountered by importers, one could say that local manufacturers of import-substitutes enjoyed a high level of administrative protection. They have consequently been operating in a sellers' market. Exporters, on the other hand, had no such protection. They were at the mercy of economic forces, and the over-valuation of the cedi worked overwhelmingly against them. It is particularly for exporters that the negative rate of protection started rising with the increase in the over-valuation of the cedi and inflation.

6.8 CONCLUSIONS

The manufacturing sector, which has been actively encouraged since independence, grew rapidly during the 1960s and continued to grow up to the mid-1970s. Different policy measures encouraging the growth of import substitution, providing various incentives to investors (tax holidays and accelerated depreciation), and protection greatly helped the growth of the manufacturing sector in Ghana. A wide variety of products started appearing on the market with the label 'Made in Ghana'. It was, however, the large-scale manufacturing sector which was the main beneficiary of manufacturing growth, while the small-scale sector, which uses chiefly local raw materials and local technology, remained more or less neglected as far as investment incentives were concerned. Even though a large number of state enterprises still dominate the manufacturing sector of Ghana, they played their most significant role in the industrialisation process during the first half of the 1960s.

The choice of technology by *sources* and *types* for medium and large-scale manufacturing has remained a weak area in Ghanaian industrial development. At times, capital-intensive technology, though not necessarily suited to local circumstances, has been imported *en masse*. Such technologies have often been of a much larger-scale than justified under normal supply and demand considerations. Inter-industry linkages have also remained poor and the dependence on imported inputs has been very high.

In the 1960s investment decisions in manufacturing, as in other areas, were often not based on serious appraisals. In the case of projects financed by suppliers' credit it was, according to Grayson, 'a completely closed deal. The equipment peddler prepared the feasibility study... where feasibility studies were prepared at all. He chose the technology, determined the size of the plant and, of course, the source and nature of the equipment, and arranged the financing. If technical advice was needed, he found that too'.[44] Unfortunately, a major source of financing projects during the 1960s, especially during the first half of the decade, was suppliers' credit.[45]

The situation improved during the 1970s when most projects were made subject to analysis using criteria based on the discounted cash flows method, and the debt service ratio (see Chapter 14). But the period, particularly since the mid-1970s, coincided with serious price distortions. This was also a period of high inflationary pressure, but the external value of the cedi remained largely unadjusted (see Chapter 10). The over-valuation of the cedi at the official exchange rate created a situation where imports were encouraged, when import licences were available, while exports were discouraged. An important consequence was low capacity utilisation in the manufacturing sector as necessary foreign exchange for investment, spare parts and other imported inputs were not available mainly because of the decline in export earnings. The end result was the rapid decline of the manufacturing sector. At 1970 prices, gross output in the medium and large scale manufacturing sector declined, on average, by about 12 per cent per annum from 1975 to 1983.

In a situation of serious price distortions it is necessary that project appraisals and policy measures take into account shadow prices (that is, social prices based on social opportunity costs) of inputs and outputs. In the absence of central policy guidelines relating to estimates and uses of shadow prices in Ghana, social profitability estimates have either been applied only partially or not at all, thus leading to a misallocation of resources (see Chapter 14). If such distortions are allowed to continue and no measures are taken to improve the basis of policy guidelines, bad investment decisions are likely to follow in manufacturing as in other sectors of the economy.

7 Mining

7.1 DEVELOPMENT OF MINING

Apart from cocoa and its derivatives and timber and its products, minerals account for the largest share of Ghana's exports (see Table 10.2). The principal minerals are gold, diamond, manganese and bauxite. Other important minerals which have potential for the future are petroleum, limestone, iron ore, industrial clays (ball clay, kaolin, feldspar, silica sand), mica, refractories and some abrasive minerals. Some of these, including petroleum, ball clay, kaolin and feldspar, are currently mined only on a small scale.

Oil was first found in 1970 at a site eight miles off the Saltpond coast. Further production tests also showed that the oil was of very high quality. It is that find which is currently being tapped. The search for oil continues, with attention now concentrated in the south-western coastal area. Ball clay, kaolin and feldspar are used in the ceramic industry.

Below we deal with the principal minerals that are currently mined on a large scale.

7.1.1 Gold

Gold has been mined and exported for over a thousand years. Long before the arrival of the Europeans in this part of Africa, gold was being extracted from alluvial materials deposited on the banks and bottoms of rivers. One of the first commodities to be traded in, on the commencement of the European contact, was gold. History records that in 1471 the Portuguese exchanged European items for gold at the estuary of the Pra River.[1]

Modern gold mining involving the use of machines and extraction of the precious metal from both alluvial deposits and hardrock dates from about 1880 when this type of mining began in the Tarkwa area.[2] Pierre Bonnat, a Frenchman, first reported the possibilities of mining gold in the Tarkwa area in 1877. A company was formed in Paris for the purpose. A gold rush or 'Jungle Rush' (as it was termed) subsequently occurred in the Tarkwa area, and many gold mining companies began operations there.

At this time the ore extracted was headloaded or transported in barrels to Tomentu, a town on the Ankobra River, where it was loaded on to barges or steam launches and taken down the river to the surf port of Axim at the mouth of the river for export. In 1898 modern mining was extended to Obuasi, where the gold deposits were found to be even richer than the Tarkwa deposits.

The major problems encountered by the mining companies in this period were high transport costs and irregular supply of labour, since southern Ghanaians were averse to working in the mines at regular hours for wages. These problems were solved by the construction of the Sekondi-Kumasi railway line (which was started from Sekondi in 1898, reaching Tarkwa in 1901, Obuasi in 1902 and Kumasi in 1903) and the inflow of abundant labour from northern Ghana.

This dynamic initial development of the gold-mining industry was very important to the Ghanaian economy in that it introduced modern capital-intensive mining techniques. It also instilled in the labour force the habit of working at regular hours for regular pay and the introduction of other techniques of modern industry. The gold-mining industry thus laid the basis for the subsequent industrialisation drive of Ghana.

After 1914 the Geological Survey Department, established in 1913, discovered other gold deposits both in the south and the north of the country. Most of the finds were exploited by mining companies until the Second World War, when the British Government ordered the closure of a number of mines in Ghana. Many of them were put under care-and-maintenance, and mineral exploration and mine development were postponed until after the war. The Nangodi mines in Northern Ghana were closed down completely at the end of the war, followed by those at Aboso, Bibiani and other localities in the south.

In 1960 the Ghana Government took over the gold mines at Prestea, Tarkwa, Konongo, Bibiani and Dunkwa (when the private companies operating them threatened to close them down) and formed the State Gold Mining Corporation (SGMC) in 1961 to operate those mines. Its functions were as follows: 'The object of the Corporation is not to run the gold mines directly. It will be a holding company for all of the shares of the mines taken over and will direct overall policy. It will be subject to the general and specific directions of the Government'.[3]

As they became highly unprofitable to operate, the SGMC closed down the Bibiani mines and put the Konongo mines under care-and-maintenance.

A number of gold mines that were closed down during the war were never reopened, and since 1945 the number of operating mines has dropped steadily from 11 to 4. The gold mines recorded a peak production of 915 317 fine ounces of gold in 1959-60.[4] Production subsequently dropped to 402 000 ounces in 1978 and further to 330 000 ounces in 1982.[5]

It has been estimated that in the period 1493-1600 Ghana produced 35.5 per cent of the world's total gold production.[6] This share has consistently dropped over the centuries and currently averages no more than 2 per cent per annum. In the period 1931-1980 Ghana produced about 995 tonnes, an average of approximately 20 tonnes per annum.

Currently two major companies, the SGMC and Ashanti Goldfields Corporation, are engaged in gold mining in the country. The Ghana Government has acquired a 55 per cent shareholding in Ashanti Goldfields. The remaining shares are owned by the London-Rhodesia Company (LONRHO), a multinational conglomerate.

7.1.2 Diamond

Another historically important mineral mined in Ghana is diamond. Both industrial and gem diamonds are found in the country. The two main areas containing diamondiferous ores in commercial quantities are the Birim and Bonsa River valleys in the Eastern and Western Regions respectively. As in other sectors of the economy, the diamond industry has exhibited the characteristics of dualism: well-organised mining companies with modern, sophisticated machinery using capital-intensive techniques, co-existing with individual African diggers whose tools are pickaxes, shovels and pans.

In the 1960s the organised mining companies dominated operations in the Birim Valley, while African diggers monopolised activities in the Bonsa Valley. The diamond companies included Consolidated African Selection Trust, Akim Concessions, Cayco (London) Limited and Holland Syndicate NV. The last company was taken over by the State Gold Mining Corporation in 1961 and renamed the Tarkowase Diamondfields.

In 1972 the Ghana Government acquired a 55 per cent shareholding in the Consolidated African Selection Trust, which was renamed Akwatia Consolidated Diamonds Limited. However, production from its Akwatia mines has been declining sharply as a result of near depletion of reserves.

The Diamond Marketing Corporation is responsible for the purchase of all diamonds won in the country from mining companies and individual winners. It arranges sale of the diamonds on the external market in co-operation with the Bank of Ghana.

7.1.3 Manganese

Another mineral that has been mined in the country for a considerable length of time is manganese. For instance, in the 1953–54 period total production of manganese ore was as high as 710 700 tonnes.[7] It is mined at Nsuta near Tarkwa in the Western Region, and until 1975 the operation was handled by the African Manganese Company, a sibsidiary of the Union Carbide Corporation of the USA.

In 1972 the Ghana National Manganese Corporation (GNMC) was established to purchase all the manganese produced by the African Manganese Company and to find external markets for it. However, in 1975 the Government acquired the African Manganese Company and entrusted its operations to the GNMC. Thus the Corporation currently mines and markets all the manganese produced in the country.

In 1982 the Fuller Company of the USA built for the GNMC a US $25 million nodulising plant capable of using 500 000 tonnes of manganese carbonate to produce 304 800 tonnes of manganese oxide nodules of high metallurgical content.[8] This is an oil-fired plant, using 50 000 gallons of oil per day. Once the kiln is fired, for efficiency the plant must work non-stop. The need for managerial efficiency is therefore high as any under-utilisation of the plant will have serious consequences for unit costs. The need for an adequate supply of residual oil is also very important as the storage tank can hold only 10 days' stock.

The transportation of manganese to the port is another area of concern. The GNMC may have to organise a fleet of tankers attached with tippers to minimise the transport cost of fuel and manganese. The tipper-tanker will bring oil from Takoradi to Nsuta and, on the return journey, will carry manganese. The load factor will remain the same in both journeys. Alternatively, Ghana Railway will have to provide similar facilities. The latter alternative would be preferable as it would reduce the pressure on the Takoradi-Tarkwa road. In view of the above, if an electric fired kiln technology is not available and, further if a lower scale plant is not a viable proposition, it would appear doubtful if GNMC was right in going ahead with the oil-fired nodulising plant.

7.1.4 Bauxite

Deposits of bauxite occur on the tops of many hills in the country. Among the largest deposits so far discovered are those on top of Mount Ejuanema and the Atewa Range near Kibi (both in the Eastern Region), the Nyinahin deposits in Ashanti and the Kanaiyerebo deposits at Awaso in the Western Region. Of all these deposits only the Awaso deposits are mined by the Bauxite Company Limited, formerly the British Aluminium Company, in which the government has a 55 per cent shareholding. All of its output is exported to Britain while the Volta Aluminium Company's (VALCO) aluminium smelter, located at Tema, uses imported alumina derived from Jamaican bauxite.

Bearing in mind the original intentions of the Volta River Project, the delay associated with the establishment of an integrated aluminium industry in the country appears bewildering.[9] The gigantic Volta River Project involved, among other things, the construction of a huge dam and hydro-electric installation on the Volta River and the exploitation of domestic bauxite deposits to feed an alumina plant whose output would be used by an aluminium smelter. In other words, the Project envisaged the establishment of an integrated aluminium industry in the country. The dam and the hydro-electric installations are in place, and so is the aluminium smelter. However, VALCO uses imported alumina as the input for its aluminium smelter at Tema, while the country's bauxite is exported in its raw form.

The aluminium industry currently enjoys a boom on the world market and indeed has a bright future. It is the most frequently used non-ferrous metal in the world. Its use is growing faster than the use of any other important metal. Again, aluminium reduction and fabrication can be used to develop an industrial base. It is forwardly and backwardly linked to the development of other industries. These include, among others, the chemical, petroleum, iron and steel industries. An integrated aluminium industry will offer employment to both highly trained and unskilled Ghanaians. It can give a boost to the development of the areas where mining of bauxite is undertaken and where the alumina factory and the aluminium smelter are sited. Finally, it will earn foreign exchange for the country.

Given the availability of resources the failure to have an integrated aluminium industry can be traced to the peculiar nature of the agreement that the Ghana Government signed with VALCO. The basis of that agreement seems to be that Ghana needed VALCO more than

VALCO needed Ghana. VALCO was able to obtain from the Ghana Government favourable terms in the Master Agreement. For instance, it was agreed that VALCO would buy power from the Volta River Authority (VRA) at a price only slightly higher than the cost price. Thus VALCO buys electric power from the VRA at a price far lower than prices paid by the Electricity Corporation of Ghana, the mines and the Republics of Togo and Benin. Again it was agreed that the government would not insist on a firm commitment on the part of VALCO to install an alumina plant in the country. This condition assured VALCO that it did not need to be the instrument through which an integrated aluminium industry would be established in the country.

This last condition is a crucial one, since the bauxite-alumina-aluminium industry has the special feature of being vertically integrated. In the western world, six companies dominate the aluminium industry. These are Kaiser (USA), Reynolds (USA), Alcoa (USA), Alcan (Canada), Pechiney (France) and Alusuisse (Switzerland). These companies are highly vertically integrated. Kaiser and Reynolds, the shareholders in VALCO, would therefore be very careful about committing themselves to the utilisation of Ghanaian alumina unless it could be absorbed by their world-wide organisations.

While the agreement with VALCO hinders the establishment of an integrated aluminium industry in the country through that company, efforts on the part of the Ghana Government to attract the Japanese and some socialist countries to the venture have not, to date, yielded any concrete results.

7.1.5 Salt

Another important mineral is salt. Unlike other West African countries, Ghana (like Togo) possesses natural features that permit better recovery of salt from sea water through solar evaporation.[10] Extraction methods used by small-scale operators are traditional and result in the extraction of both sodium chloride and other compounds.

Enterprises from the organised sector have also undertaken heavy investment for large-scale production. Such facilities exist in Accra, Ada, Apam and Elmina. Production basically depends on the length of the dry season. Brine salt mining also has potential in the north.

The output is wholly consumed in the country apart from a limited quantity which is exported. Recently an agreement has been signed with Burkina Faso for the export of salt. Potential for large exports to other West African countries is very considerable, since most of those countries depend on imports from as far away as Europe.

Again, the establishment of an integrated aluminium industry in the country will lead directly to a great demand for caustic soda in refining bauxite into alumina. Caustic soda in turn is derived from salt. Thus salt extraction has a great potential for the future.

7.2 OUTPUT AND FACTOR INTENSITY

Detailed statistics on the mining sector are not available for the earlier years. The only exception is gold, on which a major study was conducted in 1980 and mainly because of this data on gold production are available for a long period. However, data as available are presented in Table 7.1 for 1970, 1975, 1980 and 1982 for the major mineral products. As may be seen, production has declined in all four sectors, The average annual rate of decline in gold was 6.1 per cent, while in manganese it was 7.3 per cent, in diamond 10.4 per cent and in bauxite 10.8 per cent, the rate of decline being particularly sharp in the last three sectors during 1980–82. The factors responsible for the decline of mining are discussed in the next section.

As may be seen from Table 7.1, employment statistics provide some disturbing features. While output declined in all the sectors, employment increased up to 1980 in gold, manganese and diamond, and in the case of bauxite there was a slight drop, but the percentage decline was in no way comparable to the fall in output. From 1980 to 1982, the fall in output has been much faster than the fall in employment in all the sectors but particularly in bauxite, manganese and diamond. There was thus a significant fall in labour productivity in all the sectors.

Measured in terms of value-added it was only the gold sector which recorded an increase in value-added per employee from 1970 to 1980, mainly because of a sharp rise in the price of gold. As far as other sectors are concerned there has been a sharp decline in value-added per employee from 1970 to 1980 (see Table 7.2).

Table 7.2 also shows capital intensity. Capital intensity, measured by the value of fixed assets per employee, shows the highest figure for diamond (₡4827) in 1970; followed by bauxite (₡3141), gold (₡682) and manganese (₡305). The inter-sectoral difference in capital intensity is wide indeed. While in 1980 the ranking in terms of capital intensity is the same as in 1970, the values are very low, at least for

Table 7.1 Employment, output, value added and fixed assets in gold, bauxite, manganese and diamond, 1970, 1975, 1980 and 1982 [1]
(value in current prices)

	Total employed* ('000)	Gross output Quantity ('000)	Gross output Value ₡ million	Net value added ₡ million	Fixed assets* ₡ million
Gold (grams)					
1970	18.34	21 890.6	27.37	16.21	12.50
1975	20.17	16 294.8	15.98	68.23	15.23
1980	20.31	10 981.0	546.16	455.45	94.72
1982	19.56	10 280.3	405.51	225.81	127.26
Bauxite (tonne)					
1970	0.54	245.9	2.46	1.91	1.70
1975	0.52	325.2	3.96	2.41	2.28
1980	0.50	225.1	9.44	7.56	2.36
1982	0.45	63.5	3.36	0.29	4.95
Manganese (tonne)					
1970	1.31	398.6	6.89	5.36	0.41
1975	1.34	415.3	19.46	16.31	0.18
1980	1.69	249.8	24.07	16.43	5.51
1982	1.53	159.9	20.18	14.55	5.16
Diamond (carat)					
1970	2.92	2 549.5	13.09	9.86	14.09
1975	1.82	2 336.2	13.81	10.24	3.84
1980	2.50	1 149.3	38.02	19.70	17.42
1982	2.29	684.2	21.06	4.42	24.79

[1] Data refer to establishments with 30 or more workers.
* Data refer to end of year.
Sources: CBS, *Industrial Statistics* 1970-72, 1975-77, 1979-81 and 1982-84; and *Economic Surveys* 1967, 1968, 1972-74, 1981 and 1982.

diamond and bauxite, and the sharp inter-sectoral differences are no longer there. It is apparent that at 1970 prices there has been a significant decline in capital intensity in all of the major mining sectors. The decline was due to two main factors. First, most companies have failed to revalue their assets properly with the passage of time, and this has serious implications particularly during the high inflationary period. Second, there has been a decline in investment which has, in some cases, gone so far as to cause negative net investment because of the inability to even meet the depreciation.

Table 7.2 Value added and capital intensity in major mining sectors, 1970 and 1980 (¢ 1970 prices)

	Value added per employee		Capital stock per employee	
	1970	1980	1970	1980
Gold	884	1700	683	387
Bauxite	3541	1146	3141	392
Manganese	4092	737	305	271
Diamond	3379	597	4827	579

Notes: Export deflator and gross fixed capital formation deflator (see Table A.8) were respectively used to convert net value added and net value of fixed assets shown at current prices in the source. Data refer to end of year.
Sources: CBS, *Industrial Statistics* 1970-72 and 1979-81.

An analysis of the percentage distribution of materials used by source (that is, domestic or foreign) shows that in gold mining the share of local materials increased from 5 per cent in 1970 to 29 per cent in 1980.[11] This may be due to the high prices of local items caused by high inflation. The percentage increase in 1980 of local materials in bauxite mining appears unusually high. The share of local materials, which was less than 1 per cent in 1970, increased to 73 per cent in 1980. The consumption of local material in manganese remains almost as low in 1980 (3 per cent) as in 1970 (2.2 per cent) while in diamond the use of local materials was nil in 1980 as against 10 per cent in 1970.

There has been almost no change in the number of establishments in any of the major mining sectors from 1970 to 1980. If anything, from the production point of view the number of establishments in gold mining declined from 5 to 4 as one of the gold mines (Konongo) has been put into maintenance-and-care by the SGMC. In 1970 the number of establishments in diamond mining was 3, while it was one each for bauxite and manganese, and this remained the same in 1980.

7.3 RECENT TRENDS

Though the Ghanaian mining industry has a long history, the sector has been on the decline for over a decade. The firms engaged in mining have been confronted with acute shortages of spare parts and equipment, essential chemicals and explosives for extracting the minerals (particularly gold), frequent breakdown of machinery and massive mineral theft and smuggling to neighbouring countries in the case of gold and diamond. Much of the mining equipment and machinery has broken down or become obsolete and there is a need to rehabilitate or replace it.[12]

The consequences of the over-valued currency and of the resultant unrealistic price structure on production and exports in general are discussed at some length in the opening Overview chapter. It is perhaps pertinent to mention here that the over-valuation of the cedi over the years has rendered exports less remunerative than they would otherwise have been. This has greatly affected the mining sector, all of whose output is exported. For instance, the sharp rise in the cost of production in recent years has, until the *de facto* devaluation of the cedi in early 1983, been very disproportionate to the low export earnings from the minerals in terms of local currency. For example, it has been estimated that in 1981 the cost of production of a carat of diamond was about 218 per cent above its cedi income at the official exchange rate.[13] Consequently, the mines operated at huge losses and faced acute financial crises.

The inability of the railway system to efficiently service the mining sector has also been a major problem. For instance, the production and export of bauxite and manganese mining operations have been seriously affected by the inability of the railways to transport output to the Takoradi harbour for export and to deliver fuel and oil from the harbour to the mines.

There has also been a massive emigration of both experienced expatriate staff and skilled Ghanaian personnel employed by the mines, ostensibly due to the deterioration in general living conditions in the country. With their departure from the industry managerial efficiency, especially at the state-owned mining enterprises, has declined precipitously. Pilfering of gold and diamond in particular is currently on a large scale because of mismanagement. In the gold mining industry particularly a major obstacle to the ready availability of regular labour has been the widespread practice of illegal panning of gold. Workers have found this practice more rewarding than regular employment in the mines. As a result, the government in 1984 had to send a contingent of the Ghana Army to the gold mining town of Obuasi to help solve the problem of theft at the mines.

Another problem relates to the choice of technology. It appears, for instance, that the choice of technology for the production of oxide nodules at Nsuta was not based on a serious appraisal and that the management was facing problems in making the plant operational.[14]

The establishment of an integrated aluminium industry in the country has yet to become a reality. Ghana's huge deposits of bauxite still lie unexploited (except for the small mine at Awaso) while the aluminium smelter at Tema uses imported alumina. The establishment of an integrated aluminium industry will help stimulate an industry which has been historically important but whose current performance is deplorable.

Another problem still unsolved is what role should be accorded to foreign investment in the mining sector. Should the government permit direct foreign investment in mining and, as has happened in some Latin American countries, thereby risk political interference in national affairs? Should the government only go in for foreign loans and thereby lose the expert management and technical know-how that often accompany foreign direct investments? These are unresolved questions that demand immediate answers if the mining sector is to be given a new lease of life and expanded within the foreseeable future. This observation is supported by the authors of the *Five-Year Development Plan* (1975-76 to 1979-80) who in their discussion of the role of foreign private capital observed that

> if these facilities are exploited exclusively by foreign private direct investors their benefits for Ghana will be minimal, but on the other hand, if such investors are excluded, the full potential of the facilities will not be achieved.[15]

Part III
Money, Banking, Trade and Employment

8 Money and Credit

Money and credit have many functions in an economy, but from the point of view of economic development their role in financing agriculture, industry, commerce and other activities is vital. Credit, as available from different institutions, is mainly of two types: short-term and long-term. Short-term financing is usually provided by the commercial banks dealing with money deposited by their customers, while long-term credit is generally provided by development banks. The next chapter is devoted to such long-term development financing.

8.1 STRUCTURE OF BANKING

With the Bank of Ghana as the Central Bank, the banking system of the country includes three large commercial banks – Ghana Commercial Bank, Barclays Bank (Ghana) Ltd. and Standard Bank (Ghana) Ltd.) – two relatively smaller commercial banks (National Savings and Credit Bank and Social Security Bank) and six specialised banks (Agricultural Development Bank, National Investment Bank, Bank for Housing and Construction, Ghana Co-operative Bank, Merchant Bank Ghana Ltd. and the Premier Bank Ltd.). The three large commercial banks are classified as *primary* banks by the Bank of Ghana, while the two smaller commercial banks together with the six specialised banks are designated as *secondary* banks. In recent years a growing number of rural banks have joined the banking system (see Chapter 9).

Bank of Ghana

Established as a central bank on 4 March 1957, two days before the country's independence, by the Bank of Ghana Ordinance (No 34) of 1957, the Bank is at the apex of the banking system of Ghana.

The main functions of the Bank, as described in the Bank of Ghana Ordinance, are 'to issue and redeem bank notes and coins; to keep and use reserves and to influence the credit situation with a view to maintaining monetary stability in Ghana and the external value of the Ghana pound; and to act as banker and financial adviser to the Government'.[1] The Ordinance also specified that there was to be a separate Issue Department which would be responsible for ensuring that the currency needs of the country were adequately provided for and that the currency in circulation was sufficiently backed by gold, foreign exchange and acceptable securities and other bills.

The Bank, whose initial period of operation concentrated mainly on the establishment of a new currency and the administration of Government's accounts and the public debt,[2] started issuing Ghana's own currency (G£) on 14 July 1958, thus replacing the West African pound issued by the West African Currency Board.

After 1960 Ghana started to experience balance of payments problems, with the resultant fall in the foreign exchange assets backing for the currency. On the other hand, the share of Ghana Government securities in the portfolio of the Bank's assets rose considerably as the Government resorted to borrowing from the banking system.

In the Bank of Ghana Act 1963, which comprehensively amended the 1957 Ordinance, the strict British-tradition-inspired separation of the Issue and the Banking Departments was abolished. In addition to the pound sterling assets specified by the Ordinance, the assets for backing the currency were extended to include government securities other than the ones issued by the Ghana Government and expressed in convertible currency such that no more than 40 per cent of the total assets of the Issue Department was so held. But such restrictions did not apply when the assets were held for the purpose of financing agricultural products or marketing of crops, and when the assets held were Treasury Bills of the Ghana Government maturing within 90 days, or other securities of the Ghana Government maturing in not more than 20 years. The fiduciary issue was fixed at 40 per cent of the currency in circulation and, at the discretion of the Minister of Finance, could be varied up to 60 per cent of the currency in circulation.[3] Very much unlike the 1957 Ordinance which put more emphasis on rules, the 1963 Act concentrated on consultations with the government. Thus, much harm could be done to the economy if the government chose to pursue irresponsible monetary policies since the rules to check them would not be available.

The country changed to a decimal currency system from 19 July 1965.[4] A Decimal Currency Committee under the chairmanship of M. C. Kessels, the then Governor of the Bank, had been studying the problem since 1960. The Committee had recommended a change and had preferred the 10 shillings-cent system. However, the government chose to introduce the 8s 4d-cent system instead (100 pence = 100 pesewas). The major unit was called the 'cedi' and the minor unit the 'pesewa'. Following the military coup of 24 February 1966, the new government decided to replace the existing currency which bore the effigy of Nkrumah with one that did not bear the effigy of any particular person. At the same time the original recommendation of the Kessels Committee for a 10s cent-system (120 pence = 100 pesewas) was reviewed and accepted. In 1967 the Act was amended to raise the Bank's holding of Ghana Government Treasury Bills, stocks, inland commercial bills of exchange and promissory notes from 60 to 75 per cent of the currency in circulation.

As the central bank, the Bank of Ghana is adviser to the government. The advice comes mainly from the Research Department of the Bank. Though the 1957 Ordinance did not mention the promotion of economic growth as one of the functions of the Bank, there developed in subsequent years an awareness that the promotion of growth and development by monetary measures is a proper and necessary function of a central bank, especially in a developing country like Ghana. Consequently, the 1963 Act charged the bank with the responsibility 'to propose to the Government measures which are likely to have a favourable effect on the balance of payments, movement of prices, the state of the public finances and the general development of the economy and monetary stability'.[5]

Thus, in addition to the traditional functions, the Bank considers it its legitimate duty to promote rapid economic development. In pursuit of this duty the Bank of Ghana has been engaging in the following activities:

(a) direct funding for the establishment of financial institutions such as the Agricultural Development Bank, the National Investment Bank, the Bank for Housing and Construction, and many Rural Banks;
(b) direct provision of finance for institutions such as the Ghana Export Promotion Company and the Ghana Tourist Development Company;
(c) provision of credit guarantees to approved commercial banks in respect of loans to small borrowers in the industrial and agricultural fields, and agricultural produce marketing bills discounted by the other banks; and
(d) direct participation in projects.

As the central bank, the Bank of Ghana also supervises the operations of the other banks. For instance, all commercial banks have to be licensed by the Bank of Ghana before commencing business.

Other financial institutions need not obtain licenses but they operate under specific approvals from the Bank of Ghana and the Ministry of Finance. If a non-commercial bank choses to have a commercial banking section and the assets and liabilities of the commercial banking section are not separated from that of the non-commercial section, then the bank is treated as a commercial bank by the Bank of Ghana.

The Bank of Ghana is also the agency through which the Ghana Government usually deals with international monetary institutions such as the International Monetary Fund and the World Bank.

Commercial banks

A commercial bank is a financial institution whose major function is to receive deposits from the public and to make short-term loans to its customers. Commercial banks lend money by creating current account overdraft facilities and retire loans by cancelling them. They also compete with other financial institutions in the provision of services such as time deposit facilities and the safekeeping of valuables.

Of the three *primary* commercial banks operating in the country the Ghana Commercial Bank is owned by the state which also has shareholdings in each of the other two banks (Barclays and Standard).

Historically, commercial banking in the country started during the closing years of the nineteenth century when the Bank of British West Africa (now Standard Bank) was established. In 1917 the Barclays Bank D.C.O. (now Barclays Bank of Ghana) was also set up in the country.[6] These banks were overseas subsidiaries of multi-national banks incorporated in Britain. The main motive for their establishment was not the mobilisation of domestic savings for domestic investment, but rather the financing of exports and imports. Their customers were thus mainly British, Lebanese, Syrian and Indian merchants engaged in export and import and distributive trade activities. Commercial banking in the country expanded slowly until the 1950s. For instance, Barclays Bank had only 9 branches in the whole of British West Africa in 1926 and only 17 in 1946. The expatriate banks preferred to follow the traditional cautious methods practised in Britain.

Most Ghanaian businessmen and nationalists were not happy that the expatriate banks dealt mainly with expatriate customers. Therefore the State-owned Ghana Commercial Bank, formerly the Bank of the Gold Coast, was established in 1953 so as to provide more banking facilities to Ghanaians.[7]

Between 1951 and 1963 the total number of bank branches in Ghana increased seven-fold and banking facilities became available to a far wider section of the Ghanaian population. By 1962 the number of commercial bank branches was 139. This subsequently increased to 187 by 1970 and further to 191 by 1982. However, with the distribution of branches concentrated in the southern part of the country, a high percentage of the credits given goes to the south.

In the first half of the 1960s the Ghana Commercial Bank was given a near monopoly of handling the banking business of public bodies such as the universities and State enterprises. In 1963, for instance, about 90 per cent of the total deposit liabilities of the Ghana Commercial Bank was owed to the public sector. The deposits of the expatriate banks, on the other hand, are owned predominantly by private firms and individuals.

The Ghana Commercial Bank is the largest of the three primary banks. As at 1975, it was holding over 50 per cent of the total deposits of public institutions and the private sector[8] while by December 1983 it was holding about 60 per cent of total deposits in the primary commercial banks.[9] Because of its expansionist policies in banking, the Ghana Commercial Bank had 96 branches and 14 agencies by June 1973. By December 1983 it had 141 branches. Barclays Bank of Ghana and the Standard Bank of Ghana, on the other hand, had only 32 and 28 branches respectively and together held 40 per cent of total deposits by December 1983. There was a rapid expansion of bank deposits, increasing at an average annual rate of 17 per cent from June 1968 to June 1973.

Although the *Big Three* continue to play an important role in the creation of credit, two *secondary* commercial banks – National Savings and Credit Bank (NSCB) and the Social Security Bank (SSB) – appeared on the scene during the 1970s.

The NSCB was originally known as the Post Office Savings Bank, which was established on 1 January 1888 as a savings-attracting unit within the Posts and Telecommunications network. Its main function was the mobilisation of savings for investment in foreign and government securities. It was renamed Ghana Savings Bank in 1972 when it was reorganised as an autonomous body to operate as a full commercial bank, and completely separated from the Posts and Telecommunications Department. In 1975 the name was changed again to the National Savings and Credit Bank in order to distinguish it completely from the new Posts and Telecommunications Corporation, and also to draw the public's attention to the fact that it was now a full commercial bank. It has opened branches outside its headquarters in Accra, but the bank still uses the services of many post offices throughout the country on an agency basis. Although the oldest institutionalised savings unit, the bank is in most respects still an infant commercial bank, having 12 branches as at December 1983.

The SSB was established in 1977. The bank is owned by the Social Security and National Insurance Trust, which administers the Social Security fund for workers throughout the country. It is a full commercial bank, with 38 branches as at December 1983. However, in order to help workers its loans are mainly consumer-oriented.

The first co-operative bank to be established in the country (indeed, in the whole of West Africa) was founded at Akpeve, in the Volta Region, in 1938. It was owned by the Central Co-operative Society of the Trans-Volta Togoland (now the Volta Region). However, it could not survive the Second World War and was closed down after only five years in operation. A new co-operative bank known as the Gold Coast Co-operative Bank was opened by the Association of Cocoa Co-operative Societies in 1946. The new bank's aim was to mobilise savings in the cocoa growing areas and advance loans to co-operative societies in those areas. The bank remained in operation until 1960, when the Cocoa Co-operative Societies which had established it were abolished and replaced by the United Ghana Farmers Co-operative Council. Upon the abolition of the Cocoa Co-operative Societies the bank was liquidated in 1961 and its assets and liabilities were taken over by the Ghana Commercial Bank. Although it was not a bank, the United Ghana Farmers Co-operative Council advanced loans to cocoa and other cash crop co-operative societies in the country. Thus, co-operative banking lay dormant until 1973, when the Ghana Co-operative Bank was established. However, initial administrative and financial difficulties prevented the bank from commencing formal banking business until 1975. The bank has now about 50 branches and agencies, mainly in the rural areas.

There are two merchant banks in the country. These are the Merchant Bank Ghana Limited and the Premier Bank Limited. The Merchant Bank Ghana Limited (MBG) was established in 1972. The MBG (originally the National Merchant and Finance Bank) is jointly owned by the Ghana Government, the National Grindlays Bank of the United Kingdom, the National Investment Bank and the State Insurance Corporation. It undertakes business involving bulk transactions as well as commercial bill discounting for corporate bodies. Its loans go mainly to corporate bodies and are in the short and medium-term range. Its liabilities are the funds it borrows from the commercial banks for re-lending to its corporate customers and its share capital. It also has a Credit Installment Division which provides hire purchase facilities for firms and other enterprises for the purchase of machinery and other items.

This division may also lease machinery to corporate customers. It also has a Corporate Finance Division which provides management, consultancy and investment services. It underwrites and floats stocks and share issues. It also extends advisory services to firms, undertakes project appraisals for customers and also determines whether or not the MBG should finance any given project. This division also advises customers on their investment portfolios. The Corporate Finance Division is assisted by two subsidiary companies of the MBG. These are the National Stock Brokers Limited and the Investment Holding Company, which act as brokers to companies in the sale or subscription of stocks and shares.

The other merchant bank, the Premier Bank, was established in 1978. It is a joint foreign/Ghanaian enterprise and is affiliated to the Bank of Credit and Commerce International of Luxembourg. The Premier Bank undertakes managing/merchant banking activities.

Various development banks also provide some commercial banking facilities, but their main role is obviously confined to development financing (see Chapter 9).

8.2 MONEY SUPPLY

Ghana began its independent life with a currency system termed the sterling exchange standard system, a carry-over from the West African Currency Board of which Ghana was a member until independence. That currency system required that the common currency of Britain's West African Colonies be backed 100 per cent by the pound sterling.

The essence of the West African Currency Board system, like its counterparts in other parts of the British Empire, was the convertibility of the currency on demand into the pound sterling and vice versa. A major feature of the system was that the issue and redemption of currency were completely automatic. In other words, the Board did not exercise any discretion as to the amount it issued or redeemed. The Board was thus unlike a money issuing central bank. It was merely a passive money changer. Currency supply in British West Africa was thus strictly dependent on the balance of payments of the area.

Moreover, although the Board was required to maintain a 100 per cent backing of its currency in the pound sterling, in practice it maintained a higher percentage, a practice commented upon by Hazelwood, Mars, Greaves and other economists and bankers.[10]

This requirement of 100 per cent backing was a safeguard against excessive printing of the West African currency, a practice that would have destroyed its parity with the pound sterling. The colonies also benefited in that the reserve backing of their currency was held in interest-bearing assets. This practice, however, may be criticised on the grounds that by investing in British securities the Board was diverting funds that could have been invested in the colonies themselves. Following the establishment of its own central bank, Ghana left the Board in 1958.

The 1957 Bank of Ghana Ordinance empowered the Bank to issue fiduciary money, but it did not use that facility until 1960. Instead it continued to back its currency 100 per cent with the pound sterling. However, beginning from 1960 Ghana quickly moved away from the rigid relationship between its currency and the balance of payments, and the fiduciary issue became increasingly important. In that year the Bank of Ghana decided to provide rediscount facilities, and for the first time the commercial banks acquired Treasury Bills issued by the government. Once the Bank of Ghana rediscounted those Treasury Bills, it could issue fiduciary money.

Thus currently if the Ghana Government needs money to spend over and above its revenue, it simply prints securities which it hands over to the Bank of Ghana for sale to other banks, firms, institutions and the general public. But if these securities are not bought by the prospective buyers (because the price is not attractive enough) the Bank of Ghana purchases the securities which it pays for by simply crediting the government's account at the central bank. The Ghana Government in turn issues cheques to pay its bills, using the amount credited to it by the Bank of Ghana. With the securities as backing the Bank prints money. Once the money enters into circulation or is credited to the current accounts of the non-bank public, it becomes part of the money supply.

Over the years the Bank of Ghana has attempted to set credit guidelines in the form of sectoral ceilings to influence the distribution of credit, so that resources may be directed into selected government-priority sectors – agriculture, exports, manufacturing, transport, storage and communications. For example, in the 1973–74 credit guidelines, the total expansion in commercial bank credit was envisaged to be 15 per cent. The priority sectors were permitted a 20 per cent increase, while the non-priority sectors (domestic commerce and utilities) were permitted a 10 per cent increase. Import trade and other sectors were allowed only a 5 per cent increase. Performances under the guidelines were, however, disturbing. Credit to priority sectors showed virtually no change, while that to the non-priority sectors rose by ₵19.3 million (20 per cent) – a substantial amount of which went to the import trade (₵7.5 million or 39 per cent).[11]

It is worth noting that credit creation has played quite a significant role in the growth of money supply in the country. The total money supply was ₵239 million in 1965, rose to ₵385 million by 1970, and expanded further to ₵1009 million by 1975. By 1980 the money supply was ₵6058 million, and expanded to ₵11 439 and ₵23 744 million by 1982 and 1984 respectively.[12]

Table 8.1 shows the periodic average annual rate of growth in money supply. There was a tremendous rise in money supply in the 1970s, a direct consequence of huge budgetary deficits as compared to the rather low growth rates in the 1960s. For the period 1961–80 the average annual growth rate was 25 per cent, despite the statutory limit of 15 per cent per annum. The average annual growth rate of money supply during the early 1980s has remained almost as high as during the second half of the 1970s.

Money supply, defined as currency in circulation and demand deposits with the banks (except the Bank of Ghana), has been rising *pari passu* with the expansion in government budgetary deficits. In

Table 8.1 Average annual growth rate of money supply, 1961–84 (percentages)

Period	In circulation	Currency Held by commercial banks	Held outside banking sector	Demand deposit (over 12 months)	Total
1961–65	7.46	4.66	7.77	17.08	11.98
1965–70	5.46	5.92	5.42	13.64	10.01
1970–75	25.70	19.51	26.34	17.35	21.25
1975–80	49.38	57.33	48.60	37.14	43.12
1980–84	42.86	46.01	42.52	38.07	40.70

Source: See Table A.14.

the early 1980s the Ghana government was running huge budgetary deficits, ₡4400 million in 1981 and ₡4299 million in 1982 (see Table 8.2). The budgetary deficit was as high as 135 per cent of the total central government revenue in 1980-81; the figure was however lower than 100 per cent in 1982.

Most of the financing has been from domestic sources, mainly from the banking sector and to a lesser extent from the Social Security Fund.[13] Net credit to the government by the banking system was ₡258.8 million in 1970, ₡964.5 million in 1975 and ₡6522.2 million in 1980. Net credit extended by the banking system to public institutions stood at ₡590.7 million in 1981 and ₡707.6 million in 1982. Credit for cocoa financing was ₡2949.7 million in 1981 and ₡5552.9 million in 1982, while the private sector took ₡1341.6 million and ₡1557.4 million of credit in 1981 and 1982 respectively. The large credit for cocoa financing was needed as the cedi earnings through the artificially low foreign exchange rate failed miserably to cover the cocoa purchases.

Table 8.2 Central government budget deficit, 1961-82 (selected years)

	1961	1965	1970	1975	1980	1981	1982
Deficit (₡ million)	72.5	159.4	88.5	356.7	1645.4	4440.2	4299.0
As % of total government expenditure	26.6	35.9	19.3	30.7	35.2	57.2	48.0
As % of total government revenue	36.2	56.1	24.0	44.3	54.4	135.4	92.0

Source: See Table A.11.

Thus the Bank of Ghana has moved increasingly towards fiduciary issue. It has failed to manage the money supply so as to avoid inflation and to maintain the external value of the currency. Acute inflation has been assailing the economy since the early 1970s (see Chapter 11). In 1983 the cedi was devalued by about 1000 per cent. As Newlyn so rightly says: 'those countries which start their monetary development with a token currency which is entirely fiduciary, are immediatedly faced with the problem of managing such a currency so as to avoid inflation, and with the need to maintain its external value.'[14]

8.3 CREDIT SUPPLY

By credit supply we mean the loans and advances made available to the different sectors of the economy. Here the Ghanaian scene may be described as dualistic. Well organised financial institutions exist side-by-side with the unorganised market for the provision of credit. In this section we deal first with the organised market (consisting of the banking institutions) and then follow with a brief discussion of the unorganised market. The role of the development banks in the supply of credit is treated in the next chapter.

During the past decade or so the Bank of Ghana has been the biggest lender to the government. Its total claims on government increased by ₡1013.4 million (66.4 per cent over the previous year) in 1977 and by ₡1682.3 million (66.0 per cent) in 1978.[15] During the currency exercise of March 1979, the currency in circulation was reduced by about ₡625 million.[16] The government used this credit to redeem part of the Bank of Ghana's holdings of government stocks. This action thus reduced the government's indebtedness to the Bank. However, government's borrowing from the Bank went up again in 1980 as a direct result of large wage and salary increases and other expenditures, at a time when government revenue was on the decline. Thus at the end of 1980 government's indebtedness to the Bank of Ghana had expanded by 33.1 per cent compared with the position at the end of 1979.

At the end of 1977 the Bank of Ghana held ₡90.0 million in cocoa bills. In 1978 credit for cocoa financing totalled ₡560 million. By 1979 the amount of cocoa bills had increased to ₡1016.0 million as a result of the increase in the producer price of the commodity. Indeed, by the end of 1980 about 85 per cent of the Bank of Ghana's assets consisted of credit extended to the government and other public institutions.

By comparison, total claims by the three primary commercial banks on the private sector were about one-quarter of their total assets at the end of 1976. This fell to about one-fifth by the end of 1978 and further to one-sixth by the end of 1980.[17]

A significant feature of the big three commercial banks' operations is the changing nature of their assets and liabilities over the years. In

1957 for instance, a large percentage of their assets was held abroad. The productive sectors such as agriculture, mining, manufacturing, logging and sawmilling received only 14 per cent of total commercial bank credit to the public for all purposes other than cocoa marketing. However, as shown in Table 8.3, there was a shift in the sectoral distribution of commercial bank credit over the years. For example, the share of credit in import trade fell sharply from the mid-1970s, while that for services dropped sharply from 1970. On the other hand, there has been in general a notable rise in credit for agriculture and export trade in recent years. The underlying reason for this shift is the direction of government policies which encouraged agriculture and exports.[18]

Loans and advances by commercial banks rose from ₵504.1 million at the end of 1976 to ₵1517.0 million by December 1980 – an average increase of 31.7 per cent per annum. It went up further to ₵3618.5 million by December 1983 – an average annual increase of 32.5 per cent from 1976 to 1983. This increase was lower than the expansion

Table 8.3 Percentage distribution of the primary commercial banks' loans and advances by sectors, 1965–82
(percentages: current prices)

Sector	1965	1970	1975	1980	1982
Agriculture	6.93	6.61	8.26	9.58	21.61
Industry	20.89	42.13	40.89	42.50	40.83
Mining and quarrying	1.73	2.57	1.52	3.27	9.42
Manufacturing	3.56	29.59	27.69	20.54	16.15
Construction	14.60	9.29	11.30	11.10	13.47
Electricity and water	1.00	0.68	0.38	7.59	1.79
Commerce and finance	48.18	30.33	32.61	32.71	24.23
Import trade	0.27	0.68	8.61	1.68	1.64
Export trade	20.53	4.22	4.85	17.19	11.08
Others	27.37	25.43	19.15	13.84	11.51
Services	24.00	20.93	18.34	15.20	13.34
Transport and communication	0.73	3.59	10.90	9.73	8.49
Services	12.23	12.49	5.10	3.55	3.63
Miscellaneous	11.04	4.85	2.34	1.92	1.22
Total	100	100	100	100	100

Sources: CBS, *Quarterly Digest of Statistics*, September 1965, December 1970 and September 1983.

in money supply during the same period, which was 43.5 per cent per annum. The general decline in economic activity over the period had the effect of depressing the demand for credit facilities. For instance, with the falling supply of raw materials, the manufacturing sector reduced substantially its borrowing from the commercial banks. But even then it continued to be the major borrower from the commercial banks, taking between one-fifth and one-quarter of total credit extended by the commercial banks between 1975 and 1980.

The second largest sector to receive commercial bank credit in the 1975-80 period was commerce and finance (excluding cocoa marketing). In 1965 this sector took about ₵93.8 million which increased to ₵203.9 million in 1980. Loans to the agricultural sector remained fairly stable, around 9 to 10 per cent between 1975 and 1980. However, the share of agriculture in total credit advanced by commercial banks rose sharply to 21.6 per cent in 1982.[19]

With respect to commercial bank loans and advances by type of ownership, in 1981 one-third of the total credit granted by the three *primary* commercial banks went to public institutions, slightly less than two-thirds was taken by the private sector and about 2.5 to 3 per cent by the joint state/private sector. In 1983 the corresponding figures were 21.5 per cent (public), 77.4 per cent (private) and 1.1 per cent (joint venture), thus showing the increased dominance of the private sector.[20] Following the April 1983 Budget and the subsequent formal devaluation of the cedi in October 1983, the credit requirements of the Cocoa Marketing Board fell significantly, a factor which has contributed greatly to the fall in the public sector's share in bank credit in 1983.[21]

Another source of institutionalised credit in Ghana is the development banks, which are discussed in the next chapter. Apart from institutionalised credit, there exists an unorganised financial market which was discussed in the context of agricultural credit (see Chapter 4). The non-institutional credit market is made up of money lenders, thrift societies and credit unions. Both licensed and unlicensed money lenders operate in the cities, towns and villages. However, their lending activities and the rate of interest they charge are shrouded in secrecy. In 1940 and 1951 Ordinances were passed to regulate the activities of money lenders who were alleged to be exploiting poor farmers.[22] In 1958 a Government Committee on Agricultural Indebtedness reported that the money-lender was of much help to the village community and that he was held in high esteem.[23] However, to date there is no serious study shedding light on the behaviour of the private money lender and his credit granting activities. Such a study would have helped in evaluating the role of money lenders in the provision of rural credit. Thrift societies and credit unions too are widespread in the country, and are especially popular among market women, church members and the working class. They mobilise small savings and provide credit facilities on liberal terms to their members.

9 Development Banking

9.1 NEED FOR DEVELOPMENT FINANCING

A major characteristic of bank loans in less developed countries (LDCs) is that they are of a short-term nature. This does not go well with the declared objectives of LDC governments to foster development by encouraging investment in projects of medium and long-term duration.

Branches of well-known commercial banks in the metropolitan countries were established in the LDCs during the colonial era, basically to finance import-export and internal trades. With time, other indigenous commercial banks sprang up to compete with these banks in the provision of short-term capital. The financial institutions providing short-term credit in Ghana have been discussed in Chapter 8. Here we deal with the institutions that are engaged in the provision of medium and long-term finance.

The Bank of Ghana, in keeping with its role of encouraging the growth of the capital market, has been promoting the establishment of development banks to engage in the medium and long-term financing of industry, agriculture and other sectors. The Bank lends heavily to the development banks on a long-term basis to enable such funds to be re-lent, mainly to agriculture and industry. Also, the Bank initially provides office space or releases its own personnel towards the establishment of these development banks.

A development bank is a financial institution whose main function is the granting of medium and long-term loans either for one specified sector or for several sectors. Funds for this purpose are obtained by borrowing from the government or other banks or from paid-up capital.

Three main theses have been advanced to explain the establishment of development banks – the Macmillan Gap Thesis, the Exigency thesis and the Catalyst Thesis.[1] The UK Macmillan Committee on Finance and Industry of 1931 discovered that in the UK the government's banking needs were catered for by the central bank and those of commerce were provided for by the commercial banks, while there was a gap as far as financing investment in agriculture and industry was concerned. Thus the Macmillan Committee recommended the establishment of specialist financial institutions that would meet requirements for medium and long-term capital.[2]

According to the Exigency Thesis, with the establishment of international financial institutions such as the International Monetary Fund and the International Bank for Reconstruction and Development (the World Bank), it was felt that these institutions, especially the World Bank, would find it difficult to operate at the local level without local development banks. The World Bank therefore encouraged the establishment of such banks. With the establishment of institutions like the Agricultural Development Bank and the National Investment Bank in Ghana it is easier for the World Bank's affiliates, the International Finance Corporation and the International Development Agency, to assist the country by extending loans through these local banks.

The Catalyst Thesis emphasises the need to establish development banks to provide easy access to finance, particularly for small enterprises, thus serving as a stimulus to rapid economic development. All these arguments are relevant to the establishment of development banks in Ghana.

9.2 SPECIALISED DEVELOPMENT BANKS

The activities of three major development banks appear highly important in the development of directly productive sectors in Ghana. In this section we discuss in some detail the activities of these three institutions – the National Investment Bank, the Agricultural Development Bank and the Rural Banks.

Another development bank which we have not selected for detailed discussion, but which has played an important role in providing the financial requirements of the building and construction sectors, is the Bank for Housing and Construction. Established in 1973, it provides assistance to clients in the form of pre-investment investigations, extension of medium and long-term loan facilities and post-finance control. After its establishment, the bank's operations expanded so rapidly that with an authorized capital of ₵10 million (out of which ₵2 million was paid-up at the beginning of operations), the bank's total assets increased to ₵180 million at the end of 1977. The shortage of building and construction materials in recent years has affected the activities of the bank. New ground which the bank has broken of late is the joint venture quarrying activities, in which it is now participating with other local institutions and individuals all over the country. In addition, the bank has turned more and more towards commercial banking.

9.2.1 National Investment Bank

The National Investment Bank (NIB) was established by an Act of the Ghana Parliament in 1963, as a joint state-private development finance company. The objectives of the bank, as set out in the 1963 Act, are to:

(a) assist in the establishment, expansion and modernisation of enterprises;
(b) encourage and facilitate the participation of internal capital in such enterprises;
(c) counsel and encourage Ghanaian business concerns;
(d) identify emerging investment opportunities; and
(e) bring together capital, capable management and technical expertise to establish viable new enterprises.[3]

In specific terms, the NIB seeks to confine itself to directly productive activities, with a policy of encouraging private enterprise and good management in all the activities it supports.

In granting its medium and long-term credit for projects, the NIB is guided by its annual programme specifying the economic priorities approved by its management. This programme conforms to the credit regulations and investment priorities set by the government. Projects likely to merit inclusion in the highest priority category are those which are export-oriented, which serve to substitute for imports, and which are expected to increase capacity utilisation and are capable of using local raw materials to a large extent. Such projects would be screened under some stringent criteria to ensure net foreign exchange savings. According to the NIB 1975 *Annual Report*, projects that would satisfy these criteria are considered to be those that have linkage effects in the economy and these include:

(a) wood, clay, mineral and fishing industries;
(b) basic industries for the manufacture of capital goods; and
(c) agro-business.[4]

The NIB insists on project appraisal before providing any financial assistance. Projects requiring loans of over ¢250 000 require the approval of the Board of Directors and need to undergo an 'economic rate of return' analysis. On the other hand, smaller projects requiring loans or equity up to ₡250 000 can be approved by the Finance Sub-Committee of the bank after a financial analysis has been prepared for them. The NIB has been relying greatly on the technical expertise of its Development Service Institute (DSI) for analysing proposed projects.

In appraising projects the DSI employs the Discounted Cash Flow (DCF) method together with the Debt Service Ratio (DSR) criterion. The latter seeks to assess the liquidity of an investment and hence its ability to repay the required loan along with interest within the stipulated periods. Until the late 1970s, when the World Bank recommended the use of shadow prices in project appraisal, the NIB used official (controlled) prices. Such use of official prices at a time of serious price distortions failed to consider the social profitability of projects. Since the beginning of 1981, on the recommendation of the World Bank, the NIB has been applying a set of 'partial' shadow prices. Prior to the 1983 devaluation a shadow exchange rate of ₡10.00 to one US dollar and a shadow wage rate of 0.75 of the unskilled labour wage were used for appraising projects in which the NIB had been involved.[5]

Reports on projects analysed for financial assistance contain the technical specifications of the production process and the required equipment. Also included are the financial and economic analyses and manpower requirements. As regards the choice of technology, the NIB has been constrained by lack of adequately trained personnel and shortages of information on alternative technologies by types and

Table 9.1 Distribution of NIB joint venture projects by ownership and NIB equity participation as at December 1979
(no. of projects)

% Share of NIB equity	Public	Private Ghanaian	Foreign	Total
Less than 25%	7	10	1	18
25% to 50%	7	10	5	22
More than 50%	2	2	0	4
Total	16	22	6	44

Note: Projects shown under 'Private Ghanaian' and 'Private' in the source have been added and shown under 'Private Ghanaian' in the table.
Source: NIB, Annual Report 1979.

sources. Thus, no explicit choice of technology has been made, although in a number of project appraisals some form of implicit technology choice has been attempted.[6] The NIB has often participated in joint ventures with either local or foreign partners (Table 9.1). Technology transfer in the NIB-funded projects is normally on the basis of turnkey contracts or suppliers' credits. In the case of a turnkey contract the entire technology is supplied and installed by the machinery supplier in the form of a package deal. With suppliers' credit, finance usually in foreign exchange for the equipment is arranged by the machinery supplier who also usually undertakes to install the machinery and equipment.

Within the context of the bank's objectives, it had, by December 1982, cumulatively committed a total of ₡271.1 million to investment of which 61 per cent was directed towards manufacturing enterprises and 21 per cent towards agriculture and fishing. Mining and quarrying took 8 per cent and the other sectors took 10 per cent. Table 9.2 shows loans and equity participation by the NIB in different sectors over the period 1963-82. The share of manufacturing in NIB's participation in equity has been very high (78.5 per cent). Thus the NIB has taken an active interest in the development of the productive sectors such as agriculture and manufacturing. An analysis of the size distribution of loans and equity participation shows that the average size of loans has been over ₡800 000 in mining and quarrying, usually because of heavy investment requirements in this sector for which, in total, 24 loans were approved over the period 1963-82. However, the highest number of loans (405) was approved for manufacturing, the average amount being ₡377 000. Also, the number of projects in which the NIB had equity was highest in manufacturing (48) with an average contribution of ₡333 000.

By the end of December 1979 the NIB had promoted 44 projects, six of which were foreign-owned. Of the six, there was none in which the NIB had more than a 50 per cent equity share, and in one it had less than a 25 per cent shareholding. In the other five, it had between 25 and 50 per cent equity participation. An explanation for the NIB's minority equity participation in foreign-owned projects may be that most of such projects were financed substantially from foreign sources. A 25 to 50 per cent equity participation by the NIB in five of the six foreign-owned projects is a likely outcome of the Investment Policy Decree of 1973, which allowed Ghanaians and Ghanaian institutions to acquire at least 40 per cent shares in all enterprises within certain selected sectors of the economy.

Table 9.2 Sectoral distribution of NIB approved loans and NIB equity investment, 1963-82 (current prices)

Sector	Loans ₡ Mil	Loans Average size ₡000	Loans % distribution	Equity ₡ Mil	Equity Average size ₡000	Equity % distribution
Agriculture and fishing	53.0	97.1	21.1	1.2	120.3	5.9
Mining and quarrying	19.2	801.5	7.7	0.2	81.5	0.8
Manufacturing	152.8	377.4	60.8	16.0	332.8	78.5
Construction	0.7	168.0	0.3	—	—	—
Commerce	3.3	181.4	1.3	3.0	429.1	14.8
Transport and storage	11.2	78.6	4.4	—	—	—
Services	11.2	154.8	4.4	—	—	—
Total	251.4	207.6	100	20.4	303.7	100

Source: NIB, Annual Report 1982, Table 4.

9.2.2 Agricultural Development Bank

The establishment of this bank resulted from the recognition by the government that institutionalised agricultural lending was necessary to promote agricultural development. In its desire to accelerate national (but partcularly agricultural) development, the Bank of Ghana saw the need to provide some credit to the small-scale farmers on whom the bulk of agricultural production depended. In April 1964, therefore, the Rural Credit Department was created within the Bank charged with the responsibility of studying the problems of agricultural credit and also with preparing the groundwork for the establishment of a fully-fledged agricultural credit bank.

On the advice of the Bank, the government passed Act 286 in April 1965 which incorporated the Agricultural Credit and Co-operative Bank. It took over the assets and liabilities of the Rural Credit Department of the Bank of Ghana and commenced operations in August 1965. In April 1967 NLC Decree 182 amended portions of the original Act to change the Bank's name to the Agricultural Development Bank (ADB).

In the broadest sense, the ADB aims at identifying and promoting agricultural enterprises either individually or jointly and also at mobilising and channelling resources (financial and human) for the development and modernisation of agriculture and allied industries in the country. Over the years the ADB has encouraged, promoted and financed projects of varying sizes in the food crop, industrial crop, livestock, fishing and agro-based sectors of the economy. Other schemes supported by the bank include the Commodity Credit Scheme, Cocoa Rehabilitation and direct participation in special projects.

In the identification of projects, the initiative is either taken by the ADB or by prospective investors, but the project must fall within the priority sectors of the bank (which conform with the government's priorities). Where a project is initiated by the bank, its Research and Project Departments work hand-in-hand. After initial studies, the Research Department comes up with proposals for the Projects Department to study. The Projects Department conducts pre-feasibility studies for examination by the bank's Board of Directors. Sometimes it engages the services of outside consultants to prepare feasibility reports.[7]

In the case of clients who go into small and medium-scale productive ventures, the bank does not usually insist on pre-feasibility studies except where the project to be undertaken is of an unfamiliar nature. Where pre-feasibility studies are required, some clients employ consultants to prepare detailed reports for them. In some cases, the Research or the Projects Department undertakes the preparation of reports for clients for a fee.

To assess profitability the ADB uses criteria based on Discounted Cash Flow, and the Debt Service Ratio. The rate of return acceptable to the ADB is not fixed, but much emphasis is put on the repayment of loans. Available information indicates that the ADB does not use shadow prices for appraising projects. All it does is to use government approved prices for such items as petroleum, fertilisers and other imported inputs, as well as the official foreign exchange rate and open market prices for inputs like labour.[8]

The ADB has occasionally received financial support from international agencies such as the African Development Bank and the International Development Association. Apart from these, the Government of Ghana in conjunction with other countries or agencies render similar useful assistance to the bank. One such programme is the Managed Input Delivery and Agricultural Services (MIDAS), sponsored by the Government of Ghana and the Government of the USA. Through this programme the ADB has been able to support an expansion in its branch network and loan portfolio to small-scale farmers in the country.[9]

The ADB's activities over the years have expanded tremendously, particularly in its special projects. It has invested in over a dozen projects, two of which are solely owned by it. In eight of these projects it has a 50 per cent share or more. By the end of 1979 the bank had invested a total of ₵258.2 million in various ventures in the agricultural sector, the greater part of which was directed to 'group farmers' (see Chapter 4). Table 9.3 shows the distribution of the ADB's loans to different sectors for 1979 which was a typical year with respect to the precentage distribution of different sectors, although compared to, say 1977, the total amount of loans granted by the ADB was lower in 1979 even at current prices.

The ADB lends heavily for food crop production. In 1977 as much as 53 per cent of the total amount of disbursed loans was for food crop production. In both 1978 and 1979 the food crops sector took 72 per cent of the total credit made available by the bank. Owing to the great need for short-term loans of small-scale farmers, ADB loans have concentrated mainly on this area, which accounted for half of the loans given to small farmers in 1977 and as much as 93 per cent

Table 9.3 Sectoral distribution of ADB loans, 1979

	Food crops	Livestock	Indus. crops	Fishing	Agro-business	Total
₡ Million	31.0	2.5	1.3	3.2	4.9	42.9
Percentage distribution	72.3	5.8	3.1	7.4	11.4	100

Source: ADB, *Annual Report 1979.*

in 1979. The shares of medium-term ($1\frac{1}{2}$ to 5 years) and long-term (over 5 years) loans have fallen respectively from 12 and 7 per cent in 1977 to 2 and 6 per cent in 1979.

9.2.3 Rural Banks

Another type of development bank which has been springing up all over the country since the late 1970s is the rural bank. The rural banking scheme is the outcome of a courageous move on the part of the Bank of Ghana to provide the rural areas of the country with necessary credit facilities. It was to serve as a catalyst institution in the economic development efforts of the rural population.

The need for rural banks became apparent in the closing years of the 1960s when it was discovered that the Agricultural Development Bank was not meeting all of the credit needs of the small-scale rural farmers and industrialists. Moreover, funds mobilised in the rural areas by the commercial banks were transferred to the cities and larger towns for investment in sectors like commerce and housing, thus leaving the rural communities short of investment funds.[10]

Nationwide research by the Bank of Ghana in 1971 into the credit needs of the small-scale farmers/industrialists resulted in the formation of the Rural Banking scheme. Owned, managed and patronised by the local people, rural banks provide basic banking facilities for the rural communities in which they are located. Since they have no branches, money mobilised in the locality is retained for investment in the locality. The local people contribute a substantial amount of the equity capital and the Bank of Ghana helps in the establishment of the Rural Bank by buying equity shares and providing loans and other assistance.

The first rural bank was established in July 1976 at Agona Nyarkrom, a food and cocoa growing area in the Central Region. A second one was opened in February 1977 at Biriwa, a fishing town, also in the Central Region.[11] By the end of 1981 the number had increased to 34;[12] and as at August 1983 the number stood at 70.[13] Table 9.4 shows the number of rural banks existing in the regions between 1976 and 1983. The location of the rural banks demonstrates the unbalanced nature of regional development. Out of 75 banks in 1983, the Northern and Upper Regions have only three. The *raison d'etre* of the rural banks is the perceived neglect of the rural communities by the other banks, but the failure to open more rural banks in the northern areas may hinder economic development in that part of the country.

Table 9.4 Cumulative total of rural banks by regions, 1976-83

Year	Gr. Accra	Eastern	Central	Western	Volta	Ashanti	Brong-Ahafo	Northern	Upper	Total
1976	—	—	1	—	—	—	—	—	—	1
1977	—	1	2	1	—	—	—	—	—	4
1978	—	2	2	1	—	—	—	—	—	5
1979	—	2	5	1	1	—	—	—	—	9
1980	1	4	5	2	3	4	1	—	—	20
1981	1	6	7	4	6	4	3	1	2	34
1982	3	12	10	4	8	10	7	1	2	57
1983	3	13	16*	7†	8	15	10	1	2	75

* Includes three banks in operation but not officially inaugurated.
† Includes one bank not yet operative.
Sources: CBS, *Economic Survey 1981*, p. 148; and Centre for Development Studies (1983), pp. 109-11.

Operationally, the rural banks are entirely in the hands of the local shareholders who are expected to be residents of the locality. The shareholders appoint a Board of Directors which manages the bank. Like all other banks, the rural banks are governed by the Banking Act of 1970 under the supervisory powers of the Bank of Ghana. They are required to submit monthly returns and reports of their activities and the general economic situation of their localities.

In general the rural banks are reported to be performing creditably, even though their share of total banking business in the country remains rather small. However, a pilot study conducted by the Centre for Development Studies, University of Cape Coast, on rural banks in 1982-83 found that

> there is still much to be accomplished by this banking scheme in its major objective of accelerating rural development within the broad national effort of developing the productive capacity of rural Ghana. Individual banks have still a long way to go in stepping up the process of rural resource mobilization and development in specific areas of their operation. The lack of detailed information and micro-data on bankable projects has handicapped the banks in promoting many primary economic activities as envisaged by the Bank of Ghana. As a result, most rural banks tend to keep investible funds in sterile assets and government stocks.[14]

9.3 CONCLUSIONS

The development banks have undoubtedly played a major role in expanding investment in Ghana. They have been particularly important in the development of manufacturing industries and agriculture. The expansion of rural banking, as pursued by the Bank of Ghana, has obviously emerged as a new source in the area of development financing.

Since the development banks are themselves beneficiaries of subsidised capital from the Bank of Ghana and other sources, the rate of interest they charge their customers is usually very low. For example, the Agricultural Development Bank and the National Investment Bank have received loans at interest rates of 2 to 3 per cent, while the Rural Banks have been provided with funds by the Bank of Ghana almost at a zero rate of interest. The ADB and the NIB have also been recipients of foreign exchange loans from the World Bank, the African

Development Bank and other international bodies at highly subsidised rates of interest.

Given the shortage of capital, it would of course be misleading to accept the view that the low rate of interest charged by these development banks can properly reflect the scarcity value of capital. Moreover, given that at the fixed exchange rate the cedi has been over-valued, one needs to take into account the shadow rate of exchange because the effect of making capital relatively cheap will be reflected in the proportionate use of the factor inputs.

As will be shown in Chapter 14, the basis of investment decisions in Ghana has remained poor. In its investment allocations the ADB has not used shadow prices, and although the NIB has used some partial estimates of the shadow foreign exchange rate and the shadow wage rate for unskilled labour, the basis of this estimation, particularly that relating to the foreign exchange rate, has not been realistic (see Chapter 14). However, it is understandable that in the absence of central guidelines relating to the use of shadow prices, investment allocation is likely to suffer, particularly if product and input prices remain highly distorted.

10 External Trade

10.1 EXCHANGE RATE POLICY

In the pre-independence era the commercial banks and a few big commercial houses constituted the foreign exchange market. It was generally free of government interventions and it closely followed movements on the foreign exchange market in London. Its controls, which were operated at the Ministry of Finance, applied only to trade relations with countries outside the Sterling Area.[1]

In July 1961 Ghana's rapidly dwindling foreign exchange reserves forced the government to pass the Exchange Control Act (Act 71) as part of a policy package to control and conserve the supply of foreign exchange. The Act made drastic changes in the existing foreign exchange market. The entire range of foreign exchange transactions was almost wholly restricted to the Bank of Ghana and to a few accredited agents.

The objectives of the Act include the conservation and control of the use of foreign exchange; ensuring the judicious allocation of scarce foreign exchange resources; and the maintenance of the exchange rate stability of the cedi. The exchange control system created by the Act is supervised by the Bank of Ghana. The three primary commercial banks are authorised to process initial applications and sales of foreign exchange to the public. Since December 1961 exchange controls have been supplemented by import controls. The administration of import controls is handled by the Ministry of Trade.

The Ghana pound, which became legal tender in July 1958, was at par with the pound sterling up to July 1965 when a decimal currency system was introduced under an 8s 4d-cent system (100 pence = 100 pesewas). The new currency had *cedi* as the major and *pesewa* as the minor unit. The name was changed to *new cedi* (N₵) with the introduction of the 10s-cent system (120 pence = 100 pesewas) in February 1967; in July of the same year the currency was devalued by 44 per cent from N₵0.71 to 1.02 per US dollar, following which import trade was liberalised. The liberalisation process continued for the next four and a half years. The currency was further devalued in December 1971 by 78 per cent from N₵1.02 to 1.82 per US dollar. The devaluation was followed by a military coup, and the new government revalued the currency in February 1972 from N₵1.82 to 1.28 per US dollar, and the letter 'N' was dropped from N₵. The new foreign exchange rate continued until December 1973, when an adjustment was made fixing the rate at ₵1.15 per US dollar.

Since January 1972 strict import controls have remained in operation. This period has also witnessed a rapid fall in the *internal* value of the currency following the high average inflation rate of 55 per cent per annum from 1972 to 1981. There has naturally been a corresponding fall in the intrinsic *external* value of the cedi, although this was not reflected in the official rate of exchange which remained fixed at ₵1.15 per US dollar from December 1973 to September 1978 and thereafter at ₵2.75 up to October 1983.

As the official supply of foreign exchange failed to satisfy demand there developed a parallel or black market in hard currencies. At times these were even encouraged by government policy measures,[2] the rates for which started rising as the imbalance in supply and demand became wider and wider. Considering the difficulties involved in getting accurate information on parallel markets, one should approach figures such as those shown in Table 10.1 with caution. However, it appears to indicate a fairly realistic picture relating to the direction of change. The black market rate for the US dollar, which was reported to be about three times as high as the official rate in September 1978, went up to fifteen times the official rate in October 1981 and further to about twenty-six times the official rate in March 1983.

The government was reluctant to deal with the problem of the sharp fall in the external value of the cedi by devaluation. As was observed by the Council of State in its Report on Economic and Fiscal Policies,

Table 10.1 Official and parallel foreign exchange rates of selected currencies, 1978-83
(₵ per unit of foreign currency)

Currency	Official rate (cedis per unit)	Sept 1978	Oct 1981	March 1983
Pound sterling	4.50-5.50 (approx)	16	75-85	110-120
US dollar	2.75	10	40-45	70-80
CFA franc (1000)	8.50-11.50 (approx)	36	120-140	200-220

Sources: Council of State, *Reviving Ghana's Economy*, p. 25; and data collected from personal interviews.

'the discussion of the value of the cedi has been characterized by undue emotion, and even fear'.[3] According to the PNDC Secretary for Finance and Economic Planning:

> Against the background of high inflation rates in the post-1978 period, the maintenance of the rigid official dollar/cedi rate has resulted in the over-valuation of the cedi against all major currencies. A number of anomalies follow from this over-valuation. Principal among these is the conferring of large and often illegal and untaxed profits on those who get access to foreign exchange.[4]

The April 1983 Budget officially recognised the over-valuation of the cedi through the introduction of surcharges and bonuses at either 7.5 or 9.9 times the face value of the official foreign exchange rate depending on the type of import or export.[5] This *de facto* devaluation of the cedi became *de jure*, although at a slightly higher rate, with the declaration by the government in October 1983 of the new foreign exchange rate at ₵30.00 per US dollar. The cedi was further devalued in March 1984 by 16.7 per cent, from ₵30 to ₵35 per US dollar. Following these devaluations the ratio of the parallel market rate to the official exchange rate fell drastically, from 25:1 in March 1982 to 4:1 in June 1984, although the cedi remained highly over-valued at the official exchange rate.

The continuing over-valuation of the cedi at the official exchange rate has deprived the economy of remittances, through official channels, by Ghanaians staying abroad. Developing countries such as Bangladesh and Pakistan have shown that exchange rate policies can be devised to raise foreign exchange earnings from this source. The example of Bangladesh is particularly relevant in that during the immediate post-independence years the local currency (Taka) was highly over-valued at the prevailing exchange rate and, as in Ghana, the booming parallel market in foreign exchange was depriving the country of remittances through the official channel. But the drastic devaluation of the Taka in 1974 and the introduction of a bonus scheme (called Wage Earners Scheme) have helped to correct the problem. According to the scheme, foreign exchange received through remittances from Bangladeshis working abroad is auctioned every day at the Central Bank and the rates (called bonus rates) are determined by the market.[6] Realistic foreign exchange rates in Pakistan have enabled that country to earn so much in remittances that the amount received through this source now exceeds that received from exports.[7]

Many countries from both Eastern Europe and the developing world have followed a two-tier foreign exchange rate policy to attract tourism. By contrast, in Ghana, the continuing over-valuation of the cedi has caused serious damage to the tourist industry. Before the cedi was devalued in 1983, the rent per night of two-star hotel room and the cost of a standard meal were around US $100 and $50 respectively at the official rate. The corresponding rates in nearby Abidjan, the 'Paris of Africa', were less than one quarter of the charges in Ghana at that time. An attractive rate of exchange for tourists could have significantly helped the tourist industry in Ghana.

The administration of multiple exchange rates, however, would put pressure on the banking system. Given the shortage of trained and experienced personnel in the Bank of Ghana, any attempt to introduce a new system demanding extra administrative work needs to be carefully examined. Indeed, in the case of Brazil the administration of a multiple exchange rate system built on foreign exchange auctioning was found extremely difficult because of the enormous complexities involved.[8]

10.2 EXPORT AND IMPORT POLICIES

The economy of Ghana used to be very open – exports comprised as much as one-quarter of GDP during the immediate pre- and post-independence periods, and the corresponding share of imports was equally high (see Table 10.2).

The country's declared policy has always been to increase the volume of exports. In an attempt to encourage exports, the existing export bonus of 20 per cent for *non-traditional* exports was raised to 30 per cent in 1977. For *traditional* exports other than cocoa, a new export bonus of 20 per cent was instituted.[9] Efforts were also made at reducing

Table 10.2 Exports and imports as percentages of GDP, 1956–82
(selected years)
(percentages: current prices)

	1956	1957	1960	1965	1970	1975	1980	1982
Exports	25	25	25	16	21	18	7	3
Imports	25	26	27	22	19	17	8	3

Sources and *Notes*: See Table A.3.

the bureaucratic barriers in the payment of the bonus to exporters. Procedures relating to exportation were also streamlined. However, following the devaluation of the cedi in 1978, which resulted in increased earnings to exporters in terms of cedis, the export bonus was reduced to 10 per cent. The country has also been trying to diversify its exports by processing some commodities before export.

In 1979 the government removed the export duty on timber and increased the value of import licenses granted to that industry, the objective being to boost timber exports. In 1980 the government introduced a policy by which the mining and timber industries were allowed to retain 2.5 per cent of their foreign exchange earnings in foreign accounts for the financing of their imports of certain vital inputs, thereby avoiding the delays associated with the normal import licensing system.

In April 1983, in an attempt to encourage exports, a highly attractive export bonus scheme was introduced in the annual budget. Also high import surcharges were imposed to discourage imports.

The control of imports has been the main plank of the country's import policy since December 1961. The main objectives of the policy are the judicious use of foreign exchange, protection of infant industries and the earning of revenue from import duties.

Throughout its history the administrative control of imports through the import licensing system has given rise to abuse, patronage and corruption.[10] Various governments attempted to take action to remedy the situation. For instance, the Nkrumah Government appointed the Abrahams Commission in 1965 to study the problem, while the Busia Government opted for import liberalisation. In 1980 the Limann Government tried to prune the number of importers to a more controllable size. Stricter adherence to the criterion of ability and adequate facilities to import on the part of the importer was used in the registration of importers. In August 1980 a new Import and Export Act was passed with the aim of controlling imports and encouraging exports. The Act revived the Open General Licence, Special Licence and Specific Licence that had been used at one period or another since 1961.

The *Open General Licence* permitted an importer to bring into the country without limit any of the goods listed under that title. The *Special Licence* allowed the importation of any goods into the country for which Ghana's own foreign exchange reserves were not used. The importer had to find his own foreign exchange. This licence had at one period or another been known as Special Unnumbered Licence

(SUL). The Act, however, listed the types of goods for whose importation this licence could be used. The *Specific Licence* permitted importers to bring into the country any goods listed under that title and Ghana's foreign exchange reserve was used for the importation of these commodities. The Bank of Ghana provided the foreign currency while the importer paid the cedi equivalent.

Inspectors have been used in the country of origin (or of consignment) of goods imported under specific licences to check that their quality, quantity and price conform to the licence granted and to the invoice of the goods. This was aimed at checking over-invoicing which could be used by unscrupulous importers. It was also aimed at preventing the import of inferior goods.

Although Ghana has multilateral and bilateral arrangements with a number of countries, wherever possible it has engaged in its trade relationships entitling it to 'the most favoured nation treatment'. For instance, along with other African, Caribbean and Pacific nations Ghana has a special association with the European Economic Community known as the Lomé Convention. It is also a member of the Economic Community of West African States (ECOWAS), established in 1975, which aims at increasing intra-West African trade and cooperation. Ghana also has special arrangements with some petroleum exporting countries such as Nigeria, the Soviet Union and Libya for the import of crude oil.

The official foreign exchange rate, as fixed by the government, is however only one item in the calculation of the effective rate of exchange faced by the exporters and importers. In the case of Ghana, other cost elements entering into the effective exchange rate for imports (EER_m) have included tariffs, purchase tax, sales tax, 100-day credit and an import licence fee, while the effective exchange rate for exports (EER_x) has included export duty. By incorporating the above into the nominal exchange rate Leith calculated EER_m and EER_x for the periods 1955-71 and 1955-69, respectively.[11]

As may be seen from Table 10.3, EER_m (₵ per US $) increased from 0.8362 in 1955 to 1.4564 in 1971, while EER_x for total exports increased from 0.4910 in 1955 to 0.7046 in 1969. The rate of increase for EER_m has been gradual, if uneven. But for exports the changes are more marked, especially in the mid-1960s. For example, EER_x was as low as 0.3481 in 1965. Moreover when EER_x is calculated separately for different export items, it appears that the increase in the rate has been generally much lower in cocoa than in diamonds and timber. For example, the EER_x of cocoa increased from 0.3946

Table 10.3 Effective exchange rate for exports (EER$_x$) and imports (EER$_m$), 1955-71
(selected years)

	1955	1960	1965	1969	1971
EER$_x$	0.4910	0.6260	0.3481	0.7046	—
EER$_m$	0.8362	0.8359	1.0566	1.4054	1.4564

Note: See text for the definition of EER$_x$ and EER$_m$.
Source: Leith (1974), pp. 11-13.

in 1955 to 0.5422 in 1969 (though in some intervening years the rates were much higher), while for wood and timber the corresponding increase was from 0.7024 to 1.008 and for diamonds from 0.6794 in 1955 to 0.7142 in 1969.[12]

No attempt has yet been made to calculate effective exchange rates for exports and imports since the early 1970s, and in the absence of such data we can only work out a rough estimate of the rates. The high inflation of the 1970s and the early 1980s and the failure to devalue the cedi and to adjust tariffs and other charges so as to make the effective exchange rate reflect changes in demand and supply must have had serious consequences on the import and export sectors. In the early 1980s a stage was reached when it became much more profitable to sell at home than to export, thus implying a very low EER$_x$; and it was much cheaper to import than to buy locally, implying a negligible EER$_m$.[13] Thus, by the early 1980s, the actual export and import policies in Ghana were not working in the way intended in the policy declarations to encourage exports and discourage imports. Rather, just the opposite of what was desired by policy occurred, and as a result the government had to impose severe restrictions on imports.

10.3 MAJOR EXPORTS

There has not been much change in the composition of exports since independence. Since the mid-1960s some cocoa beans have been processed into cocoa paste and cocoa butter prior to export. The aim is to increase the value of earnings from cocoa. Thus the quantity processed locally prior to export varied inversely with the world market price of beans.

In terms of quantity, cocoa and its products maintained their historically dominant position in the export sector. However, as Table 10.4 shows clearly, there has been a marked decline in the quantity of cocoa exported since the peak of 502 000 tonnes achieved in 1965. In quantity terms timber exports have also declined sharply since the mid-1970s. The quantities of gold, diamonds and bauxite exported show the same downward trend.

As Table 10.5 shows, in value terms the cocoa group has been dominant in the export sector. In 1960 the group contributed 59 per cent to the total value of exports. By 1970 its share had risen to 72 per cent and by 1980 it accounted for about two-thirds of total export earnings. The timber group contributed 14 per cent to total exports in 1960, but only 3.3 per cent in 1980. The share of gold, a historically important export item, was 10 per cent in 1960, but in 1970 gold contributed only 6 per cent to the value of exports. In 1980, because of a sudden rise in the world price of gold, the share of gold went up to 17 per cent of total export earnings. In value terms, the shares of the other minerals – bauxite, diamonds and manganese – have not been very significant, not exceeding one per cent in each case.

At 1970 prices exports as a whole declined at the rate of 2.0 per cent per year between 1960 and 1980. For the period 1975–80 the rate of decline was as high as 8.2 per cent per year. Ghana's terms of trade

Table 10.4 Exports by quantity, 1960–80 (selected years)
(quantity in '000)

	Unit	1960	1965	1970	1975	1980
Cocoa beans	Tonne	308	502	368	322	195
Cocoa paste and cake[1]	Tonne	3	21	17	19	7
Cocoa butter	Tonne	—	21	17	19	7
Timber (logs)	Cubic metre	1 077	559	600	430	83
Timber (sawn)[2]	Cubic metre	233	228	238	185	66
Bauxite	Tonne	226	288	214	320	195
Manganese ore	Tonne	556	577	403	373	148
Diamond	Carat	3 292	3 084	2 872	2 372	930
Gold	Grams	27 775	24 354	21 679	14 593	7896
Aluminium	Tonne	—	—	—	123	169
Others		—	—	—	—	—

Notes: 1. Includes cocoa powders; 2. Includes plywood and veneer boards. Data for 1960, 1965 and 1970 shown in the source in hop. ft. have been converted into cub. metre.
Sources: Economic Survey (various issues).

Table 10.5 Exports by value and percentage distribution at current prices, 1960–80 (selected years)

	1960 ₵ mil.	%	1965 ₵ mil.	%	1970 ₵ mil.	%	1975 ₵ mil.	%	1980 ₵ mil.	%
Cocoa beans	133	58	137	61	300	65	552	55	1804	58
Cocoa paste*	2	1	1	0.4	4	1	11	1	49	2
Cocoa butter	—	—	11	5	27	6	75	7	150	5
Timber (log)†	32	14	25	11	37	8	84	8	99	3
Bauxite	1	0.4	1	0.4	1	0.2	4	0.4	9	0.3
Manganese	13	6	10	4	7	2	17	2	25	1
Diamond	20	9	14	6	15	3	13	1	28	1
Gold	22	10	19	8	26	6	87	9	540	17
Aluminium	—	—	—	—	—	—	100	10	370	12
Others	6	3	7	3	43	9	55	6	31	1
Total	229	100	225	100	460	100	998	100	3104	100

* Includes cocoa cake.
† Includes sawn timber.
Note: Due to rounding the total of percentage distribution may not always add to 100.
Sources: Economic Survey (various issues).

deteriorated during the period 1960 to 1980, except in the case of cocoa beans and gold.

The country is clearly over-dependent on one commodity, cocoa, and its derivatives. However, in quantity terms, as mentioned above, almost all exports have been on the decline over the past decade or so. As mentioned in Chapter 5, factors responsible for the remarkable decline in the quantity of cocoa exports have been the unattractive prices paid to farmers over the years, the tapering off of government assistance in programmes to control diseases that affect the cocoa trees and the erratic supply of pesticides.

The unattractive prices paid to cocoa producers have contributed substantially to the declining output over the years. These low prices have induced farmers to tend and harvest existing farms less vigorously and made them less enthusiastic about expanding acreage. Low cocoa prices have also given rise to substantial smuggling of cocoa into the neighbouring Ivory Coast and Togo, where prices (at both the official and the parallel-market exchange rates) have been consistently higher. Some farmers have also turned to the cultivation of comparatively more profitable crops, such as maize and other foodstuffs, even cutting down cocoa trees in order to free the land for the purpose of planting food crops.

Transportation from the farm-gate to the ports for export has been another major bottleneck. All these factors have militated against expansion of cocoa output. Ghana's share of the world market has consequently fallen and she is no longer the leading producer of cocoa.

For other export commodities, such as minerals and timber, the major factor in the decline of output has been the over-valuation of the cedi over the years. The cedi equivalent paid to producers for the foreign exchange earned has been too low to induce them to expand output. For timber, the domestic market prices have been more attractive than export prices (in cedi terms). Lack of vital imported inputs and transportation difficulties have also been among the major causes of decline. With low export prices, in cedi terms, it is not surprising that, unlike the neighbouring Ivory Coast, Ghana has not developed new exports in the form of, for instance, canned and fresh fruits.

10.4 MAJOR IMPORTS

Ghana's major attempt at industrialisation after independence has had some impact on the composition of imports. Table 10.6 shows

Table 10.6 Imports by value and percentage distribution at current prices, 1960-80 (selected years)

Items	1960 ₵ mil.	%	1965 ₵ mil.	%	1970 ₵ mil.	%	1975 ₵ mil.	%	1980 ₵ mil.	%
Food and animals	42.0	16.3	35.3	11.0	79.5	19.0	104.9	11.5	241.5	7.8
Beverages and tobacco	7.6	2.9	2.2	0.7	3.9	0.9	6.7	0.7	43.0	1.4
Crude materials inedible except fuel	0.6	0.2	3.1	1.0	9.4	2.2	27.9	3.1	56.9	1.8
Mineral fuels	13.6	5.3	13.2	4.1	24.4	5.8	150.9	16.6	827.4	26.7
Animal and vegetable oils and fats	0.4	0.2	3.0	0.9	3.8	0.9	9.9	1.1	28.4	0.9
Chemicals	19.0	7.4	20.2	6.3	66.9	16.0	126.6	13.9	482.7	15.6
Manufacturing goods classified chiefly by material	79.6	30.8	108.5	33.9	100.8	24.1	207.9	22.9	380.5	12.3
Machinery and transport equipment	66.8	25.9	105.9	33.1	108.1	25.8	228.1	25.1	922.8	29.7
Miscellaneous manufactured articles	25.2	9.8	23.2	7.3	16.4	3.9	28.9	3.2	64.5	2.1
Miscellaneous commodity N.E.S.	3.6	1.4	5.4	1.7	5.8	1.4	17.5	1.9	55.8	1.8
Total	258.4	100	320.0	100	419.0	100	909.3	100	3103.6	100

Note: Due to rounding, the total of percentage distribution may not always add to 100.
Sources: CBS; *Economic Surveys* 1960, 1967, 1972-74 and 1982.

imports by value and percentage share from 1960 to 1980. The historically dominant item is machinery and transport equipment. Other important groups include mineral fuels, manufactures, chemicals food and animals. Imports of manufactured goods (both consumer and intermediate) were around one-third of the total value of imports in 1960 and 1965, and have gradually, though unevenly, declined since the mid-1960s, amounting to only 12.3 per cent in 1980. The import of miscellaneous manufactured articles declined throughout the 1960s, falling from 9.8 per cent of the total value of imports in 1960 to 3.9 per cent in 1970 and 2.1 per cent in 1980. A declining trend is also observed for imports of food and animals, particularly marked since the early 1970s. On the other hand chemicals, which as a group took only 6 to 7 per cent of the total value of imports in 1960 and 1965, showed a significant increase (to 16 per cent) in 1970; its corresponding share was 13.9 per cent in 1975 and 15.6 per cent in 1980.

In quantity terms, imports of crude oil for refining in the country for domestic use have declined slightly, but in value terms the rise has been steep, following the upward revisions in oil prices after the 1973 oil crisis (see Table 10.6). Measured at 1970 prices, mineral fuels form the only major item which recorded significant increases, from ₡24.4 million in 1970 to ₡63.4 million in 1975, and further to ₡73.0 million in 1980.

Table 10.7 shows that imports of capital goods constituted 21.6 per cent of all imports by end-use in 1960. This rose to 30.2 per cent in 1965, but declined to 26.5 per cent in 1980. The percentage share of imports of raw materials rose from 23.8 per cent in 1960 to 41.0 per cent in 1975. But it has fallen since then; the corresponding figure for

Table 10.7 Composition of imports by end-use, 1957–80 (selected years) (percentages: current prices)

Item	1957	1960	1965	1970	1975	1978	1980
Imports of raw materials	24	23.8	31.5	39.7	41.0	34.4	29.8
Capital goods	13	21.6	30.2	24.4	22.1	28.7	26.5
Consumer goods	56	50.0	34.3	30.1	20.3	23.2	17.1
Fuel/lubricants	6	4.6	4.1	5.7	17.0	13.7	26.6
Total	100	100	100	100	100	100	100

Note: Due to rounding off the total may not always add to 100.
Sources: 1957 Birmingham (1966), p. 334; CDS, *Statistical Year Book*, 1965–66; and CBS, *Economic Surveys* 1967, 1972–74, 1977–80, 1981 and 1982.

1980 was 29.8 per cent. While raw material imports have been on the increase, imports of consumer goods are on the decline, mainly due to strict controls on some imported goods and out-right bans on others. The percentage share of fuel and lubricants has been rising sharply in recent years, due partly to increasing prices of petroleum since 1973 and partly to the liberal import policy for this item.

At 1970 prices, imports as a whole declined at the rate of 2.6 per cent per annum between 1960 and 1980. The rate of decline was particularly sharp in the period 1975–80, when it was 6.5 per cent per annum. Thus the country's declining export performance has forced it to tighten its belt with respect to imports, since it is mainly exports that pay for imports.

10.5 DIRECTION OF TRADE

As a developing country Ghana's imports have largely consisted of manufactured products from the industrialised nations. However, the origin of imports is currently much more diversified than in the period prior to independence, when the Sterling Area was the main source providing about half of the country's imports. The share of Britain alone ranged between 42 and 56 per cent of total imports in the early 1950s.[14]

As for exports, the share of the Sterling Area in the country's exports fell from 36 per cent in 1960 to 23.8 per cent in 1965, increased to 28.4 per cent in 1970, and declined again to 14.8 per cent in 1979 (see Table 10.8). Exports to the Dollar Area, which increased from 15.9 per cent in 1960 to 19.2 per cent in 1970, fell to 16.4 per cent in 1975, and further to about 7 per cent in 1979. The share of the EEC (excluding UK) fell from 35.1 per cent in 1960 to 21.8 per cent in 1975, but went up to 27.1 per cent in 1979. The share of Japan in Ghanaian exports has also increased significantly, from less than 1 per cent in 1960 to 8.3 per cent in 1979. A similar increase is observed for countries grouped as others, their share increasing from 3.4 per cent in 1960 to 17.3 per cent in 1979.

The African countries took a very small percentage of Ghana's exports. In 1978 exports to other African countries amounted to only ₵23.0 million, or 1.5 per cent of Ghana's exports.[15] In 1979 total exports to the rest of Africa were only ₵25.2 million or 1 per cent of total exports. Egypt and Nigeria together took about 40 per cent of the country's exports to other African countries in 1978, while in 1970

Table 10.8 Direction of exports, 1960–79 (selected years) (percentages: current prices)

Direction	1960	1965	1970	1975	1979
Sterling area	36.0	23.8	28.4	17.7	14.8
(a) Of which United Kingdom	31.3	20.8	23.2	13.4	14.7
(b) Of which African Countries	2.0	1.0	0.3	0.7	0.1
EEC excluding United Kingdom	35.1	27.8	23.8	21.8	27.1
Dollar area	15.9	18.6	19.2	16.4	7.0
USSR, China and European Countries (excluding EEC)	7.2	21.3	16.3	20.4	24.7
African countries excluding those in sterling area	1.5	0.9	0.6	1.8	0.8
Japan	0.8	2.3	6.5	7.2	8.3
Others	3.4	5.3	5.3	14.7	17.3
Parcel post	Negl.	Negl.	Negl.	Negl.	Negl.
Total	100	100	100	100	100

Note: Due to rounding the total may show slight discrepancy at times.
Source: CBS, Economic Survey 1964, 1967, 1972–74 and 1981.

Togo received as much as 46 per cent of Ghana's official exports to the rest of Africa. It is, however, believed quite strongly (and there is much evidence for the fact) that large quantities of Ghana's products, such as cocoa, timber, some manufactured items, and foodstuffs are smuggled across the borders to neighbouring Ivory Coast, Togo and Burkina Faso.

Although the Economic Community of West African States (ECOWAS) was formed in 1975 with the major objective of promoting intra-West African trade, official data do not show any appreciable increase in Ghana's trade with the ECOWAS countries since the establishment of that organisation.[16] It is true that trade with Nigeria is on the increase, but this is due mainly to that country's position as a major exporter of oil. Ghana's imports from the other ECOWAS members are negligible.

Ghana's exports to the ECOWAS members are also insignificant. The share of Egypt in Ghana's exports to the rest of Africa is even more important than the share of most members of ECOWAS. Only Togo officially imports an appreciable quantity of goods from Ghana.

The absence of a buoyant intra-ECOWAS trade may be due to the fact that they all produce similar export commodities, and that Ghana has generally been reducing its imports in recent years. Table 10.9 shows the origin of imports in the 1960-79 period. The share of the Sterling Area, Ghana's historically important trading partner nations, in the country's total import trade was as high as 41.7 per cent in 1960 but has declined since then, with some fluctuations, to 33.5 per cent in 1979. The African countries in the Sterling Area have recorded important increases since the mid-1970s, their share being as high as 13.3 per cent in 1979, compared to less than 2 per cent in 1960, while the share of the United Kingdom was down from 36.7 per cent in 1960 to 15.1 per cent in 1975, recovering slightly in 1979 (20.2 per cent). The share of the Dollar Area's exports to Ghana increased from 8.3 per cent in 1960 to 21.4 per cent in 1970, but has declined since then to 10.7 per cent in 1979.

The group shown under USSR, China and European countries showed a sharp increase in their exports to Ghana, their share rising from 4.3 per cent in 1960 to 26.3 per cent in 1965, a result of close

Table 10.9 Origin of imports, 1960-79 (selected years)
(percentages: current prices)

Origin	1960	1965	1970	1975	1979
Sterling area	41.7	28.8	28.7	38.9	33.5
(a) Of which United Kingdom	36.7	25.8	23.7	15.1	20.2
(b) Of which African Countries	1.6	1.3	0.8	12.7	13.3
EEC (excluding United Kingdom)	25.6	21.4	21.8	23.1	15.2
Dollar area	8.3	10.6	21.4	18.9	10.7
USSR, China and European countries (excluding EEC)	4.3	26.3	8.5	5.5	3.9
African Countries (excluding those in sterling area)	5.2	2.8	4.4	2.5	4.9
Japan	8.3	4.3	6.2	6.5	4.9
Others	5.2	5.4	8.5	4.7	16.8
Parcel post	1.4	0.3	0.5	0.4	0.3
Total	100	100	100	100	100

Sources and Notes: See Table 10.8.

trade links with the socialist bloc at that time. However, since the mid-1960s their share has declined to about 4 per cent in 1979. Other countries (not included in any of the major groups in Table 10.9) have recorded significant increases from the mid-1970s - 4.7 per cent in 1975 to 16.8 per cent in 1979, thus showing some signs of import source diversification on the part of Ghana.

10.6 TERMS OF TRADE

As a general concept, the terms of trade shows a relationship between the price a producer pays for the product he buys and the price he receives for his own product. The producer is in a better position if his selling price rises to a greater extent (or falls to a lesser extent) than the prices of the products he buys.

The concept is most frequently used in international trade, where it involves a comparison of the price indices of a country's exports and imports. It is defined as the amount of import a country gets per unit of its export. Thus for a country the terms of trade are unfavourable or adverse if import prices rise in comparison with export prices. The terms of trade are generally calculated by dividing the index of export prices by the index of import prices and then multiplying the quotient by 100 to put the figure in percentage terms. Thus a figure above 100 indicates that the terms of trade are favourable to a country; any figure below 100 indicates an unfavourable turn in the terms of trade. Alternatively, the index of import prices may serve as the numerator and the index of export prices the donominator, in which case a figure above 100 indicates unfavourable terms of trade and vice versa.

An earlier analysis of the country's terms of trade for the period 1950-62 showed that Ghana experienced a significant deterioration in its terms of trade throughout the period. Commenting on the data Killick observed:

> It is evident, if we compare the later years with the 1954 base year, that Ghana has indeed suffered a deterioration in her terms of trade. In 1962 Ghana would have had to export 98 per cent more in order to buy the same volume of imports that she bought in 1954.[17]

However, he was of the view that due to the weights assigned to the various export commodities, especially cocoa beans whose prices on the world market fluctuate violently, the official terms of trade figures tended to exaggerate the deterioration.

External Trade 211

Table 10.10 Terms of trade, 1969-82
(1968 = 100)

Year	Price index of imports	Price index of exports	Terms of trade
1969	111.9	118.0	105.5
1970	122.6	128.6	104.9
1971	134.6	109.3	81.2
1972	175.8	126.1	71.7
1973	197.2	176.7	89.6
1974	272.1	274.1	100.7
1975	311.8	297.7	95.5
1976	359.2	286.1	79.6
1977	419.4	433.7	103.4
1978	729.0	688.3	94.4
1979	1349.5	1283.1	95.1
1980*	1641.0	1558.8	95.0
1981*	1575.4	1270.0	80.7
1982*	1512.4	1003.3	66.5

* Provisional.
Sources: *Economic Survey* 1977-80, 1981 and 1982. Data for the later years shown in the source at 1975 = 100 have been converted so as to make these comparable with the data for the earlier years at 1968 = 100.

Data on the terms of trade covering the period 1969-82 are shown in Table 10.10. The table, which presents the ratio of the export prices index to the import prices index, shows that the terms of trade have been unfavourable in ten out of the fourteen years.

However, the upward movement of the prices of primary commodities in 1974 which enabled the country to realise a favourable terms of trade index for that year also helped Ghana to absorb the initial shock emanating from the oil crisis of the period. Since then the terms of trade have been favourable only in 1977.

It may be noted that the figures for most of the adverse years up to 1980 did not fall far below 100. Thus the decline in the terms of trade from 1975 to 1980 may at best explain a very small fall in the annual export revenue – a situation that could have been ameliorated by a small increase in the volume of exports. But as may be seen from Table 10.4, in terms of quantity, the export of cocoa, gold, timber and most other export commodities declined substantially in the period.

As far as the last two years (1981 and 1982) in Table 10.10 are concerned, the unit value of exports fell at a faster rate than that of

Table 10.11 Balance of payments, 1970–84
(mil. ₡: current prices)

Year	Trade balance (1)	Net invisible (2)	Current account (3) = (1) + (2)	Net non-monetary capital (4)	Overall balance (5) = (3) + (4)
1970	+36.8	−126.8	−90.0	+100.2	+10.2
1971	−84.9	−117.8	−202.7	+190.0	−12.7
1972	+164.6	−38.7	+125.9	−7.3	+118.6
1973	+176.8	−18.9	+157.9	−31.3	+126.6
1974	−128.3	−60.0	−188.3	+24.1	+164.2
1975	+89.2	−81.9	+7.3	+66.5	+73.8
1976	+26.6	−96.8	−70.2	+116.0	+45.8
1977	+84.1	−252.0	−167.9	+222.8	+54.9
1978	+149.9	−661.1	−511.2	+687.3	+176.1
1979	+452.0	−577.1	−125.1	−145.5	−270.6
1980	+231.7	−704.4	−472.7	+83.8	−388.9
1981	−669.9	−718.0	−1 387.9	+602.2	−785.7
1982	+144.7	−575.9	−431.2	+416.6	−14.6
1983	−1 840.0	−5 590.0	−7 430.0	+2 876.0	−4 554.0
1984	−1 018.6	−9 202.6	−10 221.2	+4 556.3	−5 664.9

Source: 1970–78 from CBS, *Economic Survey*, 1972–74, 1975–76 and 1977–80. 1979–83 from CBS, *Quarterly Digest of Statistics*, September 1985.

imports. As a result there was a further fall in the terms of trade, down to 66.5 in 1982.

10.7 BALANCE OF PAYMENTS

Earlier studies of Ghana's balance of payments cover the periods 1950-62 and 1960-71.[18] They reveal that the early 1950s were years of assured comfort, with hefty balance of trade surpluses caused partly by the Korean War boom of the early part of that decade.

Ghana thus began her independent life with external reserves estimated to be more than the equivalent of one year's import requirements (the result of pre-independence balance of payments surpluses). However, between 1960 and 1965 the international liquidity was drawn down from as much as ₵544 million to ₵59 million, in other words to the equivalent of about a tenth of the value of imports for 1965. The country had during the same period accumulated external debts estimated at approximately ₵640 million by February 1966.[19] The worsening of the balance of payments was not the result of an import boom (except during 1969-71, when a liberal trade policy was pursued) as the country had introduced stringent foreign exchange and import controls in 1961. The deficit in the current account was largely due to the increase in the deficit on invisibles like freight, insurance and net transfers abroad. Exports were generally stagnant throughout the decade and that to a large extent was the cause of the unfavourable balance of payments of the 1960s.

As Table 10.11 shows, during the fifteen-year period (1970-84) there had been current account deficits in the balance of payments for twelve years, and surpluses in only three years. The balance of trade had been negative in only five years out of fifteen, while the invisible balance was negative for all the years. The deficit in net invisibles was large enough to wipe out the surplus in trade balance in all except three years (1972, 1973 and 1975). The long-term capital account was favourable throughout the period with the exception of three years (1972, 1973 and 1979).

With the country registering deficits in the current account of the balance of payments in all the years since 1976, imports have had to be restricted. This has consequently had harmful effects, particularly on sectors such as mining, manufacturing and transportation which have depended heavily on imports.

11 Prices and Internal Trade

11.1 INFLATION

Ghana experienced her first serious bout of inflation in the mid-1960s. In terms of the national consumer price index the annual rate of inflation in 1964 was about 10 per cent and jumped to 26.4 per cent the next year. The rate of inflation slowed down a bit in 1966, and there followed three years of low inflation (below 10 per cent). The inflation rate was lowest in 1970 (3.9 per cent), but has generally increased since then and was as high as 116.3 per cent in 1977. The rate of inflation was 73.3 per cent in 1978 and hovered around 50 per cent during the next two years, but jumped again in 1981 to 117.1 per cent. It fell sharply to 22.3 per cent in 1982, but rose to 122.8 per cent in 1983. It was 40.2 per cent in 1984.

Table 11.1 shows annual rates of inflation from 1961 to 1984. One major factor behind the development of the inflationary pressure was the government's policy of budgetary deficits, and the financing of deficits mainly by borrowing from the central bank and the commercial banks, with the result that more money was pumped into the economy than was warranted by real growth in GDP. In addition, in 1964–65 there was a sharp increase in total payments made to cocoa farmers for the record 1964–65 cocoa crop. In 1963–64 the total amount paid to cocoa farmers was ₵43 million, while in 1964–65 the total amount paid to them was ₵57 million – an increase of 33 per cent. With this high level of income, farmers entered a consumer goods market in which the supply of goods was on the decline due to shortages and import restrictions. Hoarding of goods accentuated the physical shortage of imports during the period 1963–1965.

Although inflation had almost disappeared by 1970, the rest of that decade and the early 1980s witnessed high inflation in the country. The NLC Government had initiated a policy of trade liberalisation in 1967, which was pursued by the succeeding Busia Government.[1] However, the import liberalisation policy and the precipitous fall in the world market price of cocoa in 1971 combined to cause a rapid decline in the country's foreign exchange reserves, leading to balance of payments difficulties (see Chapter 10). The government responded by introducing stiffer restrictions on imports and foreign exchange transfers and the cedi was devalued in December 1971.

Table 11.1 Annual rates of inflation, 1961–84
(percentages)

Year	Inflation rate	Year	Inflation rate
1961	3.6	1973	17.5
1962	1.7	1974	18.4
1963	6.8	1975	29.7
1964	9.6	1976	56.4
1965	26.4	1977	116.3
1966	13.3	1978	73.3
1967	−9.0	1979	54.2
1968	8.8	1980	49.7
1969	7.1	1981	117.1
1970	3.9	1982	22.3
1971	9.3	1983	122.8
1972	10.1	1984	40.2

Notes: 1961–63: Implicit inflation rates in the GDP deflator. 1964–84: Implicit inflation rates in the national consumer price index.
Sources: See Tables A.8 and A.15.

The military government which took over in 1972 revalued the cedi and imposed rigorous price controls. But for all that, the government failed to control inflation. Indeed, from that year onwards inflation gathered momentum. The situation worsened in 1976 and in subsequent years Ghana's inflation could truly be termed as galloping.

In discussing the causes of inflation during the late 1970s and early 1980s, the Central Bureau of Statistics put the blame squarely on the 'huge borrowings by the Government from the central bank, which continued to increase year after year'.[2] Total money supply by the end of 1981 was about 29 times the level at the end of 1971. The Bureau cites the worsening supply position of domestic agricultural produce over the decade as another major contributory factor generating the runaway inflation.

Foreign exchange shortages resulting from the decline in the quantum and value of exports and the drying up of the inflow of external capital, the consequent import restrictions and the inability to import raw materials for import-substituting manufacturing industries, have all been working to push up prices.

However, the rate of inflation has not been the same for different sectors. At the official exchange rate, the rate of inflation for *imported* food has generally been low, and particularly so in times of severe food shortages in the country (for example in 1963–65 and 1975–77

Table 11.2 Average annual inflation rates of selected items, 1960-80
(percentages)

Period	Food Local	Food Imported	Transport and communication	Imports	Exports
1960-65	23.46*	10.02*	8.26*	1.70	-2.93
1965-70	3.99	3.11	3.18	10.50	15.76
1970-75	17.66	19.90	15.11	18.94	18.92
1975-80	107.23†	42.94†	31.87	49.57	37.52

* For 1963-1965.
† For 1975-1977.

Sources: Inflation rates for food, and transport and communication are computed from CBS, *Consumer Price Index Numbers* 1978 and *Statistical News Letter* No 18/83 (23 November 1983), while those for total imports and exports are calculated from imports and exports deflators (see Table A.8).

when the rates of inflation for *local* food were much higher). Low official prices for petrol, low fares for government transport on road and railway, and low telecommunication fees have helped to keep the rate of inflation in transport and communication generally low. The over-valuation of the cedi has also kept the rate of inflation relatively low for exports and imports in three out of the four periods shown in Table 11.2. The only exception is the period from 1965 to 1970 when the effective exchange rate was kept higher for both the exporters and importers.

11.2 PRICE CONTROLS

Price controls are a means of containing the impact on prices of excess demand over supply. The main policy objective of price controls is to reduce the harmful effects of inflation on the distribution of income.

In the case of Ghana price controls were introduced in 1962 with a view to protecting consumers by preventing importers and local manufacturers of import-substitutes from earning monopoly rents or 'super-normal' profits arising from import restrictions and exchange controls.[3] It was also regarded by the government as an 'egalitarian' device for safeguarding the interest of the poor. Price controls came into being along with the promulgation of quantitative restrictions on imports and foreign exchange controls. Indeed, at the time of the imposition of price controls and thereafter, the price control list was confined mainly to imported consumer goods and import-substitutes with high import content. The aim of price controls was thus to keep prices of goods low enough so that the poor could afford them.

By 1969 the government was seeking to administer a complex set of price controls. For instance, the number of separately specified items totalled 725, and for many of these products several prices were given depending on quantity and locality of sale.[4] At the time, 5920 separate prices were itemised. The magnitude of such a task demanded much strenuous effort on the part of the enforcement agencies. From 1962 to 1964 the Ministry of Trade used the services of voluntary, unpaid price inspectors. In 1964 the Ministry employed 20 full-time price inspectors, and by 1966 it was employing 48. But the price inspectors were alleged, by the Abrahams Commission of Enquiry into Trade Malpractices, to be in the pay of profiteers or were themselves engaged in buying at the controlled prices and selling far in excess of those prices. According to the Abrahams Report, since the cost of bribing the inspectors was passed on to the consumer, 'if there were no voluntary price inspectors, the prices of commodities could be expected to be just that much lower'.[5]

The government therefore lost interest in employing price inspectors. By 1970 the staff consisted of only one Chief Price Inspector. The task of enforcement was delegated to the police. However, price controls were honoured more in the breach than in the observance.[6] Only the large expatriate trading companies in the urban centres, such as UAC, UTC, PZ, SCOA and CFAO, and the state-owned GNTC observed the price controls to any significant extent.[7] For market-women price controls only existed on paper. Most of them had no respect for the regulations and ignored them with impunity.

Thus by 1970 it was fairly evident that price controls were ineffectual. The Busia Government, therefore, put less emphasis on controls as a government policy, and considerably reduced the number of goods covered by the price controls.

The new military government which came into power in 1972, however, extended the list considerably. The penalty for contravention was also increased. The military were added to the police as authorised price inspectors. In addition, the government established the Prices and Incomes Board in 1972 as a merger of three government departments – the Incomes Commission, the Public Services and the Pay Research Unit. The new Board was charged with the following responsibilities:

(a) to formulate appropriate incomes and prices policy for the successful development of the national economy;
(b) to conduct periodic reviews of incomes and prices with a view to making any modifications that may be necessary;
(c) to recommend specific policies on wages, salaries, interest, profits, dividends, rents and prices in the Ghanaian economy; and
(d) to advise the government generally on all aspects of wage and price policies.

But the new military régime did not fare any better than the previous governments in enforcing price controls. The civilian government that succeeded the military in 1979 also failed to stamp out the widespread practice of ignoring controlled prices. It opted for a policy of trade liberalisation, generally allowed prices to find their own dizzy levels without doing away with the low official prices and continued the over-valuation of the cedi at the fixed exchange rate of ₵2.75 per US dollar.

Following the military take-over in December 1981, the policy of price controls was revived on a grand scale. In April 1982 official price lists, published in two volumes, contained about 1900 prices, many items having separate prices depending on size, make and quality. However, in 1984, realising the futility of price controls the government opted for reducing the wide gap between official and parallel market prices by increasing the official prices.

Table 11.3 shows average percentage increases in the official and parallel market prices of five selected commodities which are commonly used and are in the PIB's price list. Parallel market prices used in the calculation are based on a sample survey.[8] As such prices are subject to significant variations we have taken an average.

As may be seen from the table, both official and parallel market prices increased sharply for all the items. Between 1976 and 1982 the average annual increase in official prices ranged from 33 to 55 per cent, while for the parallel market increases ranged from 62 to 85 per cent. The average annual rate of inflation, measured by the consumer price index, was 69 per cent from 1976 to 1982; the price increases in the parallel market were therefore broadly in conformity with inflation.

From 1982 to mid-1984 the increase in official prices of goods was sharp, ranging from 59 to 158 per cent per year. This is against the background of an average annual rate of inflation of about 77 per cent from 1982 to 1984. There were also increases in the prices of

Table 11.3 Average annual increases in official and parallel market prices of selected commodities, 1976-84
(percentages: current prices)

	1976-82		1982-84*	
Item	Official price	Parallel market price	Official price	Parallel market price
Sugar (kg)	33	85	59	12
Ideal milk (170 g)	55	76	91	15
Beer (large bottle)	42	68	112	25
Soap (Guardian per cake)	45	71	158	34
Cooking oil (per gal.)	44	62	89	49

* 1984 data refer to June 1984.
Sources: Prices and Incomes Board, and Sample Survey.

goods on the parallel market from 1982 to 1984, but the annual rate of increase was lower (12 to 49 per cent) than the official rates.

The insistence, prior to 1983, on keeping official prices down, was based on the belief that this would

(a) check inflation by keeping prices of goods low;
(b) safeguard the interests of the poor by providing them with goods at low official prices; and
(c) control monopoly rents.

That inflation was not checked does not need any explanation. The ratio of parallel market to official prices of government controlled goods was as high as 50:1 for certain commodities such as soap. As the parallel market prices charged contained 'monopoly rent' as well as 'risk premium', it is difficult to accept that in the absence of low official prices the parallel market prices would have been much higher.[9] Anyway, 'monopoly rent' remained uncontrolled, although the beneficiaries of such 'rent' probably comprised a chain of middlemen.

'Safeguarding the interest of the poor' was an objective that existed only on paper. Indeed, it was not the poor, nor even the rural middle class, who were the beneficiaries of the official controlled prices. The customers who benefited most from the control system were institutions such as schools, universities and hospitals, and individuals in the upper and middle income groups who generally belonged to organised bodies in the public and semi-public sectors.

11.3 INTERNAL TRADE

In Ghana, the unorganised sector of internal trade is the traditional preserve of women. The 1960 *Population Census* revealed that of the 369 000 people engaged in commerce as many as 74.5 per cent were women. In 1970 the share of women in the distribution trade was 85.2 per cent. In addition, many farmers' and fishermen's wives who do not regard themselves as *bona fide* traders do undertake trading in farm products and fish at certain times of the year. Thus the number of people engaged in the unorganised sector of internal trading fluctuates quite widely, especially for those who participate in seasonal food and fish distribution.

In 1965 Rowena Lawson found that about 86 per cent of the traders were self-employed.[10] They operated on a small scale, with low turnovers and low incomes and, generally, in a highly competitive environment, especially outside the cities and large towns. However, in the cities and large towns, a few pockets of controlling 'rings' existed restricting the effectiveness of market competition. These price rings were constantly being accused by politicians of hoarding goods, or of otherwise manipulating supply to their commercial advantage.

The unorganised trading sector is still characterised by low labour costs, few overheads apart from transport costs, well-developed entrepreneurship and managerial ability. The trader in local foodstuffs may buy directly from the producer and sell directly to the consumer. She may be a wholesaler or an agent of a wholesaler; she may also be a retailer or a combination of any or all of these. She may visit a village and buy maize from several farmers until she obtains a lorry load, which she then conveys to the city or large town to sell to other middlewomen. Farmers' wives, too, easily become traders if they deem the effort worthwhile.

There are also thousands of itinerant traders who visit several markets each week, moving goods from one to another. They watch each market carefully to determine which goods are in high demand, and they try to meet that demand.

Although new market places have been springing up in the suburbs of the cities and large towns, the geographically limited central markets which cannot grow physically in response to population growth have become the breeding grounds of monopoly, price rings and restrictive trade practices of all kinds. Traders in each major commodity, such as yam, maize or cassava, have their elected Queen Mothers. The functions of the Queen Mother include the protection of the sectional interests of the retailers of the particular commodity; providing bargaining strength; serving as guarantor where credits are extended by other traders or lorry drivers; and generally settling quarrels and misunderstandings among traders, or with their suppliers and drivers, in order to achieve smooth and mutually beneficial trading relations. The Queen Mothers ensure high selling prices by encouraging the formation of price rings, a policy which is easily implemented since in the cities and large towns the central markets are physically limited in size and congested, thus enabling a relatively small number of traders to wield immense economic power.

Statistics on the distributive trade are not available for recent years. However, the 1962 census of distributive trade found that as much as 95 per cent of the trade establishments engaged four or fewer persons.[11] The share of wholesale trade in the total distributive system was only 2.3 per cent, while retail trade had a share of 84.8 per cent and hotels and restaurants took 12.9 per cent. The respective shares of these three groups of trading activities in employment were 10.7, 74.6 and 14.7 per cent.

The organised sector of internal trade was, up to 1961, dominated by the large expatriate trading firms such as UAC, UTC, SCOA, CFAO and PZ, many Levantine and Indian stores and a few large privately-owned indigenous trading companies. These companies had their headquarters in Accra and branches in the other cities and large towns. In 1961 the Ghana Government established the Ghana National Trading Corporation (GNTC) by buying and merging the British-owned Commonwealth Trust Limited and the Greek-owned A. G. Leventis. The new firm was, up to 1966, favoured in the allocation of import licences and it quickly expanded to become a giant among giants. In addition, other state-owned commercial enterprises such as the Ghana Food Distribution Corporation (GFDC) have been established. However, these state enterprises are found to operate inefficiently, many of them earning meagre profits or losing heavily, and most of them suffering from over-manning (see Chapter 13).

Along with these big commercial houses there are smaller trading firms which are usually sole proprietorships or partnerships and which usually get their supplies from the big commercial houses with whom they compete keenly. The Kwahus (one of the Akan tribes) dominate the organised small-scale trading sector in Accra and many other large towns.

With Ghana's continuing balance of payments difficulties, which have substantially reduced her ability to import consumer goods and raw material for the import-substituting manufacturing industries, sales in the organised market have shrunk drastically (see Table 11.4). For instance, sales by the big seven – GNTC, UAC, UTC, CFAO, PZ, S. D. Karam, and SCOA – at 1970 prices stood at ₵203.4 million in 1975, ₵87.6 million in 1978, ₵33.0 million in 1980 and were down to ₵13.6 million in 1982.

Table 11.4 Distributive trade: sales by the big seven retail stores, 1975–82 (sales in ₵ mil.)

Year	Current prices	1970 prices
1975	484.0	203.4
1976	504.5	148.2
1977	668.8	136.5
1978	885.7	87.6
1979	970.8	69.9
1980	867.2	33.0
1981	1311.7	19.4
1982	1204.6	13.6

Sources and Notes: Economic Surveys 1977–80 (p. 313) and 1982 (p. 227). Trade Deflator (see Table A.10) has been used to convert current prices.

Because demand for goods subject to price controls far exceeded the supply, sales of these goods have not always been kept 'open' to the public. It has been a common practice for store managers to sell goods under price control mainly to people known to them and, as a result, a type of 'barter' system seems to have developed – the beer dealer selling to the filling station manager, the pharmacist to the beer dealer and so on, in exchange for the return of the favour from the customer. Those who were not able to provide return services or offer some form of favour were usually deprived of the benefit of obtaining goods at controlled prices. Visiting the stores during 1980 to 1983, one could see only empty shelves. At times one might find some local goods such as yam, wooden furniture, or cane baskets, the prices of which were comparable to open market prices and which did not strictly come under the jurisdiction of the Prices and Incomes Board.

224   *Money, Banking, Trade and Employment*

Some corrupt officials and management personnel in the distributive system also engaged in the corrupt practice of diverting goods from the controlled to the parallel market, in collusion with middlemen and market 'mummies' (as they are called in Ghana). This gave rise to what is now commonly known as the *kalabule* economy, a topic treated in some detail in Appendix B.

12 Employment

In Ghana, as in many other developing countries, employment expansion has been an important objective of government policy. One reason for this is that a strategy of employment expansion, if addressed to effective employment, would enhance the prospect for economic growth, thus narrowing the scope of unemployment and waste of human resources. A second reason is that given the rapidly growing size of the labour force, the adoption of a strategy for employment expansion has the advantage of containing any social and political tensions and outbursts due to overt unemployment.

Underlying the first argument for employment expansion is the question of efficiency and, more particularly, the extent to which labour intensive strategies are more efficient than capital intensive ones. The issue is discussed at some length in Chapter 14, in the context of technology choice. The second argument for employment expansion, however, appears to command a strong appeal in many developing countries. In Ghana, the strength of this appeal is reflected by the growing role of the government as a major employer. Since employment creation in the public sector can be motivated by non-economic factors, the economy can disguise the seriousness of the unemployment problem through over-manning, but only at the cost of undermining the prospect for economic growth.

According to the 1970 Census, the average annual increase in the labour force between 1960 and 1970 was 2.1 per cent. On this basis, the total labour force in 1982 is estimated at 4.2 million as against 3.3 million in 1970. The growth of the labour force and its distribution by sectors was discussed in Chapter 1. In this chapter, we concentrate on the use of manpower by the public and the private sectors and also review the growth of real wages and salaries.

12.1 EMPLOYMENT IN PUBLIC AND PRIVATE SECTORS

Data on total recorded employment and its breakdown into public and private sectors are shown in Table 12.1. The data shown in the table should, however, be interpreted with caution. First, the coverage of establishments in respect of which information could be obtained

Table 12.1 Recorded employment in establishments employing ten or more persons, 1957–79 (number in '000: December) (selected years)

	Public sector						Private sector						Total					
	1957	1960	1965	1970	1975	1979	1957	1960	1965	1970	1975	1979	1957	1960	1965	1970	1975	1979
Agriculture, hunting, forestry and fishing	28.7	43.3	49.6	43.6	59.4	71.0	12.2	13.9	6.3	5.3	3.9	2.9	40.9	57.7	55.9	48.9	63.3	73.9
Mining and quarrying	Neg.	Neg.	12.6	11.7	10.8	10.7	33.2	29.0	14.1	13.5	13.4	13.3	33.2	29.0	26.7	25.2	24.2	24.0
Manufacturing	2.5	3.2	8.2	15.0	15.9	19.2	16.7	20.9	23.7	37.7	61.1	60.6	19.2	24.1	31.9	52.7	77.0	79.8
Electricity, gas and water	8.5	14.0	14.0	14.8	14.0	15.9	0.2	0.3	—	—	—	—	8.7	14.3	14.0	14.8	14.0	15.9
Construction	27.4	33.4	52.7	35.2	25.7	17.1	20.2	28.4	20.1	14.8	17.6	11.4	47.6	61.8	72.8	50.0	43.3	28.5
Wholesale, retail trade and restaurants	0.6	1.7	7.3	20.0	19.0	16.1	28.8	29.5	24.5	15.9	14.8	15.5	29.4	31.2	31.8	35.9	33.8	31.6
Transport, storage and communication	20.0	23.4	23.8	31.3	23.0	16.9	6.4	7.8	2.5	1.3	2.3	1.9	26.4	31.2	26.3	32.6	25.3	18.8
Finance, insurance, real estate and business services	*	*	*	*	11.2	8.0	*	*	*	*	3.4	4.7	*	*	*	*	14.6	12.7
Community, social and personal services	54.7	64.9	106.5	116.7	138.9	184.4	17.2	18.8	19.5	21.5	20.6	12.5	71.9	83.7	126.0	137.7	159.8	196.9
All industries	142.5	184.4	274.7	287.8	317.9	359.3	134.9	148.6	110.7	110.0	137.4	122.8	277.5	333.0	385.4	397.0	455.0	482.1

* Included in community, social and personal services.
Source: 1957 from A. Seidman (1978); 1960, 1965, 1975 and 1979 from CBS, *Quarterly Digest of Statistics*, December 1982, December 1966 and September 1985; and 1970 from CBS, *Labour Statistics 1974*.

was limited. Secondly, the information relates to employment in the organised sector of the economy. Thus, those who are self-employed (for example, in cocoa farming) and those employed casually (for example, in the unorganised sectors such as agriculture and allied activities) have not been included in the estimates of employment shown in the table.[1]

However, the table provides some useful information. The recorded total employment increased from 277 400 in 1957 to 482 100 in 1979, an average annual increase of 2.5 per cent. During the same period, recorded employment in the private sector declined from 134 900 to 122 800, while that in the public sector increased from 142 500 to 359 300. The average annual growth rate in public sector employment was as high as 4.3 per cent, against the corresponding figure of −0.4 per cent in the private sector.

An analysis based on the percentage breakdown of total employment between private and public sectors reveals that the share of the former in total employment decreased from about 49 per cent in 1957 to about 28 per cent in 1970, after which there was a slight increase in 1975. However, by the end of the decade, the private sector accounted for only one-quarter of total employment. In other words, the public sector was predominant by 1965 and by the end of the 1970s was providing jobs for three out of four persons in establishments with ten or more employees.

Whether such a high increase of public sector employment was matched by growth in public sector production is, of course, another matter. Indeed, the increase in public sector employment since the mid-1970s can hardly be justified, as this was the period when both public and private sectors witnessed a decline in production in almost all sectors of the economy. Not surprisingly, the private sector responded to the situation by maintaining and, in certain sectors such as agriculture, manufacturing, construction and services, actually reducing its level of employment. On the other hand, public sector behaviour did not reflect the realities of the situation of falling output levels. In many sectors, employment provided by the government in 1979 was higher than that in 1975. For example, in agriculture, forestry and fishing, public employment increased by 20 per cent from 1975 to 1979. The corresponding increases in manufacturing and public utilities were 21 and 14 per cent respectively. The increase of public employment in services (which included, among others, civil service and local government) was simply phenomenal, from 138 900 in 1975 to 184 400 in 1979.[2]

Data available on a different classification system by ownership enable us to gain some additional insights. According to this classification, as shown in Table 12.2, data are provided separately for four categories of ownership: State-owned, Joint state-private, Private and Co-operative. Employment in the last category (co-operative) was nil until 1980. Employment in pure state-owned enterprises increased rapidly from 1970 to 1975 and further to 1980. Similar increases are observed for employment in the joint state-private sector. On the other hand, employment in the purely private sector declined from 1970 to 1975 and further to 1980. Indeed, the average annual increase in employment in the pure state-owned enterprises was faster in the second half of the 1970s – 9.5 per cent as against 5.2 per cent in the first half.

Table 12.2 Growth of employment by ownership in medium and large-scale establishments, 1970-80 (average no. of persons engaged)

	1970 No.	1975 No.	% Annual change from 1970	1980 No.	% Annual change from 1975
State	46 660	60 130	5.2	94 706	9.5
Joint State-Private	28 075	57 398	15.4	68 650	3.6
Private	94 342	91 580	−2.9	78 554	−3.0
Co-operative	—	—	—	646	—

Sources: CBS, *Industrial Statistics*, 1970-72, 1975-76, and 1979-81.

It is therefore difficult to escape the conclusion that the government has assumed the responsibility for providing employment to all Ghanaians or at least those who worked for wages or salaries. The following observation by the Council of State, made in 1981, in its *Report on Economic Recovery* submitted to the Limann Government, bears testimony to the above view.

Broadly speaking successive Governments of Ghana have behaved as if they were solely responsible for the employment of all persons and that such employment does not have to be related to performances. The result is over-employment and inefficiencies in the public sector. In a mixed economy the Government does not have to undertake directly the employment of all labour. In such an economy the Government's full employment objective can be achieved through stimulation of the private sector, and in our Ghanaian context the private agricultural sector appears to be an area where return on investment is high.[3]

As a consequence of over-manning in the public sector, the government was caught in a trap of its own making. Although most of the State enterprises were making losses (see Chapter 13) the employees had to be paid and, in a situation of falling government revenue, this simply meant borrowing money from the banking sector and adding further to the money supply (see Chapter 8). And in an inflationary situation when real wages and salaries were falling sharply, the government had to borrow more and more to meet the increased wage commitments (see Section 12.2). Thus a vicious circle was created.

Following the decline in economic activity in the productive sector, private entrepreneurs appear to have succeeded in reducing their surplus labour force despite government controls on lay-offs by employers.[4] But this has not happened in the public sector. The increase in public sector employment in the face of declining output has inevitably led to a sharp fall in labour productivity. In the early 1980s, although many of the State-owned factories were non-operational or partly operational because of non-availability of raw materials or spare parts, the total workforce remained more or less constant. By 1983 over-manning in the Cocoa Marketing Board became so high that one-third of the total employment was considered unnecessary.[5]

The above observation needs to be qualified, since there was a shortage of qualified and skilled people in many departments. As real wages and salaries declined sharply (see Section 12.2), those with skills left the country in large numbers. As observed by the CBS in its 1982 *Economic Survey*, 'the exodus of professional and technical personnel in search of more remunerative employment abroad continued unabated'.[6]

Thus, on the one hand, there was significant over-manning so far as unskilled labour was concerned and, on the other, there was a serious shortage of skilled and qualified manpower. By the mid-1980s the country probably lost between half and two-thirds of its top-level experienced professional manpower,[7] who went to international organisations, to Nigeria and other neighbouring countries and further afield, thus depleting this special manpower resource of the country.[8]

The above estimate is, however, only a guess as no statistical data are available 'to indicate the number of emigrants, their levels of education, expertise or experience. But the large numbers of posts which have fallen vacant in many government offices, public organisations, universities and private sector undertakings indicate that a substantial number of professional and technical personnel have joined in the exodus in search of better living conditions elsewhere'.[9]

In the past, the shortage of professional people was partly remedied by recruiting highly-trained manpower from abroad, but the severe foreign exchange situation recently being felt has in fact meant parting company with some of the existing expatriate staff, many of whom left voluntarily because of harsh economic conditions.

12.2 WAGES AND SALARIES

In Ghana a minimum wage was first introduced in 1960, when the government fixed the daily minimum wage at 65 pesewas. Since then the minimum wage has been raised a number of times.[10] In the mid-1970s the minimum wage was ₵2 and it was raised to ₵4 in June 1977.[11] A big jump in the minimum wage took place in October 1980 when it was raised to ₵12. The rate was again raised to ₵25 in March 1983 and further to ₵40 in March 1984.

The concept of minimum wage however is misleading in Ghana, at least in the way it has been used in recent years. The basic aim of the minimum wage regulation is to protect labour from exploitation. But the market wage rate in Ghana, particularly in the early 1980s, has been much higher than the statutory minimum wage rate. In the agricultural sector the prevailing wage rate in 1984 was reported to be ₵200 per day, against the government minimum of ₵40. The market wage rate cited above is, however, subject to fluctuations depending on demand and supply of labour, while in regular employment such as in manufacturing or in government employment an employee receives extra benefits on top of minimum wages and also enjoys holidays.

Available data for the industrial sector are shown in Table 12.3, in the form of average annual wages and salaries received by industrial employees during 1967-84. As the figures are averages for the entire industrial sector they do not reveal the disparity in income between high and low wage earners. They also fail to show inter-sectoral wage differences – between, say, manufacturing and mining – but such differences are not very wide. As may be seen from the table, in nominal terms there have been increases in the average wages from ₵527 per annum in 1967 to ₵24 450 in 1984. But at 1970 prices, industrial wages have declined since 1970 and particularly rapidly since the mid-1970s. The index of average monthly earnings (including salaries, wages and all other remunerations) with 1974 = 100, prepared by the Central Bureau of Statistics, shows that there has been a rapid decline – the index was only 16 in 1982.[12] In the early 1980s the real income of a salaried person deteriorated so badly that the total monthly wage packet could barely meet a week's living costs.

The main cause of the decline in real wages has been the high rate of inflation since the early 1970s. Increases in nominal wages have failed to keep pace with the rate of inflation. The result is that though in nominal terms average wages rose at the rate of 28.4 per cent per annum from 1970 to 1984, at constant prices average wages actually declined at the rate of 13.3 per cent per annum.

Table 12.3 Average annual wages in the medium and large-scale industrial sector, 1967–84
(in ₵)

	1967	1970	1975	1980	1982	1984
Current prices	527	740	1401	5757	10 585	24 450
1970 prices	636	740	646	196	136	101

Sources: As in Table 11.2 and CBS/Statistical Service, *Industrial Statistics* (various issues).

It is not only in industry that real wages have been falling. White-collar workers, especially in the public sector, have experienced similar declines, because nominal wage increases have not kept pace with inflation.

Apart from inflation another cause of declining real wages has been low productivity. With low morale and lack of material inputs, productivity in most sectors of the economy has fallen drastically over the years. It has long been government policy to retain redundant labour on the payroll of public enterprises and organisations – a policy that leads to much under-utilisation of labour.

The immediate impact of low and falling real wages has been migration, especially to neighbouring countries and also to more profitable sectors such as trading and, more recently, agriculture. The impact of large-scale migration is severely felt in professions demanding high skill and training. For example, by 1984 the universities reached a situation where one or two departments did not have a single full-time teaching staff member. The situation in secondary schools, research centres and many government departments was equally serious. Even those who have remained in the country are finding life very difficult. This has caused low morale among employees and has severely affected productivity. As the monthly income from wages and salaries can hardly buy a week's subsistence, most people have to work outside their main job, unless they are fortunate enough to have an outside income. Given the severity of the situation, it is not surprising that many trained people such as typists, accountants and nurses have left their professions for farming and trading in order to survive.

The regulation of wages and salaries in order to maintain some form of parity in inter and intra-industry income is obviously a difficult exercise, and the way it has been conducted in Ghana is described below, based on information collected in 1984.[13] The Prices and Incomes Board (PIB) has been entrusted with this responsibility. The PIB also exercises control over the autonomous bodies and state enterprises, particularly those dependent on government grants and subventions, if they try to raise wages and salaries to a very high level. However, the PIB has to work under certain restrictions. It has to follow government guidelines on wages and salaries issued by the Ministry of Finance and Economic Planning and abide by government decisions on minimum wages. It does not have any 'escalator clauses' so that it has never applied any index-linking of wages and salaries. After the trade union and the employer in an establishment have agreed on a wage settlement, the proposal is submitted to the PIB which examines it according to the guidelines mentioned above. However, if the wage settlement is less than 2.5 per cent of the existing level, authorisation from the PIB was not required.[14]

Wages and salaries have remained a very difficult area of government policy. For example any attempt to raise the *real* wage level of 1984 to the 1970 level (which for the sake of argument may be viewed as a normal year) would involve something like a seven-fold increase in 1984 *current* wages and salaries. Such a large increase would have a disastrous effect on the attempt to reduce inflation. There is also the problem of raising the real incomes of trained and experienced people on the one hand, and encouraging the movement of people from the labour surplus sectors on the other.

It would, however, not be easy to raise the income of the skilled and qualified manpower, without at the same time raising the income of other employees, since such a measure might easily be misunderstood as being directed against the relatively poor employees. A number of indirect ways are, of course, available to raise the real income of the professional and skilled people who are presently in very short supply, without directly raising their nominal salary. First, they can be provided with extra benefits in the form of liberal allowances such as medical and transport. The government seems to have adopted this course to boost the income of the skilled and qualified manpower proportionately more than that of others. Secondly, tax exemption may be provided to all salaried people. This would benefit all employees, but the increase in income of those with higher salaries would be more because they pay more taxes.

Such a measure would obviously reduce government tax revenue. In the late 1970s as much as 84 per cent of the total income tax was paid by wage earners, as the amount of tax collected from the non-wage sector, especially from those earning very high profits during this inflationary period, remained very low, since they were able to evade income tax (see Chapter 13). In 1980 income tax accounted for about 10 per cent of total government revenue and this provides an idea of the extent of extra income the government is likely to forgo because of the removal of personal income tax. However, there is an urgent need to increase real wages and salaries across the board to raise the morale of employees. The above measure appears to provide an important mechanism by which to achieve this objective, at least for a short period. Though these measures will undoubtedly have an inflationary impact on the economy, it is not likely to be as serious as it would be in the case of a five-fold increase in salaries, and the impact may also be softened by the higher productivity that improved morale may generate.

One is tempted to suggest that the government should get rid of the surplus labour, thus enabling it to pay more to those who are retained without raising the total wage bill. However, declaring some employees redundant is not an easy matter, especially in Ghana where, as already mentioned, successive governments have behaved as if they were 'solely responsible for the employment of all persons'. In the face of the economic crisis, the present PNDC government has started taking action in this regard. On the basis of a survey conducted by the Manpower Utilisation Committee, which was set up by the government in 1983, 'it has been determined that 31 700 persons could be redeployed in the first instance, of which 5500 are in the civil service and 26 200 are in the state enterprises, (including 20 000 in Cocoa Marketing Board)...The redeployment exercise is expected to identify about 10 000 additional persons for redeployment in a second phase in 1987'.[15]

The redundancy payments are being financed with help from the World Bank. However, such a pruning exercise is potentially explosive. It is, therefore, essential that the issue should be openly presented to the public, emphasising the harm caused by over-manning in the form of low productivity and its direct implications for deficit financing. Moreover, the policy declaration should be accompanied first by an assurance of a genuine increase in the real wages of those retained and, second, by a clear prospect of an increase of employment opportunities in the private sector.[16] The first aim has to be taken seriously if only to raise the morale of employees who need to be paid a living wage. As far as the second objective is concerned it should not be difficult to achieve it once the private sector picks up following the implementation of the *Economic Recovery Programme* now being carried out by the government.[17]

Part IV
Savings, Investment and Technology

13 Finance for Investment

The role of savings in economic development was discussed in Chapter 2. In this chapter we shall deal with different sources of finance for investment.

In Ghana investment as a percentage of GDP was high in the late 1950s and early 1960s, but has generally declined since then. At constant prices, total gross investment was only around 8 per cent in the early 1980s, as against over 20 per cent in 1960. Table 13.1 shows the contribution of domestic and foreign savings to total investment in Ghana.

Table 13.1 Domestic and foreign savings at 1970 prices, 1957-84 (selected years)

Type of savings	1957	1960	1965	1970	1975	1980	1984
Domestic savings							
Million cedis	195	297	285	304	303	136	160
As % of GDP	13.6	17.3	14.2	13.5	13.7	5.8	7.3
As % of total investment	96.5	76.9	69.8	95.2	107.3	65.7	96.4
Foreign savings							
Million cedis	8	89	123	16	−21	71	6
As % of GDP	0.5	5.2	6.1	0.7	−0.9	3.0	0.3
As % of total investment	3.6	23.2	30.2	4.9	−7.3	34.3	3.6

Sources: See Tables A.1 and A.4.

13.1 DOMESTIC SAVINGS

Gross domestic savings at constant prices increased from 13.6 per cent of GDP in 1957 to 17.3 per cent in 1960. The average rate of savings has generally been on the decline in subsequent years and dropped as low as 5.8 per cent in 1980. Table 13.2 shows public and private savings at 1970 prices for selected years.

Table 13.2 Public and private savings at 1970 prices, 1961–84 (selected years)

Type of savings	1961	1965	1970	1975	1980	1984
Public savings						
Million cedis	71	63	−5	−28	−68	−28
As % of GDP	4.0	3.1	−0.1	−1.3	−2.9	−1.3
As % of total domestic savings	29.3	22.0	−1.6	−9.8	−50.0	−17.5
Private savings						
Million cedis	169	222	309	331	204	188
As % of GDP	9.6	11.1	13.7	15.0	8.7	8.7
As % of total domestic savings	70.7	78.0	101.6	109.8	150.0	117.5

Sources: See Tables A.1 and A.12.

13.1.1 Public savings

Public saving, as represented by central government current revenue less non-development (recurrent) expenditure, was generally positive in the 1960s, although as a percentage of GDP it has been declining since the early 1960s (see Table 13.2). Since the mid-1970s, current revenue has continuously failed to meet non-development expenditure and the extent of the deficit, as a percentage of GDP (which had been down to 1.3 per cent in 1975) increased to 2.9 per cent in 1980. In 1982 current revenue was only 57 per cent of recurrent expenditure, as may be seen from Table 13.3.[1]

Total tax revenue increased from 13.1 per cent of GDP in 1961 to 17.1 per cent in 1965. Both direct and indirect taxes contributed to this increase, each increasing at an average annual rate of over 8 per cent. During the next five years, direct taxes fell at an average rate of 2.7 per cent per annum, while indirect taxes remained almost the same. However, as a percentage of GDP total tax revenue stood at only 14.8 per cent in 1970. There was a slight increase in the tax revenue in 1975 (16.0 per cent of GDP), both direct and indirect taxes contributing to the increase. But since then tax revenue has fallen sharply and in 1980 it was only 6.6 per cent of GDP, direct and indirect taxes falling at an average annual rate of 15.7 per cent and 15.4 per cent respectively.

In 1982 and in 1984, the respective figures for tax revenue as percentages of GDP were 4.8 and 3.1.

Non-development expenditure as a percentage of GDP increased from 16.5 per cent in 1961 to 17.6 per cent in 1965 and remained at that level in 1970, but has generally fallen since then. In 1982 non-development expenditure was only 8.8 per cent of GDP. Considering that, as a percentage of GDP, non-development expenditure has fallen sharply (being, in the early 1980s, three-fifths of what it was on average during the 1960s) it would be difficult to blame non-development expenditure for the poor performance of public savings. One can even argue that in Ghana recurrent expenditure is on the low side and that there is indeed a case for raising its level. It is the failure of the government to maintain the tax-GDP ratio (about 17 per cent) achieved in the mid-1960s which has been responsible for the decline, and the subsequent negative performance, of public savings.

With the exception of the cocoa export duty, the relative contribution of different taxes has been highly unsatisfactory (see Table 13.4). Import duties collected in 1980 were only 14 per cent of what they were in 1965, while the amount of personal and company taxes in 1980 was less than half the amount received in 1975. The fall in import duties was due to factors such as sluggish growth in non-oil imports (caused mainly by a fall in export earnings and an increase in import

Table 13.3   Central government current revenue, and recurrent and development expenditure, 1961–84
(current prices)

	1961	1965	1970	1975	1980	1982	1984
Current revenue							
Million ₵	200.2	284.0	369.2	804.8	3026.1	4545.0	10 241.0
As % of GDP	20.0	19.4	16.3	15.2	7.1	5.1	3.8
Recurrent expenditure							
Million ₵	168.6	257.3	373.7	875.4	4076.8	7917.0	13 400.9
As % of GDP	16.5	17.6	17.5	16.6	9.5	8.8	5.0
Development expenditure							
Million ₵	104.1	186.1	84.0	286.1	594.7	1927.0	1 354.4
As % of GDP	10.2	12.7	3.7	5.4	1.4	1.0	0.5

Note: Revenue and expenditure data are for financial year ending June.
Sources: See Tables A.1 and A.11.

Table 13.4 Structure of tax revenue at 1970 prices, 1961–84

Direct tax	1961	1965	1970	1975	1980	1982	1984
Direct tax	83.1	79.2	69.1	78.2	33.3	32.4	13.9
Personal income tax		36.2	29.4	32.7	16.8	16.3	7.0
Company tax		37.3	34.9	40.5	13.9	15.0	6.1
Others		5.8	4.8	5.0	2.6	1.1	0.8
Indirect tax	178.8	263.7	265.0	277.7	120.3	68.8	53.7
Import duties		147.3	67.5	52.6	19.6	16.1	11.3
Cocoa export duty		27.1	113.5	136.6	60.3	—	22.4
Excise and local duties		29.5	38.2	41.5	32.5	46.2	14.1
Purchase tax		2.3	3.4	5.9	0.3	0.9	0.4
Others		57.6	22.3	41.3	7.6	5.6	5.5

Sources: CBS, *Economic Survey*, 1968 and 1981 and Bank of Ghana, *Annual Report* 1973 and 1977; and Statistical Service, *Quarterly Digest of Statistics*, September, 1986.

expenditure on oil following the 1973 oil crisis) and the artificially low value of imports in terms of the cedi at the official exchange rate, especially in an inflationary situation when open market prices were rising rapidly (estimated at 68 per cent on average per annum from 1974 to 1982; 40 per cent per annum during the 1970s). Corruption and tax evasion probably played their parts as well in the decline of import duties as a source of revenue.

The estimated average elasticity of import duties with respect to imports (other than oil) in Ghana was found to be 0.49 for the period of 1970–71 to 1980–81. Low elasticities were also found for the consumption taxes – sales and purchases tax (0.33) and excise (0.34), mainly because of the use of an unrealistic tax base (controlled ex-factory wholesale prices) in a situation of high inflation.[2] Of the two groups, wage and non-wage income earners paying income taxes, the main burden has been borne by the fomer who paid as much as 84 per cent of the total income tax during 1977–80. But in a highly inflationary situation, as experienced in Ghana since the mid-1970s, one would have expected a large share from the non-wage sector because of an expected larger income originating in this sector, as profits of the trading and commercial sectors are likely to keep pace with inflation. Undeclared income of the non-wage sector has probably played an important part in their low contribution to income tax, but the use of official controlled prices in calculating profits and taxable income and the gap between assessment and collection of taxes were undoubtedly two other important factors accounting for the low level of tax revenue.

The estimated average elasticity of non-tax revenue with respect to GDP for the period 1970/71–1980/81 was 0.37.[3] At 1970 prices, the amount of non-tax revenue in 1982 was less than one-fifth of what it was in 1965 (see Table 13.5). Both the main components of the non-tax revenue – (a) sales of goods and services and (b) interest and profits – contributed to this decline. The relative share of revenue from the sale of goods and services fell from 80 per cent of the total non-tax revenue in 1965 to 12 per cent in 1984, mainly because the controlled prices of these items failed to rise with inflation.

At 1970 prices, the share of interest and profits from state investment declined from ₡13.2 million in 1968 to ₡5.2 million in 1981.[4] Considering that the state is heavily involved in various ventures,[5] the total number of pure state-owned units alone standing at 284 in 1980,[6] the amount received by the state in interest and profits is undoubtedly too low. Given the large involvement of the state in productive ventures such as manufacturing, mining, farming and fishing, one would normally expect the state enterprises to generate surplus in the form of payment of interest, taxes, dividends and surplus payments for financing further investment.

There are 113 state enterprises listed under the State Enterprises Commission, but this list is not all-inclusive.[7] For example, Ghana Railways Corporation is not included. Data available show that most of the non-financial state enterprises have been making substantial losses or earning meagre profits (see Table 13.6). The only major exception is the Volta River Authority, which has been paying interest and dividend regularly because of its international obligations. It has also revalued its assets regularly.

Table 13.5 Non-tax revenue at 1970 prices, 1961–84
(selected years)

Non-tax revenue	1961	1965	1970	1975	1980	1982	1984
Million cedis	35.1	42.0	35.3	31.9	11.2	7.5	14.0
As % of total tax revenue	10.1	10.9	9.6	8.2	7.3	7.4	20.8
As % of GDP	2.0	2.1	1.6	1.4	0.5	0.4	0.7

Source: See Table 13.4.

As far as most other state enterprises are concerned, they have failed to revalue their assets, and many of them have even stopped fulfilling contractual obligations by not paying interest on capital borrowed. A number of state enterprises are not paying social security contributions to the state, and some are not even remitting to the Income Tax Commission personal income taxes collected from employees. On the contrary, big loss-making state enterprises such as the State Farms Corporation, the State Fishing Corporation and the Ghana Surgar Estates Limited have continued to be a drain on the government exchequer. Thus the contribution of state enterprises to investment financing has been far from satisfactory.

The Ghana Industrial Holding Corporation (GIHOC) is the largest organisation controlling, in 1984, 26 individual enterprises/divisions, 16 of which were originally owned by the Ministry of Industries. Distilleries, canneries, metal industry and pharmaceuticals have usually been the profit-earning divisions for GIHOC, while bricks and tiles and vegetable oil mills have always made losses (at times heavily). Another division which has been a big loss-maker is GIHOC farms, although its figures were not available. Because of the way GIHOC operates as a group the loss-making divisions can be easily shielded at the expense of those making profit. Moreover, the profit and loss statements of the individual GIHOC enterprises do not show any tax payment as it is the entire corporate profit which is calculated for tax purposes by GIHOC. Indeed, in common with most other state enterprises, GIHOC has at times defaulted in tax payment, making only part of the payment due. Social security contributions are also in arrears for the loss-making GIHOC enterprises because of their serious liquidity problems.

There are many reasons for the poor performance of the state enterprises in Ghana. First, the level of capacity utilisation has been very low due to lack of spare parts and raw materials and other technical problems. For example, capacity utilisation in GIHOC was 40.4 per cent in 1978 and it declined further to 25.5 per cent in 1980. Most other state enterprises have probably done worse.

Second, the implicit assurance that state enterprises will be supported even in the face of continual losses has provided a sense of security, thus failing to encourage them to improve performance. This has also indirectly discouraged the profit-making organisations from improving their performance, although information collected from GIHOC shows that it has taken measures to encourage the profit-making divisions by giving bonuses to their employees.

Table 13.6 Profit and loss account of selected state enterprises, 1965-80 (selected years) (₡ '000)

Corporations	1965	1970	1978	1980
1. Ghana National Trading Corporation	+6 515	+2 668	+4 068	+5 160
2. State Farms Corporation	-12 733	-1 361	-7 861	-10 658
3. State Fishing Corporation	-240	-338	+77	-1 774
4. State Construction Corporation	+354*	-615	+2 512	+3 221
5. State Gold Mining Corporation	-2 689	-6 754	N.A.	N.A.
6. State Hotels and Tourism Corporation	-137	+52	+2 624†	+3 916†
7. Ghana Airways Corporation	-3 573	-2 857	-368	-7 824
8. Omnibus Service Authority	‡	‡	-3 473	-282
9. Ghana Sugar Estates Ltd.	N.A.	N.A.	-9 149	-16 227
10. Volta River Authority	N.A.	N.A.	N.A.	+144 660
11. State Transport Corporation	N.A.	N.A.	+1 381	+3 771
12. GIHOC Enterprises	-1 479	-176	+5 580	+8 790

* 1963 figures.
† Tax payment is not mentioned.
‡ Included in GIHOC Enterprises.

Sources: 1965 and 1970 from Killick (1978), p. 219. (Data in the source are shown for financial years, 1964/65 and 1969/70, respectively.) 1978 and 1980 – Data collected from State Enterprises Commission and GIHOC.

Third, by fixing prices at levels which are significantly lower than open market prices, the price fixing bodies have caused the state enterprises to lose substantial amounts in revenue. For example, in 1984, the Cannery Division of GIHOC was finding it difficult to persuade the Prices and Incomes Board to accept the price of ₵250.00 per crate of tomatoes and fix the price of their tomato products accordingly. The Prices and Incomes Board was insisting on the price of ₵200.00 per crate of tomatoes although the open market price was in the region of ₵600.00 to ₵700.00.[8] As far as the state enterprises in the export sector are concerned, until recently the official exchange rate which kept the *cedi* highly over-valued caused very low export earnings in *cedis*. The bonus allowed was not even adequate to make the revenue cover expenses.

Fourth, financial management has been particularly poor. This is largely because of the lack of insistence on accountability, partly arising from slackness on the part of the controlling authorities. No serious pressure has been put on the state enterprises to pay back the capital borrowed, or even to pay interest regularly. According to one top official of a state enterprise, 'the Auditor-General probably does not know how much is owed by different state enterprises'.[9]

Fifth, there is also the problem of over-manning. Many of the state enterprises have high surplus labour in employment, as has now been demonstrated from the redundancy exercise conducted for the Cocoa Marketing Board in 1983. For some, like the State Insurance Corporation, half of the present labour force should be able to manage operations without loss in output. As the provision of employment has often been viewed as a social function by the state, it has often been difficult to attack this problem objectively (see Chapter 12).

13.1.2 Private savings

At current prices, private savings experienced wide fluctuations from about 5 to 12 per cent of GDP between 1961 and 1969. The variations were even wider during the 1970s – from about 3 to 16 per cent of GDP. On the other hand, public savings were negative in 1969, 1970 and 1973 and have been so since 1975. In the event, the need for deficit financing through borrowing from the private sector and increased money supply meant that resources could be mobilised domestically in the form of voluntary and forced private savings to bridge the budgetary gap.

Although the contribution of voluntary private savings (through purchase of securities and government bonds by the public and the banking sector) might have been significant during the 1960s, it was definitely not so after the early 1970s. Forced private savings through extra money supply has remained the main source of government consumption and investment, especially since the mid-1970s. Forcing the private sector to save as much as 9 to 23 per cent of GDP, at a time when GDP per capita was declining at an average rate of 3.2 per cent per year (from 1975 to 1982), must have proved extremely painful to the population. However, it shows that the private sector can be forced to save in a situation of declining per capita income.

One would hardly expect the marginal rate of savings in the private sector to be high in a situation of declining per capita income. It is, of course, possible to assume that because of inflation, income distribution could have become skewed, allowing the group enjoying growth of income to save enough to compensate for the fall in savings due to the average decline in per capita income. However, this has not happened for the following reasons.

First, there has been a great decline in confidence, resulting from political upheavals. Second, 'the fear of accountability generated by recent political events'[10] also discouraged people from saving with the banks. Third, as the foreign exchange allocations for most private sector enterprises were reduced, private sector activity remained depressed. There was also a reduction of foreign exchange allocation for the public sector, but it was not as severe as for the private sector. Fourth, the 'more lucrative returns on speculative investment on the black market'[11] induced people to invest in other currencies like the Naira, the US Dollar or the Pound Sterling. Fifth, owing to the over-valuation of the *cedi*, exports were greatly discouraged and this consequently had serious effects on the growth of the private export sector, an area where the private sector traditionally played an active role. Sixth, the poor administration of import control regulations also proved detrimental to the growth of private savings. As the private sector was deprived of machinery, spare parts and raw materials, incentives for investment were significantly reduced.

The role of the interest rate is probably worth elaborating. Table 13.7 shows nominal and real interest rates and, as may be seen, the real rates were generally negative.

In fixing the nominal interest rates no account is taken by the Bank of Ghana of the demand for and supply of money. There is some doubt as to whether an interest rate policy can be effective in attracting

Table 13.7 Nominal and real interest rates, 1961–82 (selected years) (percentages)

Year	Nominal rates Bank	Nominal rates Deposit	Real rates Bank	Real rates Deposit
1961	—	3.50	—	−0.07
1965	4.50	3.50	−21.90	−22.90
1970	5.50	3.75	1.64	−0.11
1975	8.00	8.00	−21.69	−21.69
1980	14.50	13.00	−35.22	−36.72
1982	10.50	9.00	−11.80	−13.30

Notes: Real interest rates have been calculated by deducting inflation rates from nominal interest rates. Deposit rates refer to deposits over 12 months.
Sources: Bank of Ghana, *Annual Reports* (various issues); CBS, *Quarterly Digest of Statistics*, September 1983 and September 1985.

savings, especially as the largest banks have at times refused to accept savings deposits. Anyway, the PNDC Government in its 1983 Budget Statement appears to have rejected the interest rate policy as a means of raising savings.[12] The role of the interest rate in influencing private investment was also thought to be ineffective because the main constraint in this area was lack of adequate foreign exchange to import capital equipment, spare parts and other required imported inputs.

Considering that real interest rates have remained negative for a long time, credit guidelines issued by the Bank of Ghana have been used to influence the volume and distribution of commercial bank credit to the private sector and public corporations. In recent years, the government has opted for selective and direct credit controls 'to keep the expansion of overall credit within acceptable limits and to channel the permissible increases into the various sectors of the economy in line with stated priorities'.[13]

13.2 EXTERNAL CAPITAL

The inflow of foreign savings has fluctuated over the years, at times even being negative. At constant prices foreign savings contributed as much as 6.12 per cent of GDP in 1965, the highest share so far recorded (see Table 13.1). The lowest recorded share was in 1972 (−5.46 per cent of GDP). Ironically, during the heyday of the pro-Western governments which ruled Ghana from February 1966 to January 1972, the contribution of foreign savings to financing investment was much lower than it was during the rule of the pro-socialist Nkrumah Government.[14]

The importance of external capital in the development of the economy has been recognised by the various governments which have ruled Ghana since its independence. Nkrumah, however, took an ambivalent attitude towards foreign investment capital, 'asking for it while simultaneously condemning it as neo-colonialist'.[15]

It is not difficult to understand the importance of external capital, first, in filling the savings-investment gap to achieve the desired growth rate and, second, in overcoming the foreign exchange gap, since shortage of foreign exchange, by constraining the import of essential raw materials, machinery and spare parts would make it difficult for domestic savings to be utilised. These two roles, individually or together, have been important to Ghana since the early 1960s, but particularly so over the last decade. The accumulated consequences of a long and considerable economic decline have been so damaging that it is difficult to imagine an early recovery without external assistance to ease the process of readjustment.[16]

External capital usually comes in two main forms – (a) aid and grant and (b) direct private investment. Grants are obviously preferred to loans, and long-term soft loans to short-term loans at high interest rates.

The Ghanaian economy was much more open during the colonial period, and one would have expected foreign capital to play a significant role in her economic development. However, according to Ewusi, 'there is evidence to suggest that foreign capital played a rather insignificant role in the growth of the economy during the colonial periods. Thus growth depended either upon the promotion of exports or upon an import substitution program; and it seems that the policy of export promotion was pursued during the colonial period'.[17]

It was in the *Seven-Year Development Plan* (1963/64–1969/70) that a big role for foreign capital in the financing of investment was envisaged for the first time.[18] Of the total planned investment of G£1016.5 million, G£240 million (which was 58 per cent of the projected public sector investment) was going to come from foreign loans and grants, and G£100 million (22 per cent of projected private sector investment) from foreign private investment. At the time of the launching of the Plan, arrangements had been made for loans of

G£100 million, but there were doubts as to the availability of the rest of the amount expected to come from foreign loans and grants. As it turned out, the government depended heavily on short-term loans at relatively high interest rates in the form of suppliers' credit. That such a type of credit might not necessarily be in the interest of developing countries like Ghana had been pointed out in an earlier official document in the following words: 'What the Government needs now are not suppliers of credits but really long-term soft loans to enable it to carry through its projects under the Seven Year Plan'.[19]

At the time of Nkrumah's overthrow, about 60 per cent of the total loan (principal) was in the form of suppliers' credit, while the share of long-term loans was as low as 15 per cent. In 1966–67, at the height of the much publicised foreign indebtedness, the debt service ratio of Ghana was of the order of 50 per cent of export earnings – a result of new foreign liabilities which were running at N₡155 million per annum three years previously. A large part of the additional liability was in the form of short-term suppliers' credits which, in the opinion of two informed observers

> had the effect of slowing down the development of the economy by: (i) tying up a large proportion of current and future foreign exchange earnings in the repayment of short-term debts, thus creating a debtor-creditor relationship which had far reaching economic and political implications; and (ii) utilizing this borrowing in non-optimal projects, i.e. with a relatively non-developmental design, producing insufficient income or foreign exchange flows for repayment of the credit in the short-term, thus re-inforcing (i).[20]

Following the rescheduling of external debts carried out in December 1966 and October 1968 by the government of the National Liberation Council which ousted Nkrumah's government, and the cancellation of unviable projects for which suppliers' credit contracts had already been signed,[21] a cautious approach has been followed in the acceptance of such credit. Subsequently, the suppliers' credit scheme was even suspended on a number of occasions.[22] For example, it was suspended in 1979 and re-introduced in May 1982. Again, the processing of applications under the credit scheme was generally suspended by the Bank of Ghana from January 1982 to September 1983.

The share of short and medium-term loans in total indebtedness has remained very high, a major part of the oustanding medium-term loans being related to pre-1966 contracts.[23] However, the government has been turning more towards long-term as against short-term loans. As shown earlier, of the total foreign indebtedness only 15 per cent was in long-term loans in the mid-1960s. Data available from 1976 to 1982 show that in the early 1980s the share of short-term loans in the total foreign loan of the country varied from one-quarter to a little over one-third (see Table 13.8). On the other hand, the share of long-term loans was over 50 per cent in 1980 – this share declined in 1982 but was still about 46 per cent. And within long-term loans, multilateral loans have generally increased their share. For example, of the total long-term foreign debt in 1976 the share of multilateral debt was only 26 per cent – the corresponding figure was about 52 per cent in 1982.[24]

There have also been drastic changes on the borrowers' side. In 1970 the government share of borrowing was 42 per cent of the total, while in recent years it has comprised as much as three-quarters of the total.

Annual inflow of foreign loans and grants for the period 1977–82 is shown in Table 13.9. During this period, at current prices the highest per capita loan was received in 1977 ($22.40) and the lowest in 1982 ($2.11). Per capita grants varied from a negligible figure of $0.16 in 1979 to $2.55 in 1981.

In keeping with the trend, the contribution of private foreign investment has not in recent years been significant in the financing of total investment. In 1976 the inflow of private foreign investment was ₵53 million, accounting for 9 per cent of total investment. Although the

Table 13.8 Foreign indebtedness (cumulative) by type of debt, 1976–82
(selected years)
(current prices)

	1976 Mil. US $	%	1978 Mil. US $	%	1980 Mil. US $	%	1982 Mil. US $	%
Short term	241.3	24.6	593.6	40.1	341.9	24.5	592.4	34.3
Medium term	378.6	38.6	373.9	25.2	316.4	22.6	343.3	19.9
Long term	360.3	36.8	514.4	34.7	738.7	52.9	793.3	45.9
Total	980.2	100	1481.9	100	1397.0	100	1729.5	100

Source: CBS, *Economic Survey* 1981 (p. 256) and 1982 (p. 224).

Table 13.9 Foreign loans and grants, 1977-82
(in current prices)

Year	Loans Mil. US $	Loans Per capita (US $)	Grants Mil. US $	Grants Per capita (US $)	Total Mil. US $	Total Per capita (US $)
1977	229.13 (630.1)	22.40	20.62 (56.6)	2.02	249.71 (686.7)	24.42
1978	119.82 (329.5)	11.42	25.42 (69.9)	2.42	145.24 (399.4)	13.84
1979	84.58 (232.6)	7.86	1.67 (4.6)	0.16	86.25 (257.2)	8.06
1980	238.78 (636.5)	21.63	21.53 (59.2)	1.95	260.25 (715.7)	23.58
1981	94.80 (260.7)	8.37	28.87 (79.4)	2.55	123.67 (340.1)	10.92
1982	24.55 (67.5)	2.11	13.93 (38.3)	1.20	38.48 (105.8)	3.32

Note: Figures shown in brackets are the cedi equivalents.
Sources: CBS, *Economic Survey* 1981 (p. 245) and 1982 (p. 216).

inflow of private foreign investment increased to ₵59 million in 1980, with wide variations in the intervening years, its share in total investment was only 3 per cent. The net contribution of private foreign investment is still further diminished if one takes into consideration the outflow on this account. For example, the outflow on this account was ₵18 million and ₵34 million in 1976 and 1980 respectively.[25]

13.3 CONCLUSIONS

From the above discussion we can safely conclude that, mainly because of the poor performance of public savings (resulting from the fall in tax and non-tax revenues) and the low and erratic supply of external capital, investment as a percentage of GDP has fallen in Ghana. Any attempt to increase investment will therefore necessitate attacks on both fronts. External finance is definitely needed immediately to help resuscitate the economy. However, once the economy has recovered and is able to earn adequate foreign exchange through the export of cocoa, timber and minerals, the role of external capital will need to be viewed critically. Self-reliance may be taken as a development objective, particularly because of unhappy past experiences with suppliers' credit and large-scale import of inappropriate technologies. Given that donors' preferences may not necessarily coincide with the aid requirements of a developing country, it will be necessary to plan ahead in order not to depend excessively on foreign aid.[26] Hence the need for raising domestic savings and increasing export earnings.

14 Investment and Technology Choice

14.1 LOW INVESTMENT LEVEL

Ghana provides a rather unique example of a developing country which, after a successful start on the course of development by raising its savings and investment ratios to high levels in the post-independence period, saw these fall to very low levels. Total saving and its division between domestic and foreign saving were discussed in Chapter 13. Here we shall concentrate on investment.

At current prices, gross investment as a percentage of GDP increased from 13.5 per cent in 1957 to 22.6 per cent in 1960, and remained at around 17 per cent up to the mid-1960s. It has fallen sharply since then, reaching a record low level of 3.4 per cent of GDP in 1982 (see Table 14.1).

A similar trend is observed at constant prices, although the annual fluctuations are less marked than at current prices. This is mainly because, as mentioned earlier, the inflation rates implicit in current prices for imported goods have not been as high as in the GDP series which gives averages for the whole economy. As the import content in capital formation has remained high, the investment-GDP ratios at constant prices are higher than the same ratios at current prices in the years with high inflation rates - particularly after the mid-1970s.

As mentioned in Chapter 2, the low level of investment can hardly help stem the economic decline of the country, given the high rate of

Table 14.1 Gross investment at current and 1970 prices, 1957-82 (selected years)

	1957	1960	1965	1970	1975	1980	1982
Current prices							
Million cedis	100	216	262	320	672	2410	2950
As % of GDP	13.5	22.6	17.9	14.1	12.7	5.6	3.4
1970 prices							
Million cedis	203	386	408	320	282	207	145
As % of GDP	14.1	22.5	20.3	14.2	12.7	8.8	6.9

Sources: Tables A.1 and A.7.

population growth (2.6 per cent per annum since 1970) and the high incremental capital-output ratio (see Section 14.4).

14.2 PUBLIC AND PRIVATE INVESTMENT

The distribution of total investment between the public and private sectors shows that the relative share of each sector has fluctuated greatly over the years. The share of public investment in total investment was over 70 per cent in 1965 (71.0 per cent) and in 1976 (76.2 per cent), and below 30 per cent in four years – 1969 (27.7 per cent), 1970 (26.3 per cent), 1974 (27.9 per cent) and 1980 (24.6 per cent). However, some interesting features can be observed if we divide the period from 1961 to 1980 into four five-year periods (see Table 14.2).

Thus viewed, the relative share of public investment was 60 per cent of total investment during the first-half of the 1960s. This is also the period when the ratio of public investment to GDP was over 10 per cent. This period coincides with the Nkrumah Government's investment push strategy.

The post-Nkrumah period up to the mid-1970s is marked by a significant fall in the relative share of public investment, down to less than 40 per cent for the period from 1966 to 1975. In terms of percentage of GDP, as viewed at five-yearly levels in Table 14.2, public investment has never been above 5.3 per cent of GDP during the post-Nkrumah era.

Ironically, there has not been any spectacular growth in private investment during the post-Nkrumah period, although the two succeeding governments vigorously encouraged private enterprises. It is true that the relative share of private investment was as high as 64.2 per cent of total investment in 1966–70, compared to 40.3 per cent in 1961–65, but the ratio of private investment to GDP was still rather low – at constant prices, 7.9 per cent of GDP during the 1966–70 period (the same as in the previous period, 1961–66) despite government emphasis on private enterprise. A similar trend is observed for the period 1971–75, a relatively higher share of private investment (65 per cent) in the total, although the total investment to GDP ratio was only 12.6 per cent, compared to 19.4 per cent during 1961–65 (at constant prices).

The second half of the 1970s was a period of low investment (10.8 per cent of GDP at constant prices) mainly because of a sharp fall in the investment ratio during the last three years (1978–80). This period,

Table 14.2 Public and private investment at current and 1970 prices, 1961–80

	1961–65 Current prices	1961–65 1970 prices	1966–70 Current prices	1966–70 1970 prices	1971–75 Current prices	1971–75 1970 prices	1976–80 Current prices	1976–80 1970 prices
Public investment								
Million cedis	644.6	1100.3	393.5	483.7	763.6	486.2	3128.0	606.0
As % of GDP	10.5	11.5	4.4	4.7	4.1	4.1	2.9	5.3
As % of total investment	59.7	59.3	36.0	37.4	35.5	35.1	41.9	49.3
Private investment								
Million cedis	435.6	754.0	701.2	809.9	1386.6	897.4	4344.0	624.0
As % of GDP	7.1	7.9	7.8	7.9	7.4	7.6	3.9	5.5
As % of total investment	40.3	40.7	64.0	62.6	64.5	64.9	58.1	50.7
Total investment								
Million cedis	1080	1853	1095	1294	2150	1384	7472	1230
As % of GDP	17.6	19.4	12.2	12.6	11.5	11.7	6.8	10.8

Sources: See Tables A.1 and A.13.

however, saw an improvement in the relative share of public investment which took an almost equal share with private investment.

14.3 SECTORAL ALLOCATION

Statistics about actual investment expenditure by economic sectors such as agriculture and manufacturing are not available in Ghana except in an *ex ante* form in the Plan documents. We are therefore unable to analyse the actual pattern of sectoral allocation of investment and the growth and development implications thereof.[1]

The information available enables us to present the allocation of investment by type of assets at both constant and current prices (see Table 14.3). The periodical averages show that the building sector has always taken the highest investment allocation, ranging from about 46 to 68 per cent of total investment.

Whilst investment in building, machinery and equipment increased from the first to the second half of the 1960s, investment in transport equipment and other constructional work (excluding land development) showed a decline from about 13 to 10 per cent and 22 to 13 per cent, respectively, at current prices.

During the 1970s, at current prices, all the sectors maintained a relatively stable percentage share of total investment. In terms of constant prices, however, the share of machinery and equipment was higher and that of building lower during the second than during the first half of the 1970s.

In the early 1980s the building sector maintained its dominant position, taking more than 50 per cent of total investment at constant prices, followed by machinery and equipment (20.8 per cent) and transport equipment (18.4 per cent).

14.4 RATIONALE OF INVESTMENT DECISIONS[2]

Since the early 1960s various controls have been imposed on both private and public investment decisions. Such government interventions have taken the form of exchange controls, approval of projects (by criteria relating to import substitution and use of local resources), and sanction of bank loans. It is true that some of these controls were liberalised at times (during the late 1960s and early 1970s, for instance)

Table 14.3 Allocation of investments by types of assets at current and 1970 prices, 1961-82 (percentages)

	Building	Other constructional work (excluding land development)	Land development	Transport equipment	Machinery and equipment	Total
1961-65						
Current prices	46.5	21.5*	—	13.1	18.9	100
1970 prices	—	—	—	—	—	—
1966-70						
Current prices	50.0†	13.0	0.8	10.2	26.0	100
1970 prices	63.7†	—	—	10.2	26.1	100
1971-75						
Current prices	15.1	13.7	1.0	16.5	17.7	100
1970 prices	53.9	15.1	1.1	14.5	15.5	100
1976-80						
Current prices	51.6	12.5	1.2	15.3	19.4	100
1970 prices	45.6	15.3	1.6	16.0	21.5	100
1981-82						
Current prices	67.8	10.0	0.9	10.0	11.3	100
1970 prices	53.4	7.7	0.7	18.4	20.8	100

* Includes land development.
† Includes other constructional work.
Sources: CBS, *Economic Survey* (various issues) and Table A.9.

but investment decisions have never been free from government controls and bureaucratic bottlenecks. The restrictions were more applicable to the private than to the public sector, while in joint-venture projects investing agencies such as the National Investment Bank and the Agricultural Development Bank proved helpful in alleviating the rigours of bureaucratic restrictions.

Direct government intervention to influence investment decisions should not, however, be taken to imply that the price mechanism has been replaced by a better investment allocation system. Indeed this is not the case, and for the private sector the maximisation of profits, based on the estimated cost and revenue calculations of the investors, has remained the main criterion of investment decisions. One cannot, of course, say this with certainty as far as public sector investment is concerned because, in Ghana, the basis of public investment decisions relating to most medium and large-scale projects particularly during the late 1950s and early 1960s was unsatisfactory, to say the least. Development projects were at times selected on an *ad hoc* basis or through political influence.[3] At times, public investment decisions were taken and projects carried out with very little groundwork and planning, as is apparent from the following observation in the *1963 Economic Survey*.

> Although the government has signed many agreements for factories with overseas companies and governments, and although in some cases the plant and machinery have arrived in the country, the civil engineering works have not been carried out and as a result the factories have not gone into production.[4]

The *Seven-Year Development Plan* (1963–64 to 1969–70) was critical of the project selection procedure prevailing prior to 1964. A number of projects were accepted in good faith and put into the annual development budget estimates before any calculations were worked out, as very little time was allotted for the pre-budget screening of projects.[5] The Plan, therefore, proposed a longer period (about six months) for the pre-budget examination of development projects to allow the decision-makers enough time to undertake better appraisal of projects. But six months is still a short time in which to conduct a serious examination, even of medium-size projects within familiar areas, let alone larger projects in unfamiliar or new sectors. The merit of the suggestion put forward by the planners needs, however, to be viewed against the prevailing practice of providing little or no time for the pre-budget examination of development projects.

But even this modest suggestion of the planners could hardly be implemented in a situation where different ministers were not willing to submit themselves to the discipline of planning.[6] Problems of implementing development plans in Ghana have been discussed at some length in the opening Overview section. Considering the direct relevance of the points particularly relating to investment decisions, we raise them below at the cost of some repetition.

The first half of the 1960s was a period of high investment, backed by large-scale financing through suppliers' credit. However, the way investment decisions were undertaken was far from satisfactory. The following comments by Rimmer (though a little exaggerated) are basically correct about the way investment decisions were actually taken by the government at that time.

New projects appeared which had never been envisaged in the [Seven-Year] Plan but were now being pushed by contractors willing to pay commissions to the persons who accepted them. Projects were begun without feasibility studies and without competitive tendering. New enterprises were distributed among party functionaries as private fiefs, enabling them to give patronage to relatives, friends, and supporters.[7]

Omaboe, one-time Acting Chairman of the National Planning Commission, is particularly critical of the role politicians played in Ghana. According to him, the apparent conflict between planners and politicians was such that politicians often put themselves ahead of planners in the general consideration and implementation of projects. Important projects were, therefore, initiated by politicians who usually committed the nation to a course of action before the technicians were consulted. For example, there were instances when decisions with economic and financial implications were taken by the cabinet, although such decisions were not properly assessed in the context of existing development plans.[8]

In Omaboe's view such undue interference with technical matters resulted in situations where there was 'no proper processing of development projects and projects have been implemented which with a little consideration would have been postponed for a number of years without much loss'.[9]

Killick, a highly informed observer on Ghana, remarked that the quality of investment decisions was as bad as biases in investment decisions in terms of their effects on the development of Ghana's economy, and he is convinced that investment decisions in the early 1960s had little economic rationale.[10]

The *Two-Year Development Plan* (1968-70) is also very critical of the way investment decisions were made during the first half of the 1960s.

> In many cases, no feasibility studies were made and often even the most basic survey of potential markets and raw material supplies was omitted. Political rather than economic considerations appear often to have determined siting of plants; much of the construction was carried out without adequate Ghanaian participation or supervision. The managerial requirements of many of the firms were not provided for and reliable operating and financial records were not kept.[11]

The beginning of the 1970s, however, marked an era during which some improvement was shown in the project selection procedure. From 1971 onwards approval for project implementation became the responsibility of an Inter-Ministerial Committee, whose secretariat was at the Ministry of Industries and had an official of the then Capital Investments Board (now Ghana Investments Centre) serving on it. Investment projects had to pass the initial test of falling within the government priority areas.[12] After passing the initial test, projects, especially those with high fixed capital requirements, were to be subjected to serious appraisal 'using market analysis to see whether there is a demand for the project'.[13] In 1977 Dr K. Donkor Fordwor, the then Chairman of the Capital Investments Board (now GIC), asserted that 'no direct investment will be approved for establishment unless a social cost-benefit analysis shows the investment to be socially profitable'.[14]

The interest shown in social cost-benefit appraisal is understandable considering the wide divergence between market and social prices prevailing, particularly from the mid-1970s. The over-riding cause of the wide divergence between official and open market prices has been the continuation of the over-valued exchange rate of the cedi and the absence of any accompanying measures, in the form of taxes or duties, to reflect scarcities in actual prices.

However, for the application of shadow prices (which would reflect social opportunity costs) in project appraisal, certain conditions need to be fulfilled. First, such prices need to be estimated at the macro level, preferably by the central planning body which is aware of the objectives of and constraints on development. Second, such prices need to be systematically applied for the appraisal of all projects undertaken in the country. Neither of the conditions was satisfied in Ghana. The following illustration, based on experience of applying shadow prices in Ghana, should reveal the difficulties of administering an alternative set of prices in the country. The findings are from investigations conducted by the author, mainly in February–March 1983 (that is, before the 1983 budget), to see whether shadow prices were used by investing agencies.

An important finding of the investigation is that there were no central guidelines for the use of a set of shadow prices; and in the absence of such a co-ordinated approach, different organisations were using different criteria in appraising projects, as is apparent from Table 14.4.

Table 14.4 Estimate and use of shadow prices by selected organisations

	Use of shadow prices	Estimate of shadow prices	Shadow prices Foreign exchange (₡ per US $)	Unskilled labour (ratio of wages)
National Investment Bank	Yes	Partial	10.00	0.75
Ghana Investments Centre	Yes	Partial	3.58	0.50
Bank of Ghana	No	None	—	—
Agricultural Development Bank	No	None	—	—
Ghana Industrial Holding Corp.	No	None	—	—

Notes: Information from the first four organisations was collected in February–March 1983, while that from GIHOC was in April–May 1984. Official foreign exchange rate per US $ was ₡2.75 from September 1978 to October 1983.
Sources: Questionnaire returns and personal interviews.

The National Investment Bank (NIB), a leading investment organisation in Ghana with direct and indirect participation in a large number of medium and large-scale projects in the country, has participated in national as well as joint venture projects with either local or foreign partners (see Chapter 9). Of the 44 projects developed or promoted by the bank up to 1979, 50 per cent were under private Ghanaian ownership, 36 per cent under public and the remaining 14 per cent under foreign ownership.[15] By percentage shareholding, there are only 6 projects in which the NIB equity share is more than 50 per cent of the total capital investment, one of which is the Nasia Rice Company at Tamale which has been selected for examination in terms of technology choice (see Section 14.5). Of the rest, the bank has less than a 25 per cent equity share in 18 projects, and 25 to 50 per cent in the remaining 22.

It is pertinent to note that the promotion of projects by the NIB has been at a low tempo in the early 1980s. As at 1982, the number of industrial projects in which NIB was involved was 48, 18 per cent of which were established in the 1960s and 75 per cent in the 1970s.[16]

The NIB has been applying a shadow exchange rate of ₵10.00 to one US dollar since the beginning of 1981. A shadow wage rate of 0.75 of the actual wage of unskilled labour has also been recommended by the World Bank.

The Ghana Investments Centre (GIC), another leading organisation having responsibility for appraising investment projects which have significant foreign involvement, used both market and social prices in its project appraisal. However, the shadow prices used by the GIC differed from those used by the NIB. The GIC used a conversion factor of 1.3 for correcting the distortion in the official foreign exchange rate and 0.5 for the unskilled labour wage rate.[17] In the view of an economist of the Research Division in the GIC, the conversion factor of 1.3 for correcting the over-valuation of the *cedi* is on the low side, but the original factor of 1.5 was not acceptable to the Project Analysis Division of this organisation.[18]

While the NIB and the GIC used some sort of shadow prices for foreign exchange rates and unskilled labour wage rates, information collected from the Agricultural Development Bank and the Bank of Ghana showed that these two organisations did not use shadow prices at all for appraising projects, although both of them have directly participated in a number of investment projects. The Agricultural Development Bank has invested money in over a dozen projects, of which two are solely owned by it, and another five in which it has a 50 per cent or more shareholding.[19]

The Agricultural Development Bank (ADB), set up to promote agricultural enterprises, has already been discussed in some detail (in Chapter 9). As a publicly-owned financial institution established to develop the agricultural sector, the ADB has principally been interested in food crop cultivation, although it has also contributed towards the development of industrial crops, livestock production and fishing.

In 1979 food crop cultivation took about 72 per cent of ADB's total loans while the fishing, livestock and industrial crop sectors took 7 per cent, 6 per cent and 3 per cent respectively. In the industrial crop area, the bank appears to have taken the lead in palm oil production. It has established three large-scale palm oil plants and intends to establish another one with the largest operating capacity in the country (4 tonnes of palm fruit per hour).[20] Among the palm oil processing plants established by the bank are the Anwia-Nkwanta Oil Mill (one of the projects selected for studying technology choice in the next section), the WAFF Limited at Nkwantanum in the Central Region and the Bogoso Mills Limited in the Western Region.[21]

The Bank of Ghana, in addition to its traditional central banking functions, has directly invested money in productive activities. The Bank is the owner of three projects – Alajo Brick Factory (Clay Products Limited), Shai Hills Cattle Ranch and Grains Warehousing Company Limited.[22] In all these projects some form of financial assistance has been provided by foreign institutions/agencies including the Societé General de Banque, S.A. of Belgium, which provided a loan for the Alajo Brick Factory (another project considered with respect to technology choice in Section 14.5). The Canadian International Development Agency (CIDA) provided a grant for the warehousing project and an Australian institution provided a commercial loan to finance the foreign exchange component of the cattle ranching project at Shai Hills.[23]

The promotion of the Clay Products project was in two phases. The first phase covered the establishment of four medium-sized clay brick factories at Ankaful (Central Region), Ho (Volta Region), Ashiaman (near Tema) and Alajo (near Accra). The second phase (to begin at a later date) will involve the establishment of similar factories at Kumasi (Ashanti Region), Tamale (Northern Region), Bolgatanga (Upper-East Region), Sunyani (Brong-Ahafo Region) and Axim (Western Region).[24] Feasibility studies were carried out by the Bank's Development Finance Department which, according to available information, has so far made no attempt to evaluate projects on the basis of shadow prices.

The Ghana Industrial Holding Corporation (GIHOC) is the largest single public industrial body, with 26 divisions and subsidiaries in 1984.[25] Established in 1968, with 20 pre-existing state enterprises passed to it by the Ministry of Industries, the organisation was initially faced with problems of rehabilitation and expansion.[26] Of the 20 divisions originally comprising GIHOC, five were separated from it in the mid-1970s and one more was added, thus giving 16 divisions at 1977.[27] In addition to these divisions, the Corporation has four joint-venture companies. The activities of GIHOC span many industrial sectors of the economy. Areas of operation include the production of building materials, drugs, canned food, drinks, leather products, textiles and packaging materials.[28]

As at December 1981, the Corporation had a labour force of about 7800, with the Fibre Products Division employing the largest number of workers (nearly 1500), followed by the Pharmaceuticals Division with over 800 workers.[29] Its total investment for 1981 was ₵24.5 million, which was about 10 per cent of the fixed capital investment in the manufacturing sector for that year. In cumulative terms, total investment by the Corporation up to 1981 was ₵140 million.[30] Its turnover, at current prices, which was a little over ₵170 million in 1980, increased to nearly ₵310 million in 1981.[31]

As investment decisions by GIHOC were initially related to existing plants its approach mainly involved a qualitative judgement based on expected operations, capacity utilisation, output expansion or even improvement of product quality. The past procedures applied in relation to expansion and rehabilitation have, however, influenced GIHOC to such an extent that in a number of major new investments (which, of course, might appear small compared to GIHOC's total investment) such as meat storage, steelworks, boatyard slipway and sawmill, no cash flow analysis has been used and the investment decisions were mainly based on operating performance statements. A cash flow approach was, however, applied by GIHOC for its newly-established foundry project although, as with investment appraisals in the Agricultural Development Bank and the Bank of Ghana, GIHOC has not yet used any alternative set of prices which would better reflect the scarcity values of products and inputs.

It thus follows that in Ghana little attempt has been made to use shadow prices in the appraisal of investment projects. Whatever has been attempted by the NIB and the GIC has been partial and uncoordinated. The dangers of applying such an uncoordinated approach can hardly be over-emphasised. As Little and Mirrlees warned in the context of developing countries:

> While accounting prices have been in use for some time, they have seldom been used in a comprehensive and systematic way, but rather haphazardly. This is dangerous. Once some important prices become badly distorted – e.g. the price of labour or foreign exchange – the repercussions are widespread.[32]

It is therefore apparent that in Ghana various policy measures worked to keep and maintain the price structure highly distorted, and no substantial attempt has been made to estimate and use shadow prices systematically so as to avoid serious misallocation of resources. For the application of an alternative set of prices which are significantly different from market prices it is also essential that (i) internally, taxes and subsidies are strictly administered so that the allocation of resources is also efficient in the sectors which are outside the direct control of the government and (ii) externally, goods and services are not diverted through smuggling. Because of serious administrative problems, partly caused by a decline in the quality and efficiency of administration and the fact that the country is surrounded by hard currency francophone zones,[33] neither of the above conditions is strictly fulfilled in Ghana.

A direct consequence of bad investment decisions is lower output and employment than could have been achieved. In other words, the resultant output per unit of capital invested has been lower. The approach used to measure output per unit of capital invested is one of estimating the incremental capital output ratio (ICOR). Table 14.5 shows the ICOR at selected periods considering investment first without any lag, then with a two-year lag. Both net ICOR and gross ICOR have been calculated. Bearing in mind the difficulties of applying the concept of ICOR, the ratios, as shown in Table 14.5, can only give a rough indication of the orders of magnitude. Of the different periods examined, the one from 1971 to 1975 has negative incremental Gross and Net Domestic Products at 1970 prices. For other periods where there was positive real income growth and some generalisations

Table 14.5 Incremental capital–output ratios (ICOR) at 1970 prices, 1957–80

	Net ICOR		Gross ICOR	
	Without lag	With two-year lag	Without lag	With two-year lag
1957–60	3.27	—	4.11	—
1961–65	2.77	2.70	7.84	7.27
1966–70	1.89	2.70	3.87	4.70
1971–75*	—	—	—	—
1976–80	2.50	3.95	6.87	8.75

* Change in GDP is negative.
Source: Huq (1984b), p. 21.

can be made. First, as expected, the 'gross' ICOR in all cases is higher than the 'net' ICOR. Second, contrary to expectations, the values of the ICOR with the time lag are greater (except 1961-65) than those without the time lag.

Due to bad investment decisions, partly influenced by the overvaluation of the cedi, and also to the almost uncritical acceptance of projects financed by suppliers' credit which encourages more capital intensive technologies, investment has usually turned out to be capital deepening in Ghana. During the late 1950s and early 1960s, heavy investment in infrastructure such as roads and ports with long gestation periods also contributed towards a high capital-output ratio. And such ratios are likely to be higher when investment decisions are bad. A high ICOR is also due to the capital stock remaining highly underutilised. This is particularly the case with a two-year lag during 1976-80, because while much investment took place two years prior to income generation during the period, subsequent foreign exchange constraints restricting the availability of imported materials and lack of other inputs including spare parts have caused significantly lower output.

14.5 CHOICE OF TECHNOLOGY

The decision to adopt any mode of production with its corresponding technology is influenced by a variety of factors such as relative factor prices, familiarity with sources and types of machinery and equipment and government policies in this regard. Assuming that alternative technologies to produce a particular commodity do exist, a careful consideration and appraisal should normally lead to better technology selection, suiting prevailing circumstances and the society's objectives.

However, inappropriate technologies can be selected because of ignorance, biases against certain technologies,[34] lack of proper appraisal or relative factor prices failing to reflect factor availabilities. In the past there was a belief that investors were faced with fixed factor-proportions dictated by the available technology. The situation may be explained with the help of Figure 14.1.

The OX-axis represents labour and the OY-axis capital. As usual, it is assumed that labour and capital are homogeneous and infinitely variable – two assumptions which are not realistic. In a situation of fixed factor-proportions, shown by the OA-expansion path, an investor can produce the output and expand it only by using factor-proportions

Investment and Technology Choice 267

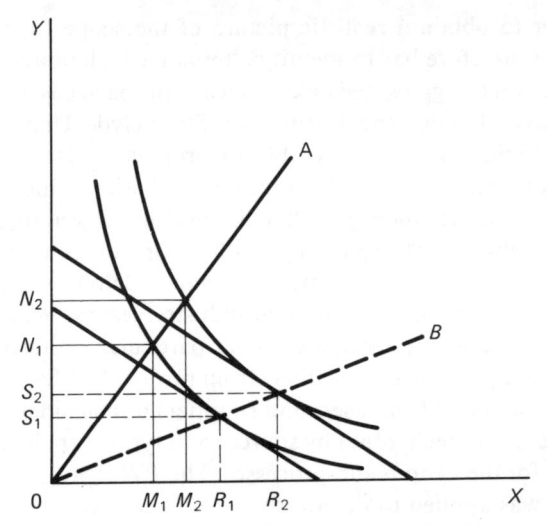

Figure 14.1 Fixed factor proportions

as dictated by the slope of OA, that is, M_1 of labour and N_1 of capital, and M_2 of labour and N_2 of capital. In other words, if existing relative factor prices demand the use of R_1 of labour and S_1 of capital in place of M_1 of labour and N_1 of capital, this cannot be done, implying lack of free movement along the isoquants and hence non-optimal input combination.

In an attempt to examine the scope of factor substitution, estimates of the elasticity of factor substitution have conventionally been made by using data on employment, value added and wages, usually from the censuses of manufacturing industries of many developing countries.[35] The approach most often used is that of the CES (Constant Elasticity of Substitution) Production Function developed by Arrow et al.[36] The estimates of elasticity of substitution have been found to vary from industry to industry in the same country, implying greater scope for technology choice in some industries than in others.

A serious limitation of the above approach is that products often vary within a particular industry, and to view each industry as having only one product is not correct. At the technology level it is also necessary to consider the different sub-processes (production stages) making up the whole technology. Indeed, the scope of factor substitution can vary from sub-process to sub-process depending on the availability of alternative techniques in each sub-process.

In order to obtain a realistic picture of the scope for technology choice one therefore has to identify alternative techniques at product level, based on disaggregated data. Such an approach has been adopted by the David Livingstone Institute of Strathclyde University in its industry studies covering a number of products.[37] These studies, as has been observed by Pickett, 'do provide a basis for moving toward a more 'rational' technology policy in developing countries'.[38]

It is probably worth expanding on the scope of technology choice as found by applying the above approach. Below we present the findings of a study on one particular industry, leather manufacturing, in which the author participated. One particular product (chrome-tanned shoe-upper finished leather) from raw cattlehides was studied, using data obtained from users and suppliers of machinery.[39] To start with, althernative techniques by sources and types were identified and examined for the various sub-processes. The PVC (present value cost) approach was applied to determine the efficiency of alternative techniques under different circumstances (discount rates, wage structures and scales of production).

Given that there are many sub-processes in leather manufacturing, for each of which there is more than one technique, it was possible to 'generate' a large number of combinations (that is, complete technologies). The scope of technology choice was further widened once LDC (less developed country) machinery and equipment were added to European sources. Indeed, data collected from Asian and African leather plants revealed a great variety in technology composition.

For purposes of comparison four technologies, shown below, were selected.

Variant	Description of technology
1	European labour-intensive technology;
2	European capital-intensive technology;
3	European least-cost technology; and
4	LDC technology.

It was found that if technology selection was guided by profitability and/or employment objectives, Variant 2 (European capital-intensive technology) would be ruled out under all the circumstances considered. Similarly, technology Variant 1 (European labour-intensive technology) would also be left out, although in situations where the technology choice was confined to European sources only, it offered the best choice in terms of employment. (The importance of Variant 1 lies in the fact that the identification of the European least-cost technology (that is, Variant 3) may not be easy considering the difficulties in obtaining information on different types of machinery, and the lack of the expertise required for the selection of techniques.) The appeal of the LDC technology turned out to be stronger at lower scales of production. Indeed, at the lowest scale of production examined (200 hides per day) both Variants 1 and 2 (European labour-intensive and European capital-intensive) showed negative Net Present Values (NPV) at a 10 per cent discount rate, although the size of the negative NPV was significantly lower for Variant 1 (−£5000) compared to Variant 2 (−£374 000).

The implications of the above findings are important for developing countries. If a country has to import technologies from Europe, and if the least-cost technology is ruled out, the labour-intensive rather than the capital-intensive technology would be selected on profitability considerations.

14.5.1 Technology choice in Ghana

In their technology selection, investing agencies in Ghana are expected to operate within the confines of the technology policy of the government. To discuss technology choice in Ghana it is appropriate, therefore, to analyse broadly the technology policy adopted by the government.

The first major attempt to introduce a technology policy was the enactment of the Investment Code Act 437 of 1981 which basically provided for the Ghana Investments Centre 'to approve and register all technology transfer contracts in Ghana'.[40] Though considered as a significant break-through, the Investment Act does not pertain 'to the assessment and development of indigenous technologies nor, as it stands, ensures that appropriate technologies are imported, imparted and assimilated by nationals'.[41]

In a rather 'solo' manner, the Nkrumah Government wanted to diversify the mono-culture economy that was considered to be a 'colonial legacy'. Besides its broadly spelt-out aims and objectives, the government pursued a policy of modernising the economy, especially the agricultural and industrial sectors. The authors of the *Seven-Year Development Plan* (1963–64 to 1969–70) remarked in their discussion of capital requirements for industrialisation that:

It is not necessary to adopt the identical capital/labour ratios of comparable manufacturing establishments in the most advanced countries. Especially in subsidiary operations it is possible to economise on the use of scarce capital resources. But there are severe limitations on the number of alternative techniques that are available for the main operations in most manufacturing processes, and if Ghana's industry is to be export-oriented then it cannot afford to employ main methods of production which are markedly less advanced – and hence in general less efficient though cheaper on capital – than those employed by its competitors in these export markets.[42]

Nkrumah's predilection for capital-intensive modes of production derives from the above objective. He equated modernisation with economic growth. His eagerness to make a 'total break with primitive methods' through the large-scale importation of foreign technology,[43] suggested an attempt to find capital-intensive substitutes for various service activities in the economy. In line with the modernisation policy, agriculture was highly mechanised in order to introduce 'ready-made' technologies especially on the State Farms, with little assistance provided to the peasant farmers. The industrialisation drive also implied the creation of new factories to complete the import-substitution process.

The modernisation attempts in agriculture and industry were, however, a flop, and in Killick's view 'much of what went wrong with industry and agriculture was a result of inappropriate technology choices, resulting in farming fiascos and in the creation of inefficient industrial enclaves'. And the choice of inappropriate technologies was considered to be the consequence of 'a deteriorating standard of investment decision-making'.[44]

If we look at successive development plans it appears that planning bodies occasionally called for the adoption of labour-intensive and domestic resource-based industrialisation. In practice, however, projects set up by the government showed no indication that these statements were meant to be taken seriously.[45] Rather foreign systems were imported wholesale consequent upon foreign advice, and with little or no regard to involving nationals in the pre-feasibility, feasibility and start-up stages.[46]

The absence of an explicit technology policy in the country has not prevented the signing of technology transfer agreements between the government, para-statal agencies, private Ghanaian firms and individuals on the one hand, and machinery suppliers (foreigners) on the other. Various investing agencies in the country have had their respective roles to play in this regard. It may therefore be appropriate to treat five such agencies in some detail in an attempt to investigate their approaches to technology choice.[47] These agencies, already mentioned in the previous section, have been selected because of their direct or indirect involvement in investment allocation in Ghana.

The GIC, established in 1963, is the only institution with the legislative authority (provided by the Investment Code Act 437 of 1981) to work out all issues relating to technology transfer contracts. It is entrusted with investigating alternative techniques of producing particular goods and considering the appropriateness of a prospective investor's technology to Ghana's factor endowments.[48] In setting the criteria (in conformity with government priorities), the GIC emphasises the calculation of the social rate of return because of the significant price distortions prevailing in the economy.[49]

Unlike the GIC, other investing agencies studied, namely the NIB, ADB, Bank of Ghana and GIHOC, undertake direct investment. There is ample evidence that the NIB considers technology choice in its project appraisal, although mainly within the context of joint-venture negotiations. In NIB-connected projects (especially where it is the consortium leader), the bank's Development Service Institute and the Board of Directors have the opportunity of directly influencing the projects' design, including the selection of the technological characteristics.[50]

In its efforts to do this, however, the bank is limited by the lack of adequate technical staff and (significantly) by the types of projects it has participated in. A number of NIB-financed projects (mostly national or joint-venture ones) involve the use of foreign capital. In such situations, technology is usually supplied on the basis of either turnkey contracts or suppliers' credits. Such contracts, which tend to leave the choice of technology largely in the hands of foreign partners, severely limit the scope for considering alternative technology.

The Bank of Ghana, the central bank, is charged with the responsibility for clearing technology agreements involving the transfer of foreign exchange. This is in accordance with the Foreign Exchange Act of 1961, regulating the flow of hard currency to Ghana. The Bank also had a department (until 1 May 1984, the Development Finance Department) which evaluated Bank of Ghana-financed projects.

Like the Bank of Ghana, the Agricultural Development Bank (ADB) claims to look for more efficient techniques both by source and type, bearing in mind the needs of the country. According to the ADB, it opens international competitive tenders giving the technical specifications of the equipment it requires. This presupposes that the bank has extensive information on machinery and equipment. However, as with the Bank of Ghana, there are limitations on the extent to which the ADB can cater for the country's needs, insofar as there is no explicit technology policy.

The GIHOC has an internal body, the Development Department, which undertakes project appraisals, although sometimes the services of external (foreign) consultants are commissioned for such purposes. According to the Corporation, the search for alternative technologies is influenced by the consideration of appropriateness of techniques to the needs and environment of the country. However, as in the case of the Bank of Ghana and the ADB, there are obvious limitations in their selection of technologies. There is no extensive identification of techniques, by sources and types, although some form of implicit choice is usually made at the time of decision-making.

The efforts of the investing agencies notwithstanding, a look at some of the projects selected in the past reveals the lack of serious search for and evaluation of alternative technologies. For our analysis we took a sample of seven selected plants located in five of the country's ten regions with a reasonably even degree of geographical spread, four from the public sector, two from the private sector, and one joint venture project. To maintain confidentiality, the projects are referred to as A, B, C and so on (see Table 14.6).

The projects have substantial capital involvement and/or sizeable output, likely indications of their importance to the economy. Three of the projects were established during the 1960s, two in the mid-1970s and the remaining two in the later part of the 1970s.

A comparison of public and private sector plants shows the heavy dependence on imported equipment by the former. In projects A, B and E (all three in the public sector) and D (which has high public investment), virtually all the machinery and equipment was imported. It was only in the case of Project G (public sector) that the scope for using local materials for making some of the simpler equipment was utilised. On the other hand, the two selected private sector plants (C and F) provide evidence that the investors made significant use of locally-made equipment. Project F (soap manufacturing) had a large percentage of its total equipment made locally, all of which is reported to be functioning well. Similarly, Project C (leather manufacturing) had its wooden drums and some other equipment locally made, at a

Table 14.6 Characteristics of technology choice in selected projects

Project (Date of establishment)	Location (region)	Ownership	Technology choice by* Source	Technology choice by* Type	Type of Evaluation Explicit	Type of Evaluation Implicit
A. Sugar (1963)	Central	Public	No	No	No	No
B. Leather (1962) 1978†	Volta	Public	No	No	No	No
C. Leather (1969)	Ashanti	Private	Yes Yes	No Yes	No No	Yes Yes
D. Rice (1973)	Northern	Joint	Yes (Partial)	Yes (Partial)	No	Yes
E. Palm Oil (1978)	Ashanti	Public	No	Yes (Partial)	No	Yes
F. Soap (1974)	Central	Private	Yes	Yes	No	Yes
G. Bricks (Mid-1970s)	Greater Accra	Public	No	Yes (Partial)	No	Yes

* For explanation of the choice of technology by sources and types see Huq (1986). See also Huq and Aragaw (1981) which deals in detail with such choices in the context of leather manufacturing.
† For Project B, a committee was set up in 1978 to advise the government on the reactivation of the project. The report of the committee contained an appraisal for reactivation.

Source: Huq (1986, p. 15) as prepared from data collected from factory visits, unpublished documents and personal interviews.

cost which was found to be significantly lower than the import cost would have been.

In the two private sector plants, the machinery and equipment was chosen by shopping around and was installed locally by the investors, while in the case of public sector plants there was heavy dependence on the supply and installation of equipment by the machinery suppliers.[51] The plants established in the 1960s are particularly noted for their lack of pre-feasibility and feasibility studies, let alone a careful examination of such reports. The lack of these vital reports made it rather difficult to ascertain the original costs of the two public sector plants, as well as what, if any, consideration was given to technology choice.

In the 1970s the situation improved, and some forms of analysis were undertaken for project selection. For example, in all the three selected projects from the 1970s, some search for technology by type was undertaken, and this was also the case for Project B's reactivation analysis carried out in 1978.

As far as technology choice by source is concerned, some search was made in the case of Projects D (rice milling), F (soap manufacturing) and C (leather manufacturing) and the reactivation committee on Project B (leather manufacturing) also considered a number of European sources for plant and equipment.

14.6 CONCLUSIONS

Improvement in investment decisions obviously demands proper appraisal of projects incorporating search and evaluation of alternative techniques by sources and types. Biases in the decision-making process can lead to non-optimal choices, that is, the selection of projects/technologies with lower profitability and/or employment than could be achieved with *obtainable* alternative projects/technologies. For example, a preference for sophisticated technologies (as Timmer found for rice marketing in Indonesia) can arise from 'a deep-felt bias on the part of Western and Western-trained technicians that identified capital-intensive with modern, and modern with good'.[52] In Nigeria, Winston found a preference for sophisticated 'high technology' so as 'to assert the power, equality and technical competence of one's people'.[53]

In developing countries, technologies have often been equated with projects, meaning that different sub-processes that go with every project/technology and the prospect of using local machinery and equipment in at least some of the sub-processes have usually been overlooked. Machinery and equipment obtained on suppliers' credit are usually associated with exceptionally high investment costs and more capital intensive techniques than required. This is mainly because the machinery suppliers have a vested interest in selling as much as they can,[54] and the investor obtaining machinery and equipment on suppliers' credit is likely to have a limited say in the selection and appraisal of techniques. Prices of machinery and equipment obtained on suppliers' credit can cost over 100 per cent more than similar goods from the cheapest available sources.[55] The experience of Ghana bears ample testimony to this fact. Harvey refers to the sale of 'inappropriate equipment, at excessive prices, on expensive insured suppliers' credit, for projects that were not only unable to generate enough profit to service their debts, but in some cases did not even produce anything at all for many years', and he blames both the foreign suppliers of machinery and the Ghanaian officials for the 'cost' of their action.[56]

A proper appraisal can be hindered by lack of adequate data, both economic and engineering, on the alternative techniques. The appraisal may also involve only market prices or the use of haphazard or partial shadow prices, as was found to be the case in Ghana. That such an appraisal in the context of a highly distorted price structure can lead to non-optimal choices is obvious.

Another area of concern is the lack of adequate analysis incorporating alternative scales of production, risk and uncertainty, and the sensitivity of the project/technology at different levels of capacity utilisation. Thus, one often finds projects/technologies in developing countries which are functioning very much below the projected capacity because of lack of adequate raw materials, difficulties of maintenance, shortage of skilled labour, and the like. It goes without saying that investment decisions in these circumstances were far from satisfactory.

We, therefore, find a strong case for technology choice in developing countries like Ghana by *explicit* evaluation, incorporating identification and evaluation of machinery and equipment by *types* and *sources* including local and other developing country sources.[57] A clearer policy frame of reference than at present can substantially help in improving investment decisions, particularly in the public sector, by making it obligatory for the decision-makers to justify their choices.

Postscript

In a work of contemporary nature such as the present one, a postscript is probably essential. Much has happened in Ghana since the first draft of the book was completed in September 1984. During the period 1983-86, the first phase (the stabilisation phase) of the Economic Recovery Programme (ERP I) was completed. Currently, the second phase (structural adjustment and development phase) of the programme (ERP II) is being implemented over the three-year period 1987-89.

The year 1983, which marked the beginning of the ERP I, has been the worst year in Ghana's history since independence. In that year a number of factors, including a severe drought and the expulsion of over 1 million Ghanaians from Nigeria, contributed towards a serious economic crisis. Real GDP per capita declined by 7.2 per cent during the year, GDP itself reaching the lowest level since 1969. However, since 1984 the economy has started picking up as a direct consequence of the various policy measures initiated by the government in 1983. It may be recalled that the April 1983 budget, for the first time since the early 1970s, announced a major *de facto* devaluation of the cedi by introducing a scheme of surcharges and bonuses on the use and earning of foreign exchange respectively. In effect, a regime of multiple exchange rates, ranging from ₵23.38 to ₵29.98 to the US dollar was introduced.[1]

The multiple exchange rate system proved unwieldy, particularly for banks and importers. The multiple exchange rate system was consequently unified at ₵30.00 to the US dollar in October 1983, replacing the old fixed rate of ₵2.75 per US dollar. The next major attempt in adjusting the rate of foreign exchange was in September 1986 when two official 'Windows' were instituted for foreign exchange transactions. The rate of foreign exchange for Window I was fixed at ₵90 = US$1. This rate was applicable to imports of crude oil and essential drugs, official debt servicing and cocoa exports. The rate in the other official Window was going to be determined by formal auctions of foreign exchange.[2] In February 1987, the first-tier exchange rate of ₵90 to the dollar was abolished, and all transactions were required to be at the rate fixed at the weekly foreign exchange auctions. Such auction rates were in the region of ₵176-179 per US dollar in early February 1988.[3]

The institution of foreign exchange auctioning has meant that the determination of the foreign exchange rate is dependent upon bids from foreign exchange users, and not set administratively. The new system has resulted in considerably narrowing the spread between the official and parallel market rates of exchange. It has also made the export business more profitable than before, thus stimulating the export sector. Economic rent on imports has also been reduced substantially. And the former vexatious red-tapeism and corruption associated with the administrative allocation of foreign exchange are now largely removed.

There have also been similar moves towards reforming the price structure in other sectors of the economy with a view to encouraging production. The primary objective of the 1987 fiscal programme was to further enhance incentives for production and at the same time to generate employment.[4] In the 1988 budget, the thinking of the government in terms of price controls has been stated as follows: 'the institution of price controls can hardly be relied upon to guarantee the prices for goods and services that consumers actually pay'.[5] The government has also unequivocally expressed its view about providing a proper incentive to savers: 'The P.N.D.C. takes the view that for the effective mobilisation of savings, the return should be made attractive. Specifically, the interest rate on savings and time deposits should be at least as high as the rate of inflation as otherwise consumption may be encouraged instead of saving'.[6]

Turning the budget deficit into a surplus has proved a difficult target for the government to achieve, although the huge budget deficits of the early 1980s are now a thing of the past. In 1986, total receipts of the government amounted to ₵73.6 billion, while total expenditure amounted to ₵73.3 billion, thus showing a modest surplus. In the 1987 budget, the estimated revenue is ₵109.7 billion and the estimated expenditure is ₵109.4 billion.[7]

The restraint in monetary expansion, following largely from the success of the government in adjusting its expenditure to its receipts, has led to the lessening of inflationary pressures. In 1986, inflation was down to 15 per cent compared with the very high level of 100 per cent or more which the economy was experiencing in some recent years. However, for 1987 the rate of inflation is estimated to be 35 per cent. This is due to some sharp rises in food prices mainly because of late and short rains.[8]

Restoring salaries and wages to what may be called 'something of a normal level' still remains a matter of serious concern to the country.

The measures taken in the 1987 budget include provision for a further increase in real income through a 25 per cent across the board increase in wages and salaries for the Civil Service with effect from 1 January of that year. In addition, transport allowances have been increased by 100 per cent against the anticipated increase of fares by 30 per cent and canteen allowance by 50 per cent. A number of tax reliefs which were introduced in 1987 have further been extended in the 1988 budget. For example, the standard relief for a single individual has been raised from the 1987 level of ₡10 000.00 to ₡24 000.00 per annum and for married people from ₡15 000.00 to ₡36 000.00. The top marginal rate of 55 per cent now applies to incomes of ₡984 000.00 and above instead of ₡310 000.00 and above. The marginal rate for salaries above ₡310 000.00 has been reduced from 55 per cent to 30 per cent. These measures are obviously likely to contribute to a demand-pull inflationary pressure.

To tackle the problem of overmanning attempts are being made to reduce the 'surplus' labour from the state sector. The first phase of the restructuring exercise has proceeded extremely well, and the Ghana Cocoa Board (old CMB) has been able to complete its exercise well ahead of schedule. A provision of ₡3 billion was made in the 1987 budget for resettling the employees made redundant from the Civil Service. In mid-1987 the Ghana Education Service of the Ministry of Education started its retrenchment exercise.

A higher inflow of foreign assistance has greatly helped the government to carry out the recovery programme. The donors have maintained their generous stand towards Ghana. The Consultative Group of donor countries for Ghana, which did not meet for a long time, had its first meeting in a decade in November 1983 in Paris. Regular meetings have taken place since then, with marked increases in total external assistance from members. Over the ERP I period, Ghana received yearly commitments averaging US$430 million which was more than double that for the four years preceding ERP I. Of the total support received for the programme from the members, the IMF provided the largest amount of US$750 million, while World Bank assistance covered four policy-based loans totalling US$274 million together with project lending of US$212 million. Indeed, the donors have remained highly sympathetic to Ghana's recovery programme. For example, in the 1987 Paris Consultative Group Meeting the members provided US$818 million worth of concessional funding although only about US$575 million had been expected by Ghana.[9]

The economy appears to have responded well to the package of measures applied since 1983. The real GDP growth rate in 1986 was estimated at 5.3 per cent, the third successive year of annual economic growth of over 5 per cent.[10] 'A significant aspect of the 1986 performance', according to the Secretary for Finance, 'was the wider spread of growth over the various sectors with the manufacturing and construction sectors registering a particularly strong recovery... Even more impressive has been the response of our cocoa farmers to the incentives provided by the steady increases in the producer price of cocoa. Cocoa exports have steadily increased from 154 000 metric tonnes in 1983–84 to over 210 000 metric tonnes in 1985–86. Our total export earnings have gone up from US$439.6 million in 1983 to US$773 million in 1986'.[11]

In brief, at the time of finalising this postscript (February 1988) it appears that the government has largely succeeded in thoroughly reforming the old price structure. The adjustment in the foreign exchange rate of the cedi has probably been taken to an extreme, with weekly auctions determining the rate. Inflation has been drastically reduced following the growth of the economy since 1984 and the successful reduction in the growth of money supply. The non-development budget has been turned into surplus, though of a modest one, after long years of deficit. The government has also been tackling the overmanning problem in the state sector. Attempts are also being made to raise the level of real wages and salaries, although the very low level of real income is still affecting the morale of fixed wage earners. The export sector has responded to the new policy measures, registering an average growth rate of 13 per cent during ERP I.[12]

Growth in manufacturing has been averaging 14 per cent per annum since 1983, a direct result of an improvement in capacity utilisation which in most industries is rapidly moving towards 50 per cent or more.[13] The performance in the agricultural sector has also been spectacular, output of most crops in 1986 being two to four times the level recorded in 1983. Indeed, with the exception of cocoyam and plantain, total output of all starchy staples in 1986 exceeded 1970 levels.[14]

Following the success of the Economic Recovery Programme, the government has been able to carry out, among other things, the repayment of some foreign debts long overdue. For example, by the end of 1986 all arrears in respect of debts incurred by the Nkrumah government had been cleared. An amount of over US$110 million owed by the Limann government to Nigeria for crude oil imports had also been rapaid. In addition, the country has been able to increase its foreign exchange reserves to the equivalent of more than two months' import cover.[15]

Thus, compared with the severe economic crisis that the country was facing in the early 1980s, there is now obviously a healthier environment in which the government can hopefully aim at achieving faster economic growth through raising the savings and investment ratios. The following targets set by the ERP II in these respects are therefore worth noting:[16]

(a) sustain economic growth at between 5 and 5.5 per cent a year over the medium term;
(b) increase the level of public investment from about 10 per cent of national income to about 25 per cent by the end of the decade; and
(c) increase domestic savings from about 7 per cent at the end of ERP I to about 15 per cent by the end of the decade.

It is gratifying to note that the measures which were frequently highlighted in the course of our analysis of Ghana's economic decline have largely contributed to the present recovery of the Ghannian economy.

Danny Safo died on 6 January 1988. He was a member of the research team which was responsible for the collection and analysis of the extensive amount of data on which the present study is based. Dan's contribution is evident throughout the book. Immediately before his death he even went through the Postscript, adding some recent information. Dan was one of the brightest academics the author has met. His satirical novel, *His Excellency the Head of State*, was published by Macmillan in 1983. Following the economic crisis, the fall in real income became so severe that Dan, like most other Ghanaians, was struggling to survive. As a Lecturer in Economics at Cape Coast University his monthly salary could barely provide a week's subsistence for him and his family. Financial and other worries ultimately led to his death. In Dan's untimely death Ghana has lost a top intellectual and the author a great personal friend. Let his death be a reminder to all concerned that economic mismanagement can indeed cause untold suffering.

Appendix A

A NOTE ON DATA USED IN THE STUDY

The main objective of this note is to explain briefly the procedure followed in collecting, arranging and interpreting data and the constraints faced during the process.

A 'data bank' was built up using information from a variety of sources. In Ghana by far the most important source of statistical information is the Central Bureau of Statistics (CBS), recently renamed as Statistical Service. The following major publications of CBS/Statistical Service have appeared fairly regularly:

Quarterly Digest of Statistics;
Economic Survey (Annually up to 1969
 and periodically since then);
Labour Statistics (Annually);
Industrial Statistics (Periodically); and
External Trade Statistics (Monthly).

Another important publication of the CBS is the *Statistical Year Book*, which unfortunately ceased appearing after 1970.

The CBS has also published the following:

1960 Population Census of Ghana (3 vols.);
1970 Population Census of Ghana (4 vols.);
1984 Population Census of Ghana: Preliminary Report;
Ghana Sample Census of Agriculture 1970;
Household Economic Survey 1974–75; and
Input–Output Table of Ghana 1968.

The Bank of Ghana is another important source of official data. It publishes regularly the following:

Quarterly Economic Bulletin and
Annual Report.

Other important official documents of Ghana include development plans and programmes, national budget statements, education statistics, annual reports from different banking and other organisations.

Appendix A

For any particular subject if we were satisfied that no official published data were available, we moved to published unofficial sources. One such source which was important in filling some data gaps for earlier years (1957-59) was T. Brown (1972). Other published unofficial sources include Birmingham *et al* (1966), Killick (1978), Leith (1974) and Steel (1977).

Unpublished information from official and unofficial sources has also been used, particularly for data which are otherwise unavailable. These include documents, mimeographed and typed scripts from various institutions and individuals. Two such important sources are ISSER, University of Ghana and CDS, University of Cape Coast.

The main period of the study is from 1957 to 1982, although more recent data as available before the book went for printing have been incorporated. Thus, the period covered in the study provides us with a reasonably long period of Ghana's experience of economic development.

We assembled national accounts data of Ghana at current and constant prices. There were no serious problems with data at current prices. But national accounts data at constant prices have so far been produced by the CBS/Statistical Service using three different base years - namely 1960, 1968 and 1975. So as to enable us to make a direct comparison of absolute figures relating to GDP and other related data at constant prices, we have reconstructed the entire time series at 1970 prices - that is with a common base year.

Once different macroeconomic variables with a common base year for the entire period were available, it obviously became easier to proceed to obtain second level calculations such as GDP and other deflators, domestic savings and foreign savings, public and private savings - all at 1970 prices.

The following definitions were used in the calculations of a number of variables and to serve as cross-checks.

$$C_t = C_g + C_p \tag{1}$$

$$S_d = GDP - C_t \tag{2}$$

$$S_f = M - X \tag{3}$$

$$I = K + \Delta q \tag{4}$$

$$I = S_d + S_f \tag{5}$$

$$I = I_g + I_p \tag{6}$$

$$I_g = G_d \tag{7}$$

$$I_P = I_{Pd} + I_{Pf} \tag{8}$$

$$S_d = S_p + S_g \tag{9}$$

$$S_g = G_v - G_{nd} \tag{10}$$

$$S_f = I_{Pf} + A \tag{11}$$

where

GDP is gross domestic product,
C_t is total consumption,
C_g is public consumption,
C_p is private consumption,
S_d is gross domestic savings,
S_p is private savings,
S_g is public savings,
S_f is foreign savings,
X is exports,
M is imports,
G_v is central government current revenue,
G_d is central government development expenditure,
G_{nd} is central government non-development expenditure,
I is total investment,
I_P is private investment,
I_g is public investment,
I_{Pd} is private domestic investment,
I_{Pf} is private foreign investment, and
A is grants and loans from abroad.
K is gross fixed capital formation in a given year, and
Δq is change in stocks in a given year.

The research was constrained not only by limited availability of data but also because of the presence of conflicting figures which appeared in different sources and, at times, even in the same source. A criterion had therefore to be followed to avoid any confusion and we decided to take in each case the latest available data as the most reliable. By depending on official sources for the basic information and also on the latest available data we experienced a number of difficulties. But the difficulties were not insurmountable and in no case did we have to use any estimate of our own. And our experience was that within the data constraints, and given the ready co-operation of people from the relevant organisations in Ghana, it was possible to make an analysis of different aspects of the Ghanaian economy. There is, of course, room for improvement – but that is for the future.

Table A.1   Gross domestic product, 1957–84
(GDP in ₵ million)

Year	Current prices	1970 prices	GDP index at 1970 prices
1957	740	1437.34	63.62
1958	780	1383.58	61.24
1959	890	1593.26	70.62
1960	956	1713.24	75.83
1961	1 022	1774.27	78.53
1962	1 094	1860.30	82.34
1963	1 208	1924.82	85.20
1964	1 357	1966.04	84.02
1965	1 466.4	2010.84	89.00
1966	1 518.4	1925.21	85.21
1967	1 504.3	1984.25	87.83
1968	1 700.2	1991.63	88.15
1969	2 000.7	2111.35	93.45
1970	2 259.3	2259.30	100.00
1971	2 500.5	2377.14	105.22
1972	2 815.4	2305.43	102.04
1973	3 501.2	2371.93	104.99
1974	4 660.1	2534.47	112.18
1975	5 283.0	2219.40	98.23
1976	6 526.2	2141.05	94.77
1977	11 163.4	2189.74	96.92
1978	20 986.1	2375.34	105.14
1979	28 221.6	2315.64	102.49
1980	42 853.5	2342.32	103.67
1981	72 626.1	2245.02	99.37
1982	86 450.8	2089.63	92.49
1983	184 038.4	1994.39	88.27
1984	270 560.6	2166.67	95.90

Note: See Appendix A for procedures followed in computing GDP at constant prices.
Sources: Brown, T. M. (1972); CBS, *Economic Survey* (various issues); and Statistical Service, *Quarterly Digest of Statistics*, September 1986.

Table A.2 Population and GDP per capita, 1957–84

Year	Population ('000)	GDP per capita (₡) Current prices	GDP per capita (₡) 1970 prices	GDP per capita index at 1970 prices (1970 = 100)
1957	6 247	118	230	87.12
1958	6 403	112	216	81.18
1959	6 563	136	243	92.05
1960	6 727	142	255	96.59
1961	6 902	148	257	97.35
1962	7 081	154	263	99.62
1963	7 251	167	266	100.76
1964	7 425	183	265	100.38
1965	7 603	193	265	100.38
1966	7 786	195	247	93.56
1967	7 972	189	249	94.43
1968	8 164	208	244	92.42
1969	8 360	239	253	95.83
1970	8 559	264	264	100.00
1971	8 780	285	271	102.65
1972	9 006	313	256	96.97
1973	9 238	379	257	97.35
1974	9 476	492	267	101.14
1975	9 720	544	228	86.36
1976	9 970	655	215	81.44
1977	10 227	1 092	214	81.06
1978	10 491	2 000	226	85.61
1979	10 761	2 623	215	81.44
1980	11 039	3 882	212	80.30
1981	11 323	6 414	198	75.11
1982	11 615	7 443	180	68.16
1983	11 914	15 447	167	63.42
1984	12 221	22 139	177	67.16

Sources: Population data (actual and estimated) from Birmingham (1966); and Population Censuses of 1960, 1970 and 1984. GDP per capita computed from Table A.1.

Table A.3 Expenditure components of the GDP at current prices, 1957-84
(₡ million current prices)

Year	Private consumption (+)	Government consumption (+)	Gross fixed capital formation (+)	Changes in stocks (+)	Exports (+)	Imports (−)	GDP
1957	596	66	112	−12	192	214	740
1958	572	70	110	−2	220	190	780
1959	650	78	154	+20	240	252	890
1960	694	96	194	+22	246	296	956
1961	804	110	210	−20	244	326	1 022
1962	830	122	184	−12	240	270	1 094
1963	916	138	218	−8	234	290	1 208
1964	987	160	232	+14	247	283	1 357
1965	1 133.0	211.9	265.7	−3.5	251.1	391.8	1 466.4
1966	1 201.0	198.0	196.5	−1.3	222.0	297.8	1 518.4
1967	1 165.4	225.0	174.0	−18.9	274.1	315.3	1 504.3
1968	1 198.5	285.3	186.8	+2.0	396.1	368.5	1 700.2
1969	1 460.6	284.7	195.3	+40.6	447.2	427.7	2 000.7
1970	1 664.7	290.3	271.4	+48.3	523.2	538.6	2 259.3
1971	1 974.8	324.5	310.7	+42.5	536.7	688.7	2 500.5
1972	2 105.7	354.9	244.4	−43.5	581.5	427.7	2 815.3
1973	2 652.1	382.3	267.8	+48.2	750.5	599.7	3 501.2
1974	3 669.5	569.3	554.5	+53.3	867.9	1 054.4	4 660.1
1975	3 873.1	688.5	613.8	+58.5	1 022.6	973.5	5 283.0
1976	5 170.5	799.0	640.8	−62.0	1 025.2	1 047.3	6 526.2
1977	8 654.8	1 409.4	1 031.9	+186.2	1 170.2	1 289.4	11 163.1
1978	17 473.0	2 370.7	1 355.3	+65.9	1 754.2	2 033.0	20 986.1
1979	23 454.9	2 902.6	1 898.6	−54.3	3 169.6	3 149.8	28 221.6
1980	35 953.3	4 784.4	2 613.1	−202.9	3 628.1	3 922.5	42 853.5
1981	63 322.9	6 384.1	3 429.9	−108.9	3 453.6	3 865.5	72 626.1
1982	77 619.3	5 602.9	3 053.0	−132.8	2 866.2	2 577.8	86 450.8
1983	172 140.1	10 787.0	6 922.1	−20.8	11 238.0	17 028.0	184 038.4
1984	233 022.8	19 640.7	18 541.7	+65.4	20 161.0	20 871.0	270 560.6

Sources: Brown, T. M. (1972); CBS, *Economic Survey* (various issues); and Statistical Service, *Quarterly Digest of Statistics*, September 1986.

Table A.4 Expenditure components of the GDP at 1970 prices, 1957–84
(₵ million)

Year	Private consumption	Government consumption	Gross fixed capital formation	Changes in stocks	Exports	Imports	GDP
1957	1108	134	224	−22	385	393	1437
1958	1063	138	219	−4	319	351	1384
1959	1176	151	294	+36	391	453	1593
1960	1244	172	348	+39	441	530	1713
1961	1348	186	358	−36	495	577	1774
1962	1272	204	326	−18	581	505	1860
1963	1333	226	387	−14	552	559	1925
1964	1285	240	396	+47	486	487	1966
1965	1377	349	412	−4	522	645	2011
1966	1339	302	294	−2	466	475	1925
1967	1479	305	228	−22	424	430	1984
1968	1404	334	219	+2	464	432	1992
1969	1557	318	211	+43	446	465	2111
1970	1665	290	271	+48	523	539	2259
1971	1703	275	311	+37	508	458	2377
1972	1591	259	203	−21	588	314	2305
1973	1750	241	204	+33	585	441	2372
1974	1958	289	305	+28	443	489	2534
1975	1627	289	258	+24	430	409	2219
1976	1560	296	242	−18	450	389	2141
1977	1563	381	285	+34	339	413	2190
1978	1732	448	246	+8	324	383	2375
1979	1690	368	230	−4	319	287	2316
1980	1795	412	217	−10	275	346	2342
1981	1635	477	191	−3	251	306	2245
1982	1421	424	148	−3	289	189	2090
1983	1427	416	145	−1	204	197	1994
1984	1645	362	165	+1	172	178	2167

Note: Due to rounding off the total may vary slightly with that shown in Table A.1.
Sources: Brown, T. M. (1972); CBS, *Economic Survey* (various issues); and Statistical Service, *Quarterly Digest of Statistics*, September 1986.

Table A.5 GDP by industrial origin at current prices, 1965–84
(₵ million: current prices)

Sector	1965	1966	1967	1968	1969	1970	1971	1972	1973	1974
1. *Agriculture*	<u>598.0</u>	<u>656.7</u>	<u>604.7</u>	<u>710.3</u>	<u>918.5</u>	<u>1051.0</u>	<u>1104.9</u>	<u>1313.2</u>	<u>1714.7</u>	<u>2383.1</u>
Agriculture and livestock production	396.8	474.9	376.9	444.0	571.6	634.4	714.3	886.5	1195.7	1662.9
Cocoa production and marketing	123.7	107.7	150.3	184.4	249.2	317.1	245.3	303.9	343.5	503.5
Forestry, logging and fishing	77.5	74.1	77.5	81.9	97.7	99.5	95.3	122.8	175.5	216.7
2. *Industry*	<u>272.4</u>	<u>278.5</u>	<u>307.9</u>	<u>345.6</u>	<u>383.7</u>	<u>412.6</u>	<u>457.1</u>	<u>498.9</u>	<u>651.2</u>	<u>845.4</u>
Mining and quarrying	35.3	41.0	42.9	42.1	40.1	38.0	40.6	63.4	81.5	100.0
Manufacturing	142.5	154.6	179.4	213.8	248.4	257.6	275.1	305.9	409.3	501.8
Electricity and Water	6.5	9.3	11.6	16.8*	18.9	23.2	23.5	25.6	29.5	30.4
Construction	88.1	73.6	74.0	72.9	76.3	93.8	117.9	104.0	130.9	213.2
3. *Services*	<u>596.2</u>	<u>583.1</u>	<u>511.6</u>	<u>644.1</u>	<u>698.4</u>	<u>796.7</u>	<u>939.5</u>	<u>1003.3</u>	<u>1135.3</u>	<u>1431.5</u>
Trade, hotels, transport and import duties	338.4	306.5	286.7	300.3	370.0	439.2	526.4	523.7	636.6	864.8
Finance, real estate, business services and other services	127.1	138.6	146.0	143.4	136.9	156.2	196.2	233.5	249.0	235.0
Government services	130.7	138.0	158.9	200.4	191.5	201.3	216.9	246.1	250.0	331.7
Gross Domestic Product	1466.6	1518.3	1504.2	1700.0	2000.6	2259.3	2501.5	2815.4	3501.2	4660.0

Table A.5 (Continued)

Sector	1975	1976	1977	1978	1979	1980	1981	1982	1983	1984
1. *Agriculture*	2518.2	3300.1	6274.4	12741.5	16924.4	24820.9	38553.2	49571.5	109927.4	133231.9
Agriculture and livestock production	1567.0	2347.1	5068.0	10010.9	12857.9	19775.5	33558.0	44002.8	92046.9	110422.1
Cocoa production and marketing	577.0	493.8	624.0	1468.5	2293.5	2269.0	1196.6	639.0	10227.5	11138.1
Forestry, logging and fishing	347.2	459.2	582.4	1262.1	1773.0	2776.4	3798.6	4929.7	7653.0	11671.7
2. *Industry*	1108.6	1254.3	1768.4	2523.8	3466.3	5095.5	6652.6	5401.0	12199.2	28630.9
Mining and quarrying	104.5	87.4	89.5	119.2	238.2	461.8	368.7	284.5	1944.2	3214.3
Manufacturing	735.9	857.5	1203.7	1813.3	2447.5	3345.8	4337.5	2116.8	7100.9	17305.5
Electricity and water	32.6	47.6	52.9	74.3	122.1	233.4	455.5	513.4	358.0	2165.9
Construction	235.6	261.8	422.3	517.0	658.5	1054.4	1490.9	1486.3	2796.1	5945.2
3. *Services*	1656.1	1971.7	3120.6	5720.8	7830.9	12937.1	27420.3	31478.3	61911.8	108697.8
Trade, hotels, transport and import duties	944.0	1211.4	1809.4	3646.0	5175.1	8854.4	22037.1	26059.0	52190.0	96488.3
Finance, real estate, business services and other services	279.1	293.8	412.0	523.8	735.0	946.3	1238.0	1366.9	1899.8	3068.5
Government services	433.0	466.5	899.2	1551.0	1920.8	3136.4	4145.2	4052.4	7822.0	9141.0
Gross Domestic Product	5283.0	6526.1	11163.4	20986.1	28221.6	42853.5	72626.1	86450.8	184038.4	270560.0

Note: *Water is included in Government Services.
Sources: Singal and Nartey (1971); CBS, *Economic Survey* (various issues); and Statistical Service, *Quarterly Digest of Statistics*, September, 1986.

Table A.6 GDP by industrial origin at 1970 prices, 1968–84
(₵ million)

Sector	1968	1969	1970	1971	1972	1973	1974	1975	1976
1. *Agriculture*	831.9	889.9	1051.0	1199.8	1238.9	1214.2	1321.7	1057.9	1040.6
Agriculture and livestock production	520.0	537.8	634.4	660.6	671.8	706.1	836.4	658.3	565.5
Cocoa production and marketing	216.1	228.7	317.1	387.5	401.8	357.9	328.0	242.4	313.7
Forestry, logging and fishing	95.8	123.3	99.5	151.7	165.3	150.2	157.3	157.2	161.4
2. *Industry*	405.0	446.8	412.6	442.6	395.5	473.0	470.1	465.9	453.7
Mining and quarrying	49.4	46.9	38.0	56.3	58.1	53.4	46.6	43.9	42.0
Manufacturing	250.5	292.2	257.6	269.1	243.2	300.5	283.4	309.2	295.7
Electricity and water	19.7*	22.6	23.2	7.4	10.1	17.6	14.4	13.8	15.4
Construction	85.4	85.1	93.8	109.8	84.1	101.5	125.7	99.0	100.6
3. *Services*	775.0	774.0	796.7	735.3	670.9	684.8	743.3	695.8	646.7
Trade, hotels and transport	351.8	390.0	439.2	464.8	408.5	428.4	467.8	396.6	355.8
Finance, real estate, business services and other services	168.4	168.4	156.2	124.3	128.1	116.5	127.0	117.2	106.8
Government services	234.8	215.6	201.3	146.2	140.2	139.9	148.5	182.0	184.1
GDP	1991.9	2111.0	2259.3	2377.7	2305.3	2372.0	2535.1	2219.6	2141.0

Table A.6 (Continued)

Sector	1977	1978	1979	1980	1981	1982	1983	1984
1. *Agriculture*	992.5	1169.8	1213.8	1240.1	1210.4	1144.4	1057.7	1167.8
Agriculture and livestock production	577.9	780.5	844.9	845.7	840.5	805.0	735.9	854.9
Cocoa production and marketing	247.6	219.6	208.7	228.5	218.9	181.3	154.6	142.5
Forestry, logging and fishing	167.0	169.7	160.2	165.9	151.0	158.1	167.2	170.4
2. *Industry*	473.7	435.5	371.6	373.9	315.9	262.3	241.9	251.7
Mining and quarrying	40.8	35.2	30.6	29.6	27.6	25.5	21.7	24.7
Manufacturing	304.0	293.5	244.2	239.9	195.0	155.0	136.8	155.5
Electricity and water	15.4	13.5	15.9	17.9	20.2	18.6	17.2	16.2
Construction	113.5	93.3	82.9	86.5	73.1	63.2	66.2	55.3
3. *Services*	723.5	769.6	727.7	728.0	718.8	683.1	694.8	747.1
Trade, hotels and transport	369.2	360.5	372.6	336.6	325.8	295.0	291.8	327.4
Finance, real estate, business services and other services	111.5	111.2	104.3	136.1	119.9	126.4	127.5	141.2
Government services	242.8	297.9	250.8	255.3	273.1	261.7	275.5	278.5
GDP	2189.7	2374.9	2315.1	2342.0	2245.0	2089.6	1994.4	2166.7

Note: *Water is included in Government Services. Due to rounding off the total may vary slightly with that shown in Table A.1.

Sources: CBS, *Economic Surveys* (various issues); and Statistical Service, *Quarterly Digest of Statistics*, September 1986.

Table A.7 Gross and net domestic capital formation at current and 1970 prices, 1957–84
(₵ million)

Year	Depreciation (Current)	Gross domestic capital formation (Current)	Net domestic capital formation (Current)	Depreciation (1970)	Gross domestic capital formation (1970)	Net domestic capital formation (1970)
1957	—	100.0	—	67.8	202.5	134.7
1958	—	108.0	—	76.4	215.0	138.6
1959	—	174.0	—	82.8	329.8	247.0
1960	—	216.0	—	95.8	387.0	291.2
1961	—	190.0	—	112.1	322.6	210.5
1962	—	172.0	—	125.9	308.2	182.3
1963	—	210.0	—	138.8	372.8	234.0
1964	—	246.0	—	156.9	442.8	285.9
1965	89.0	262.2	173.2	128.3	408.0	279.7
1966	91.2	195.2	104.0	134.3	292.1	157.8
1967	101.0	155.1	54.1	131.4	206.0	74.6
1968	109.6	188.8	79.2	128.5	221.1	92.6
1969	129.4	235.9	106.5	138.5	254.6	116.1
1970	134.3	319.4	185.4	134.3	319.4	185.1
1971	145.2	353.2	208.0	136.9	348.0	211.1
1972	170.7	200.9	30.2	137.6	182.0	44.4
1973	216.4	316.0	99.6	156.3	237.0	80.7
1974	256.1	607.8	351.7	152.8	333.0	180.2
1975	322.9	672.3	349.4	135.6	283.0	147.4
1976	378.2	578.8	200.6	137.9	224.0	86.1
1977	524.6	1 218.1	693.5	140.6	319.0	178.4
1978	724.4	1 421.2	696.8	131.0	254.0	123.0
1979	1 052.0	1 844.3	792.3	128.5	226.0	97.5
1980	1 511.7	2 410.2	898.5	131.2	207.0	75.8
1981	2 218.0	3 321.0	1103.0	133.1	188.0	54.9
1982	2 628.4	2 290.2	291.8	138.5	145.0	6.5
1983	4 196.9	6 901.3	2704.4	137.8	144.0	6.2
1984	11 207.5	18 607.1	7401.6	110.7	166.0	55.3

Sources: Tables A.3 and A.4; CBS, *Economic Survey* (various issues); and Statistical Service, *Quarterly Digest of Statistics*, September 1986.

296

Table A.8 Implicit deflators of the components of GDP by uses, 1957–84 (1970 = 100)

Year	Private consumption	Government consumption	Gross fixed capital formation	Changes in stocks	Exports	Imports	GDP	Depreciation
1957	53.8	49.3	50.0	−54.5	49.9	54.5	51.5	—
1958	53.8	50.7	50.2	−69.0	69.9	54.1	56.4	—
1959	55.3	51.7	52.4	+55.6	61.4	55.6	55.9	—
1960	55.8	55.8	55.7	+56.4	55.8	55.8	55.8	—
1961	59.6	59.1	58.7	−55.6	49.3	56.5	57.6	—
1962	65.3	59.8	56.4	−66.7	41.3	53.5	58.8	—
1963	68.7	61.1	56.3	−57.1	42.4	51.9	62.8	—
1964	76.8	66.7	58.6	+29.8	50.8	58.1	69.0	—
1965	56.3	60.7	64.5	−87.5	48.1	60.7	72.9	69.4
1966	89.7	65.6	66.8	−65.0	47.6	62.7	78.9	67.9
1967	78.8	73.8	76.3	−85.9	64.6	73.3	75.8	76.9
1968	85.4	85.4	85.3	+100.0	85.4	85.3	85.4	85.3
1969	93.8	89.5	92.6	94.4	100.3	92.0	94.8	93.4
1970	100.0	100.0	100.0	+100.0	100.0	100.0	100.0	100.0
1971	116.0	118.0	99.9	114.9	105.6	150.4	105.2	106.1
1972	132.4	137.0	120.4	−207.1	98.9	136.3	122.2	124.1
1973	151.5	158.6	131.3	+146.1	128.3	136.0	147.6	138.5
1974	186.5	197.0	181.8	+190.4	159.9	211.3	183.9	167.6
1975	238.1	238.2	237.9	+243.8	237.8	238.0	238.1	238.1
1976	314.0	269.9	264.8	−344.4	227.8	269.2	304.8	274.5
1977	553.7	369.9	362.1	+547.6	345.2	312.2	509.8	373.1
1978	1 008.8	529.2	550.9	+823.8	541.4	530.8	883.5	553.0
1979	1 387.9	788.8	832.5	+1 357.5	993.6	1 097.5	1 218.7	818.7
1980	2 003.0	1 161.3	1 204.2	+2 029.0	1 319.3	1 133.7	1 829.5	1 152.2
1981	3 873.0	1 338.4	1 795.8	+3 630.0	1 375.9	1 263.2	3 235.0	1 666.4
1982	5 462.3	1 321.4	2 062.8	+4 426.7	991.8	1 363.9	4 136.4	1 897.8
1983	12 063.1	2 593.0	4 773.9	+2 080.0	5 508.8	8 643.7	9 229.6	3 045.6
1984	14 165.5	5 425.6	11 237.4	+6 540.0	11 721.5	11 725.3	12 485.5	10 122.4

Sources: Tables A.1, A.3 and A.4.

Table A.9 Implicit deflators of the types of assets in gross fixed capital formation, 1965-84
(1970 = 100)

Year	Building	Other construction work (except land improvement)	Land improvement and plantation	Transport equipment	Machinery and equipment	Gross fixed capital formation
1965	65.0*	—	—	63.9	64.0	64.5
1966	68.6*	—	—	63.7	63.6	66.8
1967	77.4*	—	—	73.7	73.8	76.3
1968	85.4*	—	—	85.4	85.4	85.3
1969	92.0*	—	—	93.6	93.5	92.6
1970	100	100	100	100	100	100
1971	92.4	92.3	92.2	114.5	113.3	99.9
1972	108.6	108.6	108.1	149.2	148.1	120.4
1973	122.6	122.7	122.0	172.5	172.1	131.3
1974	170.8	170.8	172.6	205.5	205.5	181.8
1975	238.0	238.0	238.4	238.1	238.0	237.9
1976	259.5	259.5	258.9	274.0	274.2	264.8
1977	368.1	340.3	365.4	368.0	368.0	362.1
1978	562.9	562.8	562.3	534.4	534.6	550.9
1979	874.8	875.8	875.0	762.9	761.9	832.5
1980	1 343.3	1 344.3	1 328.4	961.1	960.7	1 204.2
1981	2 308.8	2 302.8	2 225.0	1040.7	1039.8	1 795.8
1982	2 685.2	2 681.7	2 718.2	1037.4	1037.8	2 062.8
1983	6 436.4	6 584.9	6 320.0	2739.3	2739.0	4 773.9
1984	12 345.6	12 373.5	13 740.0	9788.7	9767.5	11 237.4

* Includes other constructional works and land improvement.
Sources: CBS, *Economic Survey* (various issues), Statistical Service, *Quarterly Digest of Statistics*, September 1986; and Table A.8.

297

Table A.10 Implicit deflators of the components of GDP
by industrial origin, 1968–84
(1970 = 100)

Year	Agriculture	Agriculture and livestock	Cocoa production and marketing	Forestry logging and fishing	Industry	Mining
1968	85.4	85.4	85.3	85.5	85.3	85.2
1969	103.2	106.3	109.0	79.2	85.9	85.5
1970	100	100	100	100	100	100
1971	87.8	108.1	63.3	62.2	103.3	72.1
1972	106.0	132.0	75.6	74.3	126.1	109.1
1973	141.2	169.3	96.0	116.8	137.7	152.6
1974	180.3	198.8	153.5	137.8	179.8	214.6
1975	238.0	238.0	238.0	238.0	237.9	238.0
1976	317.1	415.0	157.4	284.5	276.5	208.1
1977	632.0	877.0	251.9	348.5	373.6	219.9
1978	1 089.2	1 282.6	668.7	743.7	580.9	338.6
1979	1 394.3	1 522.2	1098.9	1106.7	929.8	778.4
1980	2 001.5	2 338.4	993.0	1 673.5	1 362.8	1 560.1
1981	3 185.2	3 992.6	546.6	2 515.6	2 105.9	1 335.9
1982	4 331.7	5 466.2	352.5	3 118.1	2 059.1	1 115.7
1983	10 393.1	12 508.1	6515.5	4 577.2	5 043.1	8 959.4
1984	11 408.8	12 916.4	7816.2	6 849.6	11 375.0	13 013.4

Table A.10 (continued)

Manufac-turing	Electricity and water	Construction	Services	Trade	Finance and business services	Government services
85.3	85.3	85.4	85.3	85.4	85.2	85.3
102.1	83.6	89.7	90.2	94.9	81.3	88.8
100	100	100	100	100	100	100
102.2	317.6	107.4	127.8	113.3	157.8	148.4
125.8	235.5	123.7	148.2	128.2	182.3	175.5
136.2	167.6	129.0	165.8	148.5	213.7	178.7
177.1	211.1	169.6	192.6	184.9	185.0	223.4
238.0	236.2	238.0	238.0	238.0	238.1	237.9
290.0	309.1	260.2	304.9	340.5	275.1	253.4
396.1	345.8	372.4	431.3	490.1	369.5	370.3
617.5	550.4	554.1	743.1	1 011.1	471.0	520.6
1 002.3	767.9	794.3	1 075.8	1 388.9	704.7	520.6
1 394.7	1 303.9	1 253.8	1 780.9	2 630.5	702.6	1228.5
2 224.4	2 255.0	2 039.5	3 814.7	6 764.0	1032.5	1517.8
2 010.8	2 760.2	2 351.7	4 608.2	8 833.6	1081.4	1548.5
5 190.7	2 081.4	4 223.7	8 910.7	17 885.5	1490.0	2839.2
11 128.9	13 369.8	10 750.8	14 549.3	29 471.1	2173.2	3282.2

Sources: Tables A.5 and A.6.

Table A.11 Central government revenue and expenditure, 1961–85*
(₵ million current prices)

		Expenditure		
Year	Revenue	Recurrent	Development	Total
1961	200.2	168.6	104.1	272.7†
1962	180.3	174.2	108.6	282.8†
1963	203.0	199.9	122.1	322.0†
1964	294.2	261.2	123.7	384.9†
1965	284.0	257.3	186.1	443.4†
1966‖	231.4	224.0	96.4	320.4
1967	241.5	226.9	75.4	302.3†
1968	300.2	298.1	72.4	370.5†
1969	291.2	317.1	65.3	382.4
1970	369.2	373.7	84.0	457.7
1971	495.6	378.3	108.4	486.7
1972	435.4	430.7‡	103.8	534.5‡
1973	391.7	449.1‡	96.0	545.1‡
1974	583.6	569.2‡	169.3	738.5‡
1975	808.8	875.4‡	286.1	1 161.5‡
1976	814.8	997.4	441.2	1 438.6
1977	1 074.6	1 308.0	637.2	1 945.2
1978	1 539.1	2 322.2	695.4	3 017.6
1979	2 187.8	3 334.5	759.8	4 094.3
1980	3 026.1	4 076.8	594.7	4 671.5
1981	3 279.1	6 329.3	1390.0	7 719.3
1982	4 545.0	7 917.0	927.0	8 844.0
1983	4 642.5	8 842.1	936.0	9 778.1
1984	10 241.0	13 400.9	1354.4	14 755.3
1985	22 641.0	22 700.0	3991.0	26 691.0

* For financial year ending June.
† Total payments exclude payments 'below-the-line' accounts.
‡ Includes portion of loan repayment.
‖ Data available for January to June have been doubled.
Sources: CBS/Statistical Service, Quarterly Digest of Statistics (various issues).

Table A.12 Public, private and total savings at current and
1970 prices, 1961-84
(¢ million)

	Current prices			1970 prices		
Year	Public	Private	Total	Public	Private	Total
1961	32	76	108	71	169	240
1962	6	136	142	16	368	384
1963	3	151	154	7	359	366
1964	33	177	210	69	372	441
1965	27	95	122	63	222	285
1966	7	112	119	17	266	283
1967	15	99	114	26	174	200
1968	2	214	216	2	252	254
1969	−26	281	255	−24	260	236
1970	−5	309	304	−5	309	304
1971	117	84	201	232	167	399
1972	5	350	355	6	448	454
1973	−57	524	467	−47	475	381
1974	14	407	421	10	277	287
1975	−67	788	721	−28	331	303
1976	−183	740	557	−94	379	285
1977	−233	1 332	1 099	−52	298	246
1978	−783	1 925	1 142	−133	328	195
1979	−1147	3 011	1 864	−158	416	258
1980	−1051	3 167	2 116	−68	204	136
1981	−3050	5 969	2 919	−127	260	133
1982	−3572	6 801	3 229	−271	516	245
1983	−4200	5 313	1 111	−571	722	151
1984	−3160	21 067	17 897	−28	188	160

Notes: The distribution between public and private savings at 1970 prices is based on their respective shares at current prices.
Sources: Tables A.3, A.4 and A.11.

Table A.13 Public, private and total investment at current and 1970 prices, 1961–84
(¢ million)

	Current prices			1970 prices		
Year	Public	Private	Total	Public	Private	Total
1961	104	86	190	176	146	322
1962	109	63	172	195	113	308
1963	122	88	210	217	156	373
1964	124	122	246	223	220	443
1965	186	76	262	290	118	408
1966	96	99	195	144	148	292
1967	75	80	155	100	106	206
1968	72	117	189	85	136	221
1969	65	171	236	70	184	254
1970	84	235	319	84	235	319
1971	108	245	353	106	242	348
1972	104	97	201	94	88	182
1973	96	220	316	72	165	237
1974	169	439	608	93	240	333
1975	286	386	672	120	162	282
1976	441	138	579	171	53	224
1977	637	581	1 218	167	152	319
1978	695	726	1 421	124	130	254
1979	760	1 084	1 844	93	133	226
1980	595	1 815	2 410	51	156	207
1981	1390	1 941	3 331	78	110	188
1982	927	2 023	2 950	46	99	145
1983	936	5 967	6 903	20	124	144
1984	1354	17 253	18 607	12	154	166

Notes: The distribution between public and private savings at 1970 prices is based on their respective shares at current prices.
Sources: Tables A.3, A.4 and A.11.

Table A.14 Money and quasi-money supply, 1961–85*
(₵ million)

	Money supply					Quasi-money supply time and savings deposit of commercial banks	Total money and quasi-money supply
	Currency						
Year	Ghana notes and coins in circulation	Held by commercial banks	Held outside banking sector	Demand deposits over 12 months	Total money supply		
1961	96	10	86	66	152	26	188
1962	108	14	94	98	192	34	236
1963	129	11	117	117	236	51	297
1964	171	12	160	150	310	65	385
1965	128	12	116	124	239	60	308
1966	127	12	115	161	276	67	352
1967	131	11	119	142	261	75	348
1968	142	16	125	152	278	93	380
1969	164	13	151	170	320	99	428
1970	167	16	151	235	385	115	511
1971	117	18	159	194	353	146	509
1972	256	17	239	222	461	205	666
1973	269	24	245	319	564	230	793
1974	368	32	336	361	697	308	1 005
1975	524	39	486	523	1 009	378	1 386
1976	753	46	707	723	1 430	472	1 902
1977	1 331	187	1 144	1 360	2 504	816	3 320
1978	2 392	312	2 080	2 148	4 278	1 294	5 572
1979	2 624	197	2 427	2 357	4 784	1 662	6 441
1980	3 897	469	3 428	2 797	6 225	2 503	8 728
1981	6 552	607	5 945	4 203	10 148	3 486	13 634
1982	7 488	760	6 728	5 569	12 297	4 833	17 130
1983	11 435	1 398	10 037	8 733	18 770	5 559	24 329
1984	16 234	2 291	13 943	12 700	26 643	7 193	33 836
1985	25 327	3 430	21 897	21 574	43 471	11 565	55 036

* For year ending December.
Sources: CBS/Statistical Service, *Quarterly Digest of Statistics* (various issues).

Table A.15 Consumer price index, 1963–84
(1970 = 100)

Year	National	Urban	Rural
1963	57.9	59.5	57.4
1964	63.5	65.8	62.8
1965	80.3	84.2	79.2
1966	90.0	91.4	90.8
1967	82.9	84.1	83.1
1968	90.0	90.5	89.9
1969	96.5	97.3	96.2
1970	100.0	100.0	100.0
1971	109.3	110.3	108.9
1972	120.3	123.0	119.0
1973	141.3	147.3	139.6
1974	167.3	177.7	164.3
1975	216.9	244.3	209.2
1976	339.2	368.2	330.8
1977	733.4	777.8	720.5
1978	1 271.0	1 333.9	1 258.7
1979	1 960.4	1 994.9	2 008.7
1980	2 935.2	5 156.5	3 229.2
1981	6 371.2	6 225.1	6 761.7
1982	7 791.9	7 603.3	8 283.3
1983	17 363.0	16 353.0	17 056.6
1984	24 336.5	23 019.1	23 601.1

Sources: CBS, *Consumer Price Index Numbers*, October 1978; *Statistical News Letter* no. 18/83 (23 November, 1983), no. 8/84 (16 April 1984) and no. 15/85 (8 November, 1985).

Table A.16 Major exports of domestic produce by value at current prices, 1960-82
(₡ million current prices)

Year	Cocoa beans	Cocoa paste*	Cocoa butter	Timber (log)	Timber (sawn)	Bauxite	Manganese ore	Diamond	Gold	Aluminium	Others	Total
1960	132.8	2.1	—	20.8	11.0	1.1	12.8	19.7	22.2	—	6.3	228.8
1961	166.3	2.5	1.0	21.7	14.9	1.1	14.5	17.2	25.8	—	6.7	271.7
1962	160.9	2.2	6.8	13.9	15.4	1.6	13.2	17.8	27.0	—	9.2	268.0
1963	136.2	0.4	6.9	14.4	11.6	1.0	8.0	6.7	22.6	—	6.0	213.8
1964	136.2	0.4	8.9	16.2	13.3	1.3	8.7	12.2	20.6	—	8.5	226.4
1965	136.5	0.9	11.4	13.3	11.4	1.3	9.6	13.5	19.1	—	6.5	223.4
1966	103.1	1.2	11.5	10.9	10.0	1.5	12.2	10.8	17.1	—	7.7	185.8
1967	130.7	2.5	22.5	12.7	9.7	1.6	9.2	12.6	21.0	—	17.2	239.6
1968	185.6	4.5	24.1	16.3	12.3	1.5	10.5	17.4	29.0	—	36.1	337.3
1969	218.6	3.3	24.0	24.1	15.0	1.4	7.0	13.9	29.2	—	53.7	390.8
1970	300.4	3.9	27.3	19.9	17.1	1.3	7.2	14.5	25.7	—	42.9	460.2
1971	209.2	3.5	24.3	20.5	12.2	2.3	6.6	11.8	28.5	—	44.2	363.1
1972	289.1	9.9	29.0	42.3	21.2	2.7	10.0	18.6	50.4	56.9	19.2	549.4
1973	344.8	8.0	44.2	88.6	41.8	2.6	7.3	13.1	70.1	46.1	30.9	697.4
1974	439.1	10.7	85.6	63.9	42.2	3.5	10.5	14.6	109.8	96.7	40.3	916.9
1975	551.5	11.2	74.5	49.1	34.6	4.3	17.1	12.7	87.3	99.7	53.3	995.0
1976	515.5	13.1	64.5	47.5	32.5	4.0	19.7	12.7	82.5	107.6	48.7	948.2
1977	679.7	35.4	82.2	62.5	29.7	2.5	18.8	15.5	72.4	156.4	54.1	1209.2
1978	988.0	47.1	75.0	59.1	37.8	7.2	24.3	30.0	91.0	159.0	46.0	1559.8
1979	1846.3	52.9	130.6	55.8	68.8	7.9	29.7	31.4	210.4	210.3	75.9	2720.0
1980	1803.8	48.9	149.5	29.8	69.6	8.5	24.9	27.5	540.0	370.1	31.3	3103.9
1981†	1091.3	12.9	79.7	41.2	57.8	7.1	22.2	22.5	435.5	424.1	929.4	2924.0
1982†	1053.3	3.2	62.1	15.5	28.5	1.9	8.6	13.9	319.8	519.2	376.0	2402.0

* Includes Cocoa Cake.
† Provisional data.
Source: CBS, *Economic Survey* (various issues).

Table A.17   Major imports classified by main commodity groups, 1960–82
(million ₡: current prices)

Year	Food and animals	Beverages and tobacco	Crude materials inedible except fuels	Mineral fuels	Animals vegetable oils and fats	Chemicals
1960	42.0	7.6	0.6	13.6	0.4	19.0
1961	52.4	7.0	1.8	13.0	1.0	20.2
1962	44.6	2.6	1.6	14.6	0.6	19.4
1963	37.0	2.4	1.8	15.2	1.6	19.2
1964	40.0	1.4	2.0	14.2	3.0	15.0
1965	35.3	2.2	3.1	13.2	3.0	20.2
1966	39.3	2.4	2.3	10.6	2.6	16.6
1967	43.2	3.4	3.7	15.5	3.4	32.8
1968	51.0	5.0	6.3	21.5	3.9	48.3
1969	55.2	1.6	5.4	22.9	5.9	55.1
1970	79.5	3.9	9.4	24.4	3.8	66.9
1971	62.6	4.6	12.4	27.0	5.2	71.6
1972	72.2	2.3	13.2	45.3	5.2	63.9
1973	111.7	4.9	22.8	46.8	6.0	91.1
1974	140.6	7.1	28.7	156.5	15.4	123.3
1975	104.9	6.7	27.9	150.9	9.9	126.6
1976	118.8	11.6	34.3	148.0	11.6	155.9
1977	95.6	11.0	45.8	206.4	11.3	159.1
1978	143.5	22.5	42.6	230.9	28.6	242.8
1979	178.4	27.7	48.3	483.6	22.5	377.7
1980	241.5	43.0	56.9	827.4	28.4	482.7
1981	247.3	29.3	63.8	1056.2	30.8	413.3
1982	276.3	11.6	29.9	990.7	9.2	305.2

Table A.17 (continued)

Manufactured goods classified chiefly by material	Machinery and transport equipment	Miscellaneous manufactured articles	Miscellaneous commodities N.E.S.	Totals
79.0	67.4	25.2	3.6	259.2
89.0	66.2	26.4	8.2	285.2
77.8	51.8	17.6	2.8	233.4
84.2	74.2	21.4	3.8	260.8
79.4	71.2	15.0	2.0	243.2
108.5	105.9	23.2	5.4	320.0
77.1	82.2	15.2	2.9	251.2
72.3	70.4	15.3	1.5	261.5
76.3	86.0	14.0	1.7	314.0
97.4	94.4	14.6	1.8	354.4
100.8	108.1	16.4	5.8	419.0
99.4	131.5	19.2	9.6	443.1
68.2	104.3	11.3	7.3	393.2
105.8	111.3	14.9	10.6	525.9
221.0	212.4	27.7	10.9	943.7
207.9	228.1	28.9	17.5	909.3
187.7	270.6	32.1	21.1	991.7
240.1	343.2	40.9	39.8	1193.2
287.1	547.6	50.4	85.9	1681.9
334.8	768.7	62.1	42.2	2346.0
380.5	922.8	64.5	55.8	3103.6
394.9	950.5	120.1	178.2	3484.3
299.9	577.5	134.2	147.2	2781.6

Sources: CBS, *Economic Survey* (various issues); and Statistical Service, *Quarterly Digest of Statistics*, September, 1986.

Appendix B

THE *KALABULE* ECONOMY

One consequence of economic chaos, and of somewhat incoherent attempts to control it, is normally the emergence of a black economy. This can contribute to economic welfare in given conditions by making the flow of goods and services in the economy higher than it would otherwise have been. It is, however, not an entirely wholesome phenomenon. Insofar as economic growth depends on consent and willing participation in national effort, the black economy can be counter-productive.

In some African countries such as Nigeria, Zaire and Uganda, the black economy has been found to operate on a large scale.[1] There are, however, practical difficulties in dealing with a topic like the black economy because much of the relevant information is shrouded in secrecy. The best one can do is to make guesses to get an idea of the situation and, as with all guesses, one is subject to errors and omissions.

In Ghana the black economy is popularly known as the *kalabule* economy. *Kalabule* is a newly-coined word in Ghana, thought to have been first used in the early 1970s.[2] It refers to any form of cheating, trade malpractice or black marketing behaviour. This is, however, a narrow way of defining the term, at least in the way it is commonly understood. Broadly viewed, the *kalabule* or the black economy covers all types of illegal transactions including smuggling of cocoa, timber and gold; diversion of goods from official to unofficial channels; tax evasion and over-invoicing of imports and under-invoicing of exports. Thus viewed, one should be able to judge the amount of income which is not properly accounted for as far as the government and official record of transactions are concerned.

Any economy – developed or underdeveloped – may have some form of operation in the black economy, the extent of which is not likely to be high in countries with well-organised administrations.[3] But in countries with serious administrative problems and where the divergence between private and public interest is wide, as has been the case in Ghana particularly since the beginning of the 1970s, the extent of the black economy can be large indeed.

It is commonly believed that the emergence of the *kalabule* economy in Ghana is an outcome of the promulgation of the Control of Prices

Act of 1962. In the previous year the government had introduced foreign exchange controls as a means of fighting the nascent balance of payments problems of the country. The government realised that, with exchange and import controls, importers and local manufacturers of import-substitutes would reap super-normal profits in the absence of price controls. The government therefore established maximum prices above which goods should not be sold. The policy was aimed mainly at protecting consumers. Another aim was to mitigate the harmful effects of the distribution of income that would result if demand and supply were brought into equilibrium through the natural upward adjustment of prices, given the low price elasticities of demand for many imported goods and import-substitutes. A maximum price is usually set below the equilibrium market price. The consequence is that while the consumers find the maximum price fixed by the government very acceptable, the suppliers regard it as highly unsatisfactory. Demand for the commodity consequently exceeds its supply.

The effects of imposing a price ceiling in a competitive market are illustrated in Figure B.1.

P_1 and Q_1 are the competitive equilibrium price and quantity respectively. P_2 is the legal maximum price which is able to attract only Q_2 of the commodity onto the market if the supply of the good is

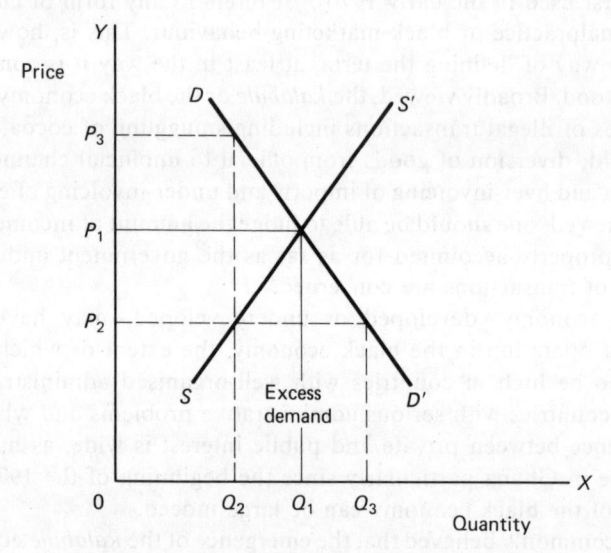

Figure B.1 Low official price and excess demand

uncontrolled. But at the price P_2 consumers would want to purchase Q_2 quantity of the commodity. Therefore demand for the commodity tends to be greater than supply at that price. As the figure shows, given the quantity Q_2, the consumers will be willing to buy that amount at the price P_3 which is above the controlled price. Thus suppliers will be able to sell above the controlled price if they so choose. Given their desire to maximize profit this is precisely what they will tend to do and thus *kalabule* comes into existence. The social interest, as perceived by the government, comes into sharp conflict with the private interest of producers, distributors and other suppliers.

Another factor which has directly contributed to the *kalabule* economy is the over-valuation of the cedi. As shown in Chapter 10, during the late 1970s and early 1980s the extent of the divergence between the official exchange rate and the parallel exchange rate became so wide that at one stage one dollar was selling in the parallel market at twenty-six times the official exchange rate. The continuation of the over-valued exchange rate and low controlled prices also contributed towards smuggling of goods. Such smuggling could have taken place both at large and small scale, tempting big producers and suppliers as well as small producers, such as cocoa farmers and maize growers.

In a sample survey conducted in June 1984 covering the capital city (Accra), one public institution (University of Cape Coast) and one village (Assin Aworabo) an attempt was made to examine, among other things, the extent of purchases by consumers through the official control source. The respondents were asked to give information for three selected years – 1980, 1982 and 1984. It was found that in 1980, 59 per cent of the total respondents had no access to the official source, while in 1982 and in the first half of 1984 the respective figures were 64 and 65 per cent (see Table B.1). This is presumably because of lower availability of goods in recent years. However, depending on the location and the status of the respondent (whether holding a job in a place like a public institution or living in an urban centre or a village) significant variations were observed in the source of purchase. About 53 to 60 per cent of the respondents from Accra did not have access to goods supplied through official channels while those holding jobs in the University of Cape Coast did much better. Only 33 to 46 per cent (depending on the year) were without any access to the official source. In Assin Aworabo (a village) there was a very high percentage of the respondents (80 to 100 per cent) without any access to the official source of goods.

Table B.1 Percentage of people buying goods from official sources and the extent of such purchases for 1980, 1982 and 1984 (percentage of people)

Goods bought from official sources as % of total purchase	Accra 1980	Accra 1982	Accra 1984	UCC* 1980	UCC* 1982	UCC* 1984	Assin–Awarabo 1980	Assin–Awarabo 1982	Assin–Awarabo 1984	Total 1980	Total 1982	Total 1984
None	53	60	58	33	46	33	80	84	100	59	64	65
Less than 5	11	0	5	22	9	25	7	8	0	12	5	9
5 to 10	12	7	0	1	9	0	13	0	0	5	5	0
More than 10	24	33	37	44	36	42	0	8	0	24	26	26
Total	100	100	100	100	100	100	100	100	100	100	100	100

* University of Cape Coast.
Source: Sample survey.

Two important conclusions emerged from the sample survey. First, the location and the status of the customers affected the availability of low priced goods (official source). People living in the city and those having jobs in public institutions had better access to the official source compared to those living in the village. Second, in general the proportion of the total purchase from the official source (low controlled price) had been low, thus indicating a high percentage of the total sale of goods in the *kalabule* economy.

As mentioned earlier, the above way of looking into the *kalabule* economy does not cover all the transactions in the economy. In an attempt to view it broadly we considered different sectors of the economy contributing to the GDP and examined them separately by looking into their nature and type of production and possibilities of their entry into the parallel economy. In this, we were able to obtain some fairly reliable information from practising economists, government officials, executives of public institutions and private corporations and many others. Some published and unpublished information also helped us trace some activities in the black economy (for instance, smuggling of cocoa).[4]

On the basis of the information so gathered we made two estimates, both on the conservative side, of the extent of *kalabule* transactions based on 1981 GDP. As agriculture (excluding cocoa) in Ghana largely takes place in a subsistence form one will not expect much of it to enter the *kalabule* economy. However, in areas having borders with the hard currency CFA zone, such as Brong-Ahafo and Volta Regions, one would expect some smuggling of agricultural goods too. We have therefore considered a range of 10 to 20 per cent of agricultural income entering into the *kalabule* economy.

On the other hand, wholesale and retail trade provides an example of a sector having a large percentage of the income likely to enter into the *kalabule* economy. So we considered a range of 50 to 80 per cent as the most likely figure for this sector. Other sectors were similarly closely examined and different values estimated for the range. In total, 25.5 to 42.4 per cent of the GDP in 1981 was estimated to be in the black economy according to the range of estimates considered.

If the extent of the *kalabule* economy is as large as 40 per cent of the economy, its impact is likely to be extremely severe. To start with, it will deprive the economy of large foreign exchange earnings through the official channel and also deprive the government of revenue in the form of export duty, import duty, income tax, sales and other taxes. It will also seriously affect the distribution of income. People engaged in *kalabule* practices will have much higher incomes, though not officially recorded. It is probably one of the reasons why it has been suggested in certain circles that in the late 1970s and early 1980s as much as 80 per cent of the national income in Ghana was in the hands of the top 10 per cent of the population. Such a figure is probably an exaggeration, as mentioned in Chapter 2, but it definitely draws attention to a serious issue.

Notes and References

Overview

1. Roemer has analysed Ghana's development, providing good insights mainly from the political angle, covering the period 1950-80. See Roemer (1984), pp. 201-25.
2. Leith (1974), p. 1.
3. World Bank (1984a), p. xv.
4. Ibid., p. xv.
5. As observed by Rado (1985, p. 9): 'The present system of everyone doing three or four jobs may be the only way for a non-swimmer to keep afloat, but it is still not swimming. The advantages of specialization, so clearly demonstrated by Adam Smith 200 years ago, are still valid. Ghana's return to normality will surely involve a return to something much nearer to one man, one job'.
6. World Bank (1984a), p. 26.
7. Ibid., p. xv.
8. Ibid., p. xvi.
9. Ibid., p. xv.
10. Rado (1985), p. 5.
11. Ibid., p. 10.
12. Government of Ghana, *Five-Year Development Plan* (1977), Part I, p. 85.
13. Omaboe (1966), p. 440.
14. Niculescu as quoted in Ewusi (1973), p. 4.
15. Omaboe (1966), p. 442.
16. Ibid., p. 442.
17. No official reasons were given in defence of this action, but Omaboe guesses it to have stemmed from a visit undertaken by a delegation of the CPP under the leadership of Nkrumah to the Soviet Union and other Eastern European countries. See Omaboe (1966), pp. 450-51.
18. Government of Ghana (1964). See also Bissue (1967).
19. Government of Ghana (1964), p. 2.
20. Ibid., p. 5.
21. Killick (1978), p. 136. For an illuminating and in-depth analysis of economic planning under Nkrumah, NLC and Busia see Killick (1978) from which we have borrowed heavily.
22. Omaboe (1966), Killick (1978), Bissue (1967) and Ewusi (1973).
23. Killick (1978), p. 139.
24. Ibid., p. 140.
25. Omaboe (1966), p. 461.
26. Rimmer as quoted in Killick (1978), p. 140.
27. Omaboe (1966), p. 461.
28. Ibid., p. 461.

29. There was an incident when Nkrumah reacted sharply by saying, 'Who decides, Mensah or me?', when told that one of his decisions ran counter to the plan. (J. H. Mensah was the Executive Secretary of the Planning Commission and chief author of the Seven-Year Plan.) See Killick (1978), p. 143.
30. Ibid., p. 143.
31. Government of Ghana, *Two-Year Development Plan* (1968), Foreword.
32. Government of Ghana (1970), p. 81.
33. Killick (1978), p. 144.
34. Quoted in above, p. 145.
35. Government of Ghana (1977), Part I, p. 30.
36. Ibid., pp. 26-29.
37. Ibid., Part II, p. 188.
38. Ibid., p. 190.
39. I am grateful to Felix Atabu, one of my Ghanaian post-graduate students, for this information.
40. Government of Ghana (1983).
41. Ibid., Part I, p. 6.
42. Ibid., p. 17.
43. Ibid., p. 17.
44. Ibid., p. 19.
45. Government of Ghana (1985b).
46. Killick (1978, pp. 153-60) has made a very good analysis of short-term economic management up to the mid-1960s.
47. Green (1971), p. 244.
48. Quoted in Seidman (1978), p. 87. The report was based on a twelve-month study conducted by the firm of G. H. Whitman, Inc, International Economic Consultants.
49. Ahmad (1970), p. 24.
50. Ibid., p. 24.
51. Killick (1978), p. 159.
52. Ibid., pp. 152-53. In 1964 the external debt amounted to £187 million of which £157 million was suppliers' credit, although in 1963 the amount through suppliers' credit was thought to be only £20 million. As observed by Dowse (1969, p. 97): 'Much of this credit had little to do with Seven-Year Plan priorities, but a great deal to do with sagging British and West German exports from whence credits were financed and the bribes to Ghanaian contacts paid'.
53. Dowse (1969), p. 96.
54. Jeffries (1982), p. 310.
55. Ibid., p. 311.
56. Esseks (1975), p. 46.
57. Jeffries (1982), p. 311.
58. According to Killick (1978, p. 303), the 1971 devaluation 'was the most decisive break with the past of any action since the 1966 coup'.
59. Rado (1985), p. 12-13.
60. Ibid., p. 15.
61. Jeffries (1982), p. 312.
62. Ibid., p. 314.
63. Rado (1985), p. 16.

Notes and References

64. Government of Ghana (1985a), p. 23.
65. Ibid., p. 23.
66. Government of Ghana (1985b), p. 19.
67. Ibid., p. 38.
68. As admitted by the PNDC Secretary for Finance and Economic Planning, 'Over the last two years (1983 and 1984), financing from multilateral agencies (particularly the IMF) has been a crucial source of our development programmes. However, the flows from these sources are likely to be substantially reduced after 1985'. Government of Ghana (1985a), p. v.
69. Observations made by the *Five-Year Development Plan* (Government of Ghana 1977, Part I, p. 85) in the context of the 1951 Ten-Year Plan.
70. Killick (1978), p. 145.
71. Omane (1980), pp. 19-22).
72. Government of Ghana (1977), *Five-Year Development Plan* (1977) Part I, p. 86.
73. Ibid., p. 85.
74. Lewis (1965), p. 31.
75. See Seidman (1978, Chapter 5) who shows the emergence and the conflicting interests of the political-economic interest groups during the period 1951-65. According to Rado (1986, p. 567), 'the over-expanded public sector has created employment and a sense of security for tens of thousands of educated Ghanaians at all levels of income well above what they would have earned under any sustainable policy. The over-taxation of the export sector is exactly mirrored by the corresponding under-taxation of everyone else. Massively protected domestic industries whose labour forces enjoy security of tenure created powerful vested interests. The withdrawal of such unwarranted privileges, however desirable in the "national interest", is bound to create stiff opposition from the interest groups concerned'.
76. See Omane (1980, Ch. 2) who discusses the role of corruption in the context of suppliers' credits contracted during 1959-65.
77. See, for example, Dumont (1966) and Dowse (1969). The working class was reported to be incited to dangerous strikes in 1961 when the government was attempting to repair the neglect of agriculture by forcing the 'privileged urban class to accept heavy taxation'. Dumont (1966), p. 294.
78. Dumont (1966), p. 294.
79. Dowse (1969), p. 65.
80. Dumont (1966), p. 294.
81. Killick (1978), p. 135.
82. Government of Ghana, *Five-Year Development Plan* (1977), Part II, p. 196.
83. World Bank (1984a), p. 2.
84. World Bank (1983), pp. 59-63. Ghana's average annual growth rate for the 1970s as estimated by the World Bank study at -0.1 per cent, however differs from our calculation which gives a figure of 0.4 per cent.
85. Government of Ghana (1982), p. 21.
86. World Bank (1981).
87. Gordon and Parker (1984).

88. Although the import of crude oil declined from 1.24 million tonnes in 1975 to 1.04 million tonnes in 1982, the expenditure on crude oil imports increased from 13.3 to 32.4 per cent of total export earnings during this period. CBS, *Economic Survey* 1982, p. 199.
89. Ewusi (1978), p. 108.
90. Seidman (1978), p. 36. The point, of course, is valid only if one could argue that had this company not existed at all, an entirely domestic one would have, and it would have made at least as much profit and would have leaked less of it. In the Ghanaian context, given the allegations of Ministers sending vast sums of money to numbered Swiss bank accounts, one would find it very difficult to make the above point.
91. Cohen and Tribe (1972), p. 535.
92. World Bank (1986), p. 204.
93. Government of Ghana, *Economic Recovery Programme 1984–86* (Review of Progress in 1984 and Goals for 1985, 1986), p. ix.
94. Nkrumah's address to the nation in a dawn broadcast, April 1961. The complaint by Nkrumah is based on some of the evidence of corruption which was rampant at the time. For example, the Cocoa Purchasing Company which was established by Nkrumah's government in the mid-1950s was reported to be syphoning off a large slice of the Cocoa Marketing Board's income for the private benefit of the party faithful. Indeed, it was far from being the only example of officially sanctioned corruption during the Nkrumah period as was pointed out by the Apaloo Report of 1967. See Austin (1964) and Dowse (1969).
95. Dumont (1966), p. 292.
96. Jeffries (1982), p. 314.
97. Council of State (1981), p. 4.
98. Ibid., pp. 3-4.
99. Killick (1978), p. 59, footnote.
100. Rado (1985, p. 11) refers to a letter he received from Lewis in which the latter wrote: 'Working with a man like Nkrumah, specific targets are a menace. I took my courage in my hands and put a target figure of 60 industrial enterprises in the draft (second five-year) plan. This was *one* thing he picked up, and he changed it to 600!'
101. Rado (ibid.), pp. 14-15.
102. Killick (1978), p. 155.
103. Ibid., p. 204.
104. Rado (1985), p. 15.
105. Mentioned to the author by Dr Stephen S. Adei who was a consultant for the preparation of the report (*Reviving Ghana's Economy: A Report by the Council of State on Economic and Fiscal Policies*). Mr William Ofori Atta was the Chairman of the Council of State.
106. I am grateful to Dr Adei for raising this point with me.
107. Gordon and Parker (1984), pp. 32-33.
108. In their analysis of the poor economic development in Sub-Saharan Africa the World Bank appears to have over-emphasized the role of internal factors, whereas the Organisation of African Unity, the Economic Commission for Africa and the African Development Bank

have over-emphasised the role of external factors. It is possible to take issue with both the stands. For a very good critical analysis of the above views see Gordon and Parker (1984).
109. Ibid., p. 8.
110. Mason (1980), pp. 252–55.
111. Hazlewood (1979), pp. 166–74.
112. Ibid., p. 173.
113. Government of Ghana (1985b), p. 9.
114. Ibid., p. 3.
115. Government of Ghana (1985b) and *1987 Budget Statement*.
116. As mentioned already, in our analysis in this section we have concentrated on the internal factors. But this in no way should suggest that we do not see the importance of external forces and, particularly of external aid. Indeed, the following observation made by the World Bank (1984b, p. 1) in the context of Sub-Saharan Africa particularly applies to Ghana that 'domestic reforms cannot be fully effective unless supported by appropriate levels and types of external assistance'.

1 Land and People

1. According to the 1960 census, the Akans comprised 44.18 per cent of the total population. The other major ethnic groups were Mole-Dagbani (15.98 per cent), Ewe (13.06 per cent) and Ga-Adangbe (8.08 per cent). See CBS, *Population Census* 1960 as reported in Central Bureau of Statistics, *Statistical Year Book* 1965–66, p. 45.
2. For more information on the history of Ghana see, for example, Agbodeka (1972); Buah (1980); Dickson (1969); Fage (1969) and Nkrumah (1957).
3. For detailed geographical data see, for example, Dickson and Benneh (1980) and CBS, *Statistical Year Book* 1955–56, pp. 1–3.
4. In subsequent discussions the Upper Region will be treated as one because of data constraints.
5. See CBS, *Statistical Year Book* (1969–70) p. 7, for information about the censuses from 1891 to 1970.
6. In Ghana, any settlement with a population of 5 000 and over is considered as an urban area.
7. The population of Tamale Urban Council, which is the fourth major urban centre in the country, however, grew at a slightly faster rate between 1970 and 1984, from 98 560 to 168 091. CBS, *1984 Population Census of Ghana: Preliminary Report*, Accra 1984, p. 59.
8. One should be careful in interpreting unemployment statistics in the context of a developing country like Ghana. As viewed by Rahim (1979), p. xiii: 'The unemployment, including under-employment and very low productivity 'full' employment, that exists in the less developed countries is different in nature – and of course, magnitude – from that which is observed from time to time in the industrialized economies'.
9. According to Ewusi, rural-to-rural migration far exceeds rural–urban migration in Ghana. See Ewusi (1977).

2 Growth and Structure of the Economy

1. Omaboe (1966), p. 18.
2. Ibid., p. 18.
3. Ibid., p. 18.
4. Leith (1974), p. 1
5. Growth rates are computed from Table A.1.
6. Omaboe, (1966), p. 18. Until July 1965 Ghanaian currency (G£) was at par with the £ Sterling.
7. Data on national accounts are published by the CBS/Statistical Service at both current and constant prices. Figures at constant prices are available for three time groups constituting the post-independence period. For purposes of a direct comparison, one needs the entire time series data at a single base year. We have constructed such a series, using 1970 as the base year. The measurement procedures and the problems relating to discrepancies in official data are discussed in Appendix A.
8. World Bank (1986), p. 180.
9. The per capita income of Nigeria is probably on the high side because of the overvalued Naira.
10. Central Bureau of Statistics, *Consumer Price Index Numbers* (October 1978), p. 13.
11. Considering that GDP at 1970 prices is calculated from GDP data at constant prices given by the CBS which used 1960, 1968 and 1975 as base years, GDP distribution at constant prices happens to coincide with that at current prices in these years. We have therefore avoided these three years in Table 2.4.
12. ₵1.15 per US dollar from December 1973 to September 1978 and thereafter ₵2.75 up to October 1983.
13. We have left out the years 1970 and 1975 (see Note 11 above).
14. Nurkse (1953); Lewis (1955); Myint (1967).
15. World Bank (1986, pp. 188–89) shows data of gross domestic savings for China, Hungary and Poland.
16. Lewis (1955); Rostow (1960 and 1971).
17. World Bank (1986).
18. Ibid., pp. 188–89.
19. Incremental capital-output ratios (ICOR) are shown in Chapter 14; net ICOR is found to vary from 2.7:1 to 3.95:1 calculated at periodic levels with two-year investment lag.
20. Chenery and Bruno (1962); Chenery and Strout (1966); and Chenery and MacEwan (1966).
21. Ahluwalia (1974), pp. 6–10.
22. Ibid.
23. Ibid., p. 11. Kuznets (1955) found that at the initial stage of development of the now developed countries, economic growth was largely associated with marked income inequality, and it was only at a much later phase of development that increased equality was observed. For a similar observation with respect to under-developed countries see Adelman and Morris (1973), who have found that economic growth in subsistence agrarian economies correlates well with inequality of income.

Notes and References

24. Killick (1978), pp. 80–3.
25. Ibid., p. 83.
26. Ibid., p. 83.
27. Ibid., p. 83.
28. Huq (1984a). See also Chapter 5.
29. Centre for Development Studies (1983).
30. According to the study (ibid., p. 21), the population share in the lower income bracket with less than ₵5000.00 annual income is over 80 per cent of the sample.
31. Government of Ghana, *Government's Economic Programme* (1981), p. 32.
32. Ghana Education Service, *Digest of Education Statistics* (Pre-University) 1974–75, p. 8.

3 Infrastructure

1. Adam Smith (1950), p. 244.
2. Ibid., p. 247.
3. Ibid., pp. 246–7.
4. Nathan Consortium (1970).
5. Ibid., p.s. i.
6. Government of Ghana, *Economic Recovery Programme 1984–86*, Vol. I, p. 5.
7. Dickson and Benneh (1980).
8. Information supplied by Ghana Highway Authority.
9. Ibid.
10. Dickson and Benneh (1980), p. 112.
11. Computed from data on government expenditure on roads and waterways, CBS, *Quarterly Digest of Statistics* (December 1966 and March 1984) and GDP (see Table A.1).
12. Nathan Consortium (1970), p.s. vi.
13. Ibid., p.s. vi.
14. Ibid., p.s. i.
15. Dickson and Benneh (1980), p. 114.
16. Nathan Consortium (1970), p.s. vii.
17. Personal communications with Ghana Railways Corporation, Takoradi.
18. Nathan, op. cit., p.s. vii.
19. CBS, *Economic Survey* (1981), p. 117.
20. Nathan Consortium (1970), p.s. ix.
21. Ibid., p.s. xi.
22. Killick (1966), pp. 391–410.
23. Government of Ghana, *Economic Recovery Programme 1984–86*, Vol. I, p. 5.
24. Ibid., p. 5.
25. Government of Ghana, *Ghana 1977*, p. 297.
26. Schultz (1960, 1961 and 1971).
27. In the early 1980s, the share of government expenditure on education (both recurrent and development expenditure) was only 2 to 3 per cent. The corresponding figure in the mid-1960s was around 5 per cent. Figures computed from government expenditure on education (CBS,

Quarterly Digest of Statistics December 1966, December 1969 and March 1984) and GDP (see Table A.1).
28. Computed from data shown in CBS, *Quarterly Digest of Statistics* December 1966 and Table A.1.
29. Computed from data on GDP (see Table A.1) and government expenditure on health (CBS, *Quarterly Digest of Statistics* December 1966 and March 1984). As a percentage of GNP government health expenditure in the mid-1970s went up to over 2 per cent (2.21 per cent in 1975-76). See Brooks (1981) p. 41.
30. Seidman (1978), p. 143.
31. Ibid., p. 142.
32. Dickinson (1982), p. 351.
33. Nathan Consortium (1970), p.s. viii.
34. Adam Smith, (1950), p. 246.
35. Huq (1984a).

4 Agriculture

1. CBS, *Economic Survey 1982*, p. 17.
2. Ministry of Agriculture, *Operation Feed Yourself* (1974). The period fortunately coincided with favourable climatic conditions which were responsible for higher output.
3. CBS, *Economic Survey 1977-80*, p. 38.
4. Ministry of Agriculture, *Ghana Sample Census of Agriculture* (1970) p. 40.
5. Ibid., p. 48. Measured at the district level it was, however, found that a number of sample districts in the Ashanti, Volta and Eastern Regions had an average number of farms per holder above 3.1. The highest recorded was 3.4.
6. Gyekye (1984), p. 7.
7. Amonoo (1978), p. 144.
8. Honny and Micah (1983), p. 52.
9. *Economic Survey 1977-80*, pp. 38 and 40.
10. Atsu (1984b), p. 13.
11. Honny and Micah, op. cit., p. 48.
12. Ibid.
13. In a study based on a number of countries in sub-Saharan Africa, including Ghana, Bond (1983, pp. 723-24) found that 'neither labour nor land presents major constraints on agricultural expansion'.
14. Ministry of Agriculture, *Ghana Sample Census of Agriculture* (1970), p. 37.
15. CBS, *Economic Survey 1981*, p. 24.
16. Atsu (1984b), p. 17.
17. Ibid. By comparison the cost of buying land is less than one hundredth of this figure.
18. Amonoo (1978), p. 180.
19. Atsu and Owusu, (1982).
20. Ibid.
21. Atsu (1984b), p. 11.

22. The figures are taken from a 1981 study by the Ministry of Agriculture, Government of Ghana, and quoted in ibid (p. 11). The increases in output and income are reported to have taken place in fairly recent years.
23. Amonoo (1978), p. 181.
24. La Anyane (1963), Chapter 2.
25. Brown, C. K., (1972), pp. 17–18.
26. Ibid., p. 17. Includes quote from 'Focus and Concentrate', prepared by Ministry of Agriculture, Accra, 1970.
27. Amonoo (1978), pp. 142–43.
28. As quoted in ADB, *Agricultural Credit Programmes in Ghana* (1975), p. 29.
29. Ibid., p. 33.
30. Ibid.
31. Ibid., p. 30.
32. Ibid., pp. 21–22.
33. Gyekye (1984), p. 6.
34. Ibid., p. 7.
35. The figure is based on a sample survey in the Twifu Prasu District (Central Region). See Honny and Micah (1983), pp. 53–54.
36. Ibid.
37. CBS, *Economic Survey 1977–80*.
38. Ibid., p. 76.
39. CBS, *Economic Survey 1977–80*, p. 60.
40. Ibid., p. 60.
41. Ibid., p. 60.

5 Cocoa

1. A detailed survey carried out between 1970 and 1975 by the Cocoa Production Division of the Ministry of Cocoa Affairs, Government of Ghana, showed that 1.65 million hectares of land were under cocoa production. This estimate, according to Nyanteng (1980, p. 29), is probably more accurate.
2. CBS, *Economic Survey 1977–80*, p. 42.
3. The implication is that the part of cocoa output that is smuggled out does not appear in the official statistics of cocoa production.
4. CBS, *Economic Survey 1977–80*, p. 48.
5. Manu (1973), p. 4.
6. Ibid., p. 5. According to Engel's law, given tastes or preferences, the proportion of income spent on food will diminish with an increase in income. The law is named after Ernst Engel, the Director of the Bureau of Statistics in Prussia, who formulated it in a paper published in 1857.
7. Ibid. (p. 5) as based on FAO, 'Agricultural Projections for 1970, *Commodity Review*, Special Supplement 1969, pp. 11–38.
8. Ibid., p. 6.
9. Blomqvist (1973), p. 11.
10. At current prices, the contribution of the cocoa sector to GDP has, however, fallen from 14.1 per cent in 1970 to 11 per cent in 1975 and further to 5.5 per cent in 1980. In 1982 it was less than 3.6 per cent of GDP.

324   *Notes and References*

11. In 1978–79 revenue from export duty on cocoa alone was ₵938 million against a total government revenue of ₵2186 million (that is, 43 per cent of the total). The export of cocoa beans alone accounted for 68 per cent of total export earnings in 1981 and including cocoa paste (1.9 per cent) the total contribution of the cocoa sector was as high as three-quarters of total export earnings. See CBS, *Economic Survey 1981*.
12. Honny and Micah (1983); See also Gyekye (1983).
13. The real income of the farmers is likely to be higher, at least for those who received benefits in the form of input subsidies.
14. Nyanteng (1980), p. 24.
15. Council of State (1981), p. 42.
16. Ibid., p. 43.
17. Nyanteng (1980).
18. Ibid., p. 99.
19. Following a recent decision by the Ghana Government, CMB has been renamed as Ghana Cocoa Board.
20. Fitch and Oppenheimer (1966), p. 44.
21. Ibid., p. 47.
22. Government of Ghana, *Ghana 1977*, p. 57. In the 1975–76 financial year the government received ₵244 million as export duty out of ₵496 million which was the total revenue realised from the sale of cocoa.
23. Nyanteng (1980).

6 Manufacturing

1. Government of Ghana, *Five-Year Development Plan* (1975–76 to 1979–80), Vol. II, p. 189.
2. CBS/Statistical Service, *Industrial Statistics* 1962–64, 1975–77, 1978–81 and 1982–84.
3. The measurement of capital intensity using the book value of capital stock is not, however, without problems. For a start, the system of book-keeping in general is rarely the same and far less efficient across firms to make inter-firm and inter-industry comparisons of such indices absolutely reliable. The use of book values of capital assets for calculating capital–labour ratios is further constrained since different kinds of equipment depreciate at different rates, maintenance conditions are likely to vary across firms, and machinery and equipment are likely to be used beyond the depreciation period in some firms and not so in others. Besides, book values are likely to be subject to irregularities arising from the price distorting effects of exchange rate policies, tariffs and subsidies, as well as invoice operations with respect to the import of capital goods. Thus, although we have used here the book value of capital stock for determining capital intensity in the absence of any better alternative, we will need to be cautious while interpreting the implications suggested by our indices of capital intensity for different industries.
4. Huq and Prendergast (1983), p. 72.
5. Data kindly provided by Ghana Investments Centre.

Notes and References

6. Data collected by the author from Suame Magazine and interviewing officials of the Ghana National Association of Garages and the Technology Consultancy Centre, University of Science and Technology, Kumasi.
7. CBS, *Industrial Statistics* 1977-79 and 1979-81.
8. Grayson (1971), p. 73.
9. Gold Coast Industrial Development Ordinance 1947, Section 3(1) as quoted in Birmingham *et al* (1966), p. 287.
10. Ibid., pp. 287-8.
11. Government of Ghana, *Second Development Plan* (1959-64) as quoted in Birmingham (1966), p. 288.
12. Nkrumah (1963), p. 112.
13. Government of Ghana, *Seven-Year Plan...* (1963-64 to 1969-70), Chapters 1 and 5.
14. Government of Ghana, *Ghana Economic Review* 1973-75, p. 91.
15. Ibid., p. 89; and discussions with the Research Division of the GIC.
16. Steel (1972), pp. 222-23.
17. Ibid., p. 220.
18. Leith (1974).
19. Government of Ghana, *PNDC's Programme for Reconstruction and Development* (1982), p. 21.
20. As the establishments employing 5 or less workers do not have to contribute 12.5 per cent of the workers' pay to the Workers's Social Security Fund, it is feared that many of those employing more than 5, but less than 29, do not contribute to the Social Security Fund. These establishments can be viewed as belonging to the informal sector.
21. The study was conducted by the CDS on behalf of the Ministry of Industries (data unpublished).
22. Steel (1977).
23. Hakam (1972).
24. Ibid.
25. CPP, *Programme for Work and Happiness.*
26. Government of Ghana, *Ghana 1977*, p. 265.
27. Ibid., p. 265.
28. Ibid., pp. 263-64.
29. CBS, *Economic Survey* 1981 (p. 84) and 1982 (p. 74).
30. Winston (1971) and Hogan (1968).
31. CBS, *Economic Survey 1982*, p. 69.
32. CBS, *Economic Survey 1981*.
33. Ibid., p. 89.
34. For first-hand information in the context of a number of East and West African countries see Huq (1986).
35. Ibid. See also Chapter 14.
36. Information obtained from the Research and Planning Department of the ADB.
37. It refers to the Anwia-Nkwanta Oil Mill.
38. Little, Scitovsky and Scott (1970).
39. For earlier estimates of ERP using data from developing countries see, for example, Soligo and Stern (1965) and Lewis and Guissinger (1968).

40. Leith (1974), pp. 68–80.
41. Steel (1972).
42. Leith (1974), p. 78.
43. Steel (1972), pp. 226–27.
44. Grayson (1973), p. 489.
45. Ibid., p. 480.

7 Mining

1. Killick (1966), p. 250.
2. Dickson and Benneh (1970), p. 71.
3. CBS, *Economic Survey 1960*, as quoted in Killick (1966), p. 251.
4. Government of Ghana, *Report of the Committee for Increased Gold Output in Ghana* (1980), p. vii.
5. CBS, *Quarterly Digest of Statistics* March 1984, p. 6.
6. *Report of the Committee for Increased Gold Output in Ghana* (1980), p. 18.
7. Killick (1966), p. 271.
8. Information collected from visits to GNMC, Nsuta.
9. Hart (1980) examines the issue in detail.
10. Personal communication with the Chairman of State Enterprises Commission.
11. CBS, *Industrial Statistics*, 1970–72 and 1979–81.
12. CBS, *Economic Survey 1981*, p. 66.
13. Ibid., p. 67.
14. The electrical fittings are reported to be non-compatible with the types used in Ghana.
15. Government of Ghana, *Five-Year Development Plan* (1975–76 to 1979–80) Part II, p. ix.

8 Money and Credit

1. Orleans-Lindsay (1967), p. 1.
2. Killick (1966), p. 295.
3. IMF (1975), p. 150.
4. Orleans-Lindsay (1967), p. 4.
5. Ibid.
6. Adjetey (1978), p. 13.
7. Ibid.
8. IMF (1975), p. 152.
9. Data collected from the Bank of Ghana.
10. de-Graft Johnson (1967).
11. IMF (1975), pp. 166–7.
12. CBS, *Quarterly Digest of Statistics*, September 1983 and September 1985.
13. IMF (1975), p. 127.
14. Newlyn (1972), p. 9.
15. CBS, *Economic Survey* 1977–80, p. 173.

16. This involved the exchanging of ₡10.00 for ₡7.00 for amounts up to ₡500.0, and ₡10.00 for ₡5.00 for amounts exceeding ₡5000. Bank deposits and government securities were not affected.
17. CBS, *Economic Survey 1977-80*, p. 184.
18. CBS, *Economic Survey 1981*, p. 158.
19. Higher share of commercial bank credit going to the agricultural sector can however be misleading in that the banks sometimes classify agro-based activities as agriculture to meet ceilings.
20. Data collected from the Research Department of the Bank of Ghana.
21. The CMB borrowed ₡255.6 million in 1981, ₡189.6 million in 1982, while only ₡13.8 million in 1983.
22. Adjetey (1978), p. 72.
23. Ibid., p. 73.

9 Development Banking

1. Nwankwo (1980), p. 95.
2. Ibid., p. 95.
3. Government of Ghana, *Ghana Economic Review* (1971-72), p. 243.
4. NIB, *Annual Report 1975*, p. 18.
5. Huq (1984b), p. 16.
6. Tribe (1983), pp. 16-19.
7. ADB, *Annual Report 1978*, p. 16.
8. Interview with an official of the Post-Finance Department of the ADB.
9. ADB, *Annual Report 1977*, p. 7.
10. The commercial banks were lending to sectors that were concentrated in the urban areas.
11. CBS, *Economic Survey 1981*, p. 149.
12. Ibid., p. 148.
13. Centre for Development Studies, University of Cape Coast, *Socio-Economic Baseline Survey of Rural Banks: A Case Study of Biriwa Rural Bank* (1983), p. 3.
14. Ibid., p. 1.

10 External Trade

1. Adjetey (1978), p. 67.
2. Importers were given special unnumbered licences (SUL) to import goods. No foreign exchange was allotted by the government for these imports, thereby encouraging licence holders to bid for scarce currencies in the black market. A scheme in this direction was introduced by the Limann Government following the 1980 Trade Liberalisation Programme (CBS, *Economic Survey* 1981, p. 258). The PNDC Government reintroduced the scheme in 1983.
3. Council of State (1981), pp. 23-24.
4. Government of Ghana, *Summary of PNDC's Budget Statement and Economic Policy for 1983*, Accra, 21 April, 1983, p. 2.
5. Ibid., pp. 3-4.

6. In 1981-82, of the total import of Tk.34 543.5 million (US$ 1 771.5 million) Bangladesh imported goods worth Tk.6 867.0 million (US$ 352.2 million) under the Wage Earners' Scheme. Total visible exports of Bangladesh in the corresponding year was Tk.12 555.4 million. Government of Bangladesh, *Bangladesh Economic Survey* 1982-83, pp. 189 and 209.
7. In 1982-83 Pakistan's foreign exchange earnings through remittances were US$2621 million as against US$2504 million through visible exports. Government of Pakistan, *Sixth Five-Year Plan*, pp. 8-83.
8. Kafka (1956). See also Bhagwati (1962).
9. CBS, *Economic Survey 1977-80*, p. 260.
10. Killick (1978), pp. 275-86.
11. Leith (1974), pp. 10-13.
12. Ibid.
13. Government of Ghana, *PNDC's Programme for Reconstruction and Development* (1982).
14. Leith (1974), pp. 10-13.
15. CBS, *Economic Survey 1977-80*, p. 285.
16. See Rimmer (1984, pp. 152-54), for a good discussion on the prospects of and constraints on the growth of trade in the ECOWAS.
17. Killick (1966), p. 345.
18. Ibid., pp. 354-56 and Killick (1978), pp. 100-102.
19. Killick (1978), p. 101.

11 Prices and Internal Trade

1. Leith (1974), Chapter 5.
2. CBS, *Economic Survey 1981*, p. 202. See also Chapter 8 of the text which shows the growth of money supply.
3. Killick (1978), p. 287.
4. Ibid., p. 287.
5. As quoted in Killick (1978), p. 291.
6. Ibid., p. 291.
7. Ibid., p. 291.
8. The survey was conducted by the author in June 1984, to examine, among other things, the reaction of the people towards official and open market prices; and also to identify the benficiaries of the controlled price system. Data were collected from the capital city (Accra), one public institution (Cape Coast University) and one village (Assin Aworabo, Central Region). Total sample size was however confined to 46 observations representing different income groups.
9. See Appendix B, which deals in detail with the parallel market in Ghana.
10. Lawson (1966).
11. 1962 Census of Distributive Trade as quoted in CBS, *Statistical Year Book* 1969-70, pp. 79-80.

12 Employment

1. CBS, *Economic Survey 1982*, p. 165.
2. CBS, *Quarterly Digest of Statistics*, September 1985, p. 44.
3. Council of State (1981), p. 4.
4. CBS, *Economic Survey 1982*, p. 165.
5. Interviews with officials of the MDPI. The figure is reported to be based on a study conducted by the World Bank.
6. CBS, *Economic Survey 1982*, p. 164. The exodus, however, has not been confined to skilled and professional people only. There has been a large-scale migration from all sections of people.
7. Rado (1986), p. 563.
8. CBS, *Economic Survey 1982*, p. 174.
9. Ibid., p. 174.
10. Baa-Nuakoh (1983), p. 11.
11. CBS, *Economic Survey 1981*, p. 188.
12. CBS, *Economic Survey 1982*, p. 166.
13. Government of Ghana, *Ghana 1977*, pp. 435-38; and interviews with the Salaries Section of the PIB.
14. Information supplied by the PIB.
15. Government of Ghana (1985b), p. 38.
16. It will also be necessary to carry out the exercise with utmost caution. As warned by the Commission on Salaries and Wages (*Report of the Commission on Salaries and Wages*, 1980-81 vol. II, p. 4), the redundancy exercise 'is a very sensitive assignment and must, therefore, be handled with maximum caution. It must take place both horizontally and vertically to ensure that no group or groups are unjustifiably spared or victimized'. So far as redeployment is concerned, the Commission took the view that the 'extent of redeployment should be determined by the proven intake requirements of those productive areas of the economy to which redeployable labour would be diverted. Since redeployable labour would first be trained or retrained and, therefore, equipped with new employable skills, self-employment must be encouraged, with possible initial government assistance for those who would be on their own'.
17. Government of Ghana (1985b).

13 Finance for Investment

1. Following the use of calendar year by CBS/Statistical Service in reporting central revenue accounts for recent years (since 1982), we have used calendar year in place of financial year here. Moreover, the deflators based on GDP and other macroeconomic indicators, as calculated from official Ghanaian sources, are in calendar year and we have used these deflators for converting current price data to constant prices. The method of conversion we have adopted is as follows: data if shown in financial year, say for 1979-80, have been converted to constant prices by using the deflator corresponding to the calendar year 1980.

Notes and References

2. The World Bank (1984a), p. 23. In both, the independent variable was private consumption.
3. Ibid.
4. Implicit industry deflator (see Table A.10) has been used to convert data at current prices shown in the source. CBS, *Economic Surveys* 1968, 1969 and 1981 and Bank of Ghana, *Annual Reports* 1973, 1976 and 1978.
5. Some ventures, such as State Gold Mining Corporation and Ghana Industrial Holding Corporation, in which the state is involved, are indeed very big.
6. CBS, *Industrial Statistics* 1979-1981, p. 12.
7. Information on state enterprises was mainly collected by interviewing State Enterprises Commission (SEC) and GIHOC.
8. Information provided by GIHOC.
9. Interview with an official of State Enterprises Commission.
10. Council of State (1981), p. 36.
11. Ibid.
12. Government of Ghana, *Summary of PNDC's Budget Statement and Economic Policy for 1983*, p. 7.
13. Ibid., p. 8.
14. One can, of course, argue that the uncritical acceptance of suppliers' credit particularly during the first half of the 1960s did not do much good to the country, although it helped increase the volume of foreign savings in financing investment.
15. Killick (1978), p. 104.
16. Huq (1984a).
17. Ewusi (1973), p. 64.
18. Ibid., pp. 70-71 and Bissue (1967), p. 25.
19. CBS, *Economic Survey* 1963, p. 38.
20. Cohen and Tribe (1972), p. 535.
21. Omaboe (1969), p. 15 and Grayson (1973), p. 481. See also Rimmer (1983), p. 399. There was a third rescheduling of external debts carried out in June 1970.
22. Ameyaw, (1983), Typescript, p. 1.
23. CBS, *Economic Survey 1982*, p. 224.
24. Ibid., p. 224 and *Economic Survey 1981*, p. 256.
25. CBS, *Economic Survey 1977-80*, p. 297.
26. Islam (1981).

14 Investment and Technology Choice

1. One can, of course, find some estimates for one or two individual sectors. For example, according to a CBS estimate, fixed capital investment in the manufacturing sector in 1980 and 1981 amounted to ₵189 million and ₵248 million respectively at current prices. CBS, *Economic Survey 1981*, p. 83.
2. This and the following section are taken, at times verbatim, from Huq (1984b).

3. Ahmad (1970); Killick (1978); and Government of Ghana, *Seven-Year Plan* (1963-64 to 1969-70).
4. CBS, *Economic Survey 1963*, p. 29.
5. Government of Ghana, *Seven-Year Plan*, op. cit., pp. 295-96.
6. Killick (1978), p. 142.
7. Rimmer as quoted in Killick, ibid., p. 140.
8. Omaboe (1966), pp. 460-61. See also Krassowski (1974, p. 58) according to whom the purchases of the VC-10 aircraft and eight merchant ships (costing in all £14 million, with credit for £9 million) 'were not foreseen in the Five-Year Plan, nor indeed was the purchase of some of the other items. Rather, the government appears to have been talked into making them by determined salesmanship; although in many cases the salesmen were met at least half-way by Nkrumah or other government ministers'.
9. Ibid., p. 461.
10. Killick (1978), p. 178.
11. Government of Ghana, *Two-Year Development Plan* (1968-70), p. 58.
12. Attoh-Okine (1983), p. 1.
13. Ibid., p. 1.
14. As quoted in Adei (1983a), p. 1.
15. NIB, *Annual Report* 1979, pp. 27-31.
16. Ibid., pp. 27-31.
17. Questionnaire returns.
18. Conversations with Dr S. Adei, Head of the Research Division of the GIC.
19. Questionnaire returns.
20. Interviews with officials of the Post-Finance Department of the ADB.
21. ADB, *Annual Report* 1976, p. 16.
22. Questionnaire Returns.
23. Ibid.
24. Bank of Ghana, *Feasibility Report on Medium-sized Brick Factories* (1979), p. 1.
25. Personal interviews with GIHOC officials.
26. Government of Ghana, *Ghana 1977*, pp. 265-8.
27. Government of Ghana, *Ghana Economic Review* 1973-75, p. 93.
28. GIHOC, *Annual Report* (Provisional) (1981), p. 1.
29. Ibid., p. 11.
30. GIHOC, *Auditors' Report* (1981).
31. GIHOC, *Annual Report*, op. cit., pp. 9-10.
32. Little and Mirrlees (1974), p. 37.
33. Botchwey (1983).
34. Pickett *et al* (1974), Timmer (1975), Winston (1979) and Hawrylyshyn 1978) confirm this.
35. See, for example, Minhas (1963). See also Morawetz (1974).
36. Arrow *et al* (1961).
37. See, for example, Forsyth (1979), Keddie and Cleghorn (1979), Huq and Aragaw (1981) and Khan and Rahim (1985).
38. Pickett (1977), p. 776.
39. Huq and Aragaw (1981).

40. Adei (1983b), p. 3.
41. Ibid., p. 4.
42. Government of Ghana, *Seven-Year Plan*, op. cit., p. 17.
43. Killick (1978), p. 192.
44. Ibid., p. 207.
45. Adei (1983b), p. 2.
46. Ibid., p. 2.
47. The only major investing agency which was not included in our investigations is the Tema Food Complex Corporation.
48. Adei (1983b), p. 5.
49. Adei (1983a), p. 1.
50. Tribe (1983), p. 19.
51. Personal interviews and data obtained from the investing agencies.
52. Timmer (1975), p. 25.
53. Winston (1979), p. 840.
54. The following observation by McBain (1977, p. 120) made in the context of footwear technology is equally applicable to many other industries: 'Although they may be guilty of recommending and over-selling inappropriate mixes of technology the suppliers are unlikely to regard this as a serious fault since they are acclaimed for services to the economies of their own countries when they obtain export orders'.
55. Harvey (1983), p. 54.
56. Ibid., p. 29.
57. For an earlier elaboration in this context based on observations from East and West Africa see Huq (1986).

Postscript

1. Bank of Ghana, *Annual Report for the Financial Year Ended 30th June 1984*, p. 5.
2. Government of Ghana, *The State of the National Economy and the 1987 Budget*, p. 5. It appears that the government recognised the parallel market as another window, the rate of exchange here being determined on informal auctions.
3. *Financial Times*, 5 and 12 February, 1988.
4. *The 1987 Budget*, p. 5.
5. Government of Ghana, *Outline of the 1988 Budget and Economic Policy for 1988*, p. 4.
6. Government of Ghana, *National Programme for Economic Development* (1987), p. 24.
7. *The 1987 Budget*, pp. 3–5.
8. *The 1988 Budget*, p. 2.
9. *National Programme for Economic Development*, pp. 4 and 9.
10. *The 1987 Budget*, p. 1. Recent figures available show that the average annual growth in real GDP over the 1984–86 period was over 6 per cent. See *The 1988 Budget*, p. 1.
11. *The 1987 Budget*, p. 1.
12. *National Programme for Economic Development*, p. 8.

13. Ibid., p. 7.
14. Ibid., p. 5.
15. Ibid., p. 8.
16. Ibid., p. 10.

Appendix B

1. For example, in the past the value of total transactions taking place in the 'black', or illegal, underground sector of the economy is believed to be about 40 per cent of Nigeria's GDP (*South: The Third World Magazine*, June 1984, p. 25). The corresponding figure for Uganda is estimated at over two-thirds of the monetary GDP, and 'one estimate for Zaire in 1971 was that 60 per cent of the state's ordinary revenues were lost or directed to other purposes than the official ones'. (MacGaffey 1983, pp. 351-2).
2. See Safo (1983) for the origin and causes of the *kalabule* economy.
3. For example, in the USA the size of the underground economy, depending on the estimates made at different times, was found to vary between 6 and 22 per cent of GNP while for the UK the corresponding range was found to vary between 2 and 7 per cent of GNP. See Tanzi (1983), pp. 12-13. One should however refrain from inter-country comparisons as the estimates were arrived at by applying different methodologies.
4. CBS, *Economic Survey* 1977-80 and 1981; Council of State, *Reviving Ghana's Economy* (1981).

Bibliography

A. Official Publications

Agricultural Development Bank, *Agricultural Credit Programmes in Ghana*. A country paper first presented at FAO/Finland Regional Credit Seminar for Africa 3-14 December 1973, prepared by the Research and Planning Division, Loans Department of ADB, Accra, 1975.
—, *A Review of Operations over a Decade 1965-1975*.
—, *Annual Report*, Accra, annually.
Bank of Ghana, *Feasibility Report on Medium-sized Brick Factories* (Alajo, Ho, Ashiaman, Ankaful and Accra) 1979.
—, *Annual Report*, Accra, annually.
—, *Quarterly Economic Bulletin*, Accra, quarterly.
Central Bureau of Statistics (CBS), *Consumer Price Index Numbers*, Accra, October 1978.
—, *Economic Survey*, Accra, originally annual series, periodically since 1971.
—, *External Trade Statistics*, Accra, monthly series.
—, *Industrial Statistics*, Accra, periodical series 1962 to 1979-81.
—, *Input-Output Table of Ghana: 1968*, Accra, October, 1973.
—, *Labour Statistics*, Accra, annual series.
—, *National Income of Ghana at Constant Prices 1965-68*, Accra, 1973.
—, *National Programme for Economic Development*, Accra, July 1987.
—, *Outline of the 1988 Budget and Economic Policy for 1988*, Accra, January 1988.
—, *1960 Population Census of Ghana*, Census Office, Accra (various volumes).
—, *1970 Population Census of Ghana*, Accra (various volumes).
—, *1984 Population Census of Ghana: Preliminary Report*, Accra December 1984 (revised printing February 1985).
—, *Quarterly Digest of Statistics*, Accra, quarterly.
—, *The State of the National Economy and the 1987 Budget*, Accra, January 1987.
—, *Statistical News Letter*, Accra (selected issues).
—, *Statistical Year Book 1965-66 and 1969-70*, Accra.
—, *Summary Report on Household Economic Survey*, 1974-75, Accra, 1979.
Council of State, *Reviving Ghana's Economy*. A Report by the Council of State on Economic and Fiscal Policies, Accra, 1981.
Ghana Commercial Bank, *Quarterly Economic Review*, Accra, January-December 1982.
Ghana Education Service, *Digest of Education Statistics* (Pre-University), Accra, annually since 1972.
Ghana Industrial Holding Corporation, *Annual Report* (Provisional) Accra, 1981.
—, *Auditor's Report*, Accra, 1981.
Government of Bangladesh, Ministry of Finance and Planning, Economic Advisers Wing, *Bangladesh Economic Survey 1982-83*, Dhaka, 1983.

Government of Ghana, *Action Programme for Agricultural Production* 1980–81, Accra, 1980.
—, *Budget Proposal 1979–80*.
—, *Economic Recovery Programme 1984–1986*, vols I and II, Accra, October 1983.
—, *Ghana Economic Review*, 1971–72 and 1973–75, Accra.
—, *Ghana 1975: An Official Handbook*, Accra, 1975.
—, *Ghana 1977: An Official Handbook*, Accra, 1977.
—, *Government's Economic Programme* (1981–82 to 1985–86), Accra, March 1981.
—, *Five-Year Development Plan*, 1975–76 to 1979–80, Parts I and II, Accra, 1977.
—, *One-Year Development Plan: Mid-1970–Mid-1971*, Accra 1970.
—, *PNDC's Budget Statement and Economic Policy for 1984*. Presented by Dr Kwesi Botchwey, PNDC Secretary for Finance and Economic Planning, 27 March 1984.
—, *The PNDC Budget Statement and Economic Policy for 1985*. Presented by Dr Kwesi Botchwey, PNDC Secretary for Finance and Economic Planning, Accra, April 1985 (Government of Ghana 1985a).
—, *PNDC's Programme for Reconstruction and Development*. Statement by the PNDC Secretary for Finance and Economic Planning on radio and television on 30 December, 1982.
—, *Progress of the Economic Recovery Programme 1984–86 and Policy Framework 1986–88*. (Report Prepared by the Government of Ghana for the third meeting of The Consultative Group for Ghana in Paris, November 1985). Accra, October 1985 (Government of Ghana 1985b).
Government of Ghana, *Report of the Commission on Salaries and Wages in the Public Services 1980–81*, in two volumes, Accra, 1980.
—, *Report of the Committee Appointed to Investigate the Health of Ghana*, Accra, 1968.
—, *Report of the Committee for Increased Gold Output in Ghana*, Accra, 1980.
—, *Second Development Plan*, 1959–64, Government Printer, Accra, 1959.
—, *The Seven-Year Plan for National Reconstruction and Development 1963–64 to 1969–70*, Accra, 1964.
—, *Summary of PNDC's Budget Statement and Economic Policy for 1983*. Presented by Dr Kwesi Botchwey, PNDC Secretary for Finance and Economic Planning, 21 April, 1983, Accra, 1983.
—, *Two-Year Development Plan: Mid-1968 to Mid-1970*, Accra, 1968.
Government of Pakistan, Planning Commission, *Sixth Five-Year Plan 1983–88*, Islamabad, 1983.
International Bank for Reconstruction and Development (IBRD). See World Bank.
International Monetary Fund, *Surveys of African Economies: Volume 6: The Gambia, Ghana, Liberia, Nigeria and Sierra Leone*, Washington DC, 1975.
Ministry of Agriculture, *Operation Feed Yourself in Figures*, Accra, 1974.
—, *Report on Ghana Sample Census of Agriculture 1970*, Accra, vol. I, 1972.
Ministry of Education, *Ghana Education Statistics 1970–71*, Accra, 1973.
National Investment Bank, *Annual Report*, Accra, 1971 to 1982.
Prices and Incomes Board, *Official Price List*, Accra, vol. 1, April 1982 and vol. 2, May 1982.

Statistical Service, *Industrial Statistics 1982-84,* Accra, July 1986.
—, *Quarterly Digest of Statistics* (September 1986), Accra, 30 September 1986.
World Bank, *Accelerated Development in Sub-Saharan Africa: An Agenda for Action,* Washington DC, 1981.
—, *Appraisal of the Volta River Hydro-electric Project,* World Bank, Washington DC, 1961.
—, *Ghana: Policies and Program for Adjustment: A World Bank Country Study,* Washington, DC, 1984 (World Bank, 1984a).
—, *Toward Sustained Development in Sub-Saharan Africa: A Joint Program for Action,* Washington DC, 1984 (World Bank, 1984b).
—, *World Development Report 1983* (New York: Oxford University Press, 1983).
—, *World Development Report 1985* (New York: Oxford University Press, 1985).
—, *World Development Report 1986* (New York: Oxford University Press, 1986).

B. Books and Articles

ADEI, S. (1983a) 'Criteria for Approving Projects in Ghana by the Ghana Investments Centre: A Note', Accra (Typescript) 1983.
ADEI, S. (1983b) 'Technology Policy and Agents in Technology Decision Making in Ghana', Centre for Development Studies, University of Cape Coast, Cape Coast 1983. (Typescript).
ADELMAN, I. and MORRIS, C. (1978) *Economic Growth and Social Equity in Developing Countries* (Stanford: Stanford University Press).
ADJETEY, S. M. A. (1978) *The Structure of the Financial System in Ghana* (A Research Memorandum) (Accra: Bank of Ghana).
AGBODEKA, F. (1972) *Ghana in the Twentieth Century* (Accra: Ghana Universities Press).
AHLUWALIA, M. S. (1974) 'Income Inequality: Some Dimensions of the Problem' in H. B. Chenery (ed.), *Redistribution with Growth* (London: Oxford University Press).
AHMAD, N. (1970) *Deficit Financing, Inflation and Capital Formation: The Ghanaian Experience 1960-65* (München: Weltforum Verlag).
AMEYAW, S. (1983) 'A Review of the Suppliers Credit Scheme' (Accra: Bank of Ghana) Typescript.
AMONOO, E. (1978) *Agricultural Production and Income Distribution in Ghana* (Cape Coast: Centre for Development Studies, University of Cape Coast).
ARROW, K., CHENERY, H. B., MINHAS, B. and SOLOW, R. (1961) 'Capital-Labour Substitution and Economic Efficiency', *Review of Economics and Statistics* (vol. 43, no. 3) August.
ATSU, S. Y. (1984a) *The Effect of Government Policies on the Increased Production of Food and Food Self-Sufficiency in Ghana* (Legon: ISSER, University of Ghana) Typescript.

ATSU, S. Y. (1984b) 'Appropriate Technology and Increased Food Output: Evaluating the Technology Alternatives for Crop Production in Ghana'. Paper presented at the ISSER Food Self-Sufficiency Conference, University of Ghana.

ATSU, S. Y. and OWUSU, P. M. (1982) 'Food Production and Resource Use in the Traditional Food Farms in the Eastern Region of Ghana' (Legon: ISSER, University of Ghana).

ATTOH-OKINE, G. (1983) 'Ghanaian Experience in Project and Technology Selection/Performance and Activities to be Completed Before a Factory goes into Commercial Production'. A paper presented at CDS/DLI Programme in Choice of Technology, University of Cape Coast.

AUSTIN, D. (1964) *Politics in Ghana 1946-60* (London: Oxford University Press).

AUSTIN, D. (1985) 'The Ghana Armed Forces and Ghanaian Society', *Third World Quarterly* (vol. 7, no. 2) January.

BAAH-NUAKOH, A. (1983) *A Reconsideration of Unemployment Estimates in Ghana* (Legon: Department of Economics, University of Ghana).

BHAGWATI, J. (1962) 'Indian Balance of Payments and Exchange Auctions', *Oxford Economic Papers* (N.S. vol 14, no. 1) February.

BIRMINGHAM, W., NEUSTADT, I. and OMABOE, E. E. (eds) (1966) *A Study of Contemporary Ghana: vol. I. The Economy of Ghana* (London: Allen and Unwin).

BISSUE, I. I. (1967) 'Ghana's Seven-Year Development Plan in Retrospect', *Economic Bulletin of Ghana* (no. 1).

BLOMQVIST, A. G. (1973) 'An Approach to an Optimal Cocoa Policy for Ghana'. Paper presented to Cocoa Economics Research Conference 9-12 April, 1973 (Legon: ISSER).

BOND, M. E. (1983) 'Agricultural Responses to Prices in Sub-Saharan African Countries', *IMF Staff Papers* (vol. 30).

BOTCHWEY, K. (1983) 'State of the Economy', *The Mirror*, Accra, 31 December.

BOAHEN, A. (1975) *Ghana: Evolution and Change in the Nineteenth and Twentieth Centuries* (London: Longman).

BROOKS, R. G. (1981) 'Ghana's Health Expenditures 1966-80: A Commentary' (Glasgow: Department of Economics, University of Strathclyde) (mimeo).

BROWN, C. K. (1972) *Some Problems of Investment and Innovation Confronting the Ghanaian Food Crop Farmer* (Legon: ISSER).

BROWN, T. M. (1972) 'Macroeconomic Data of Ghana', in *Economic Bulletin of Ghana*, Second Series (vol. 2, nos. 1 and 2).

BUAH, F. K. (1980) *A History of Ghana* (London: Macmillan).

CENTRE FOR DEVELOPMENT STUDIES (1983) *Final Report on the Socio-Economic Baseline Survey of Rural Banks, A Case Study of Biriwa Rural Bank* (Cape Coast: University of Cape Coast).

CHENERY, H. B. and BRUNO, M. (1962) 'Development Alternatives in an Open Economy: The Case of Israel', *Economic Journal* (vol. 72).

CHENERY, H. B. and MACEWAN, A. (1966) 'Optimal Patterns of Growth and Aid: The Case of Pakistan', *Pakistan Development Review* (vol. 6, no. 2) Summer.

CHENERY, H. B. and STROUT, A. M. (1966) 'Foreign Assistance and Economic Development', *American Economic Review* (vol. 56).
COHEN, D. L. and TRIBE, M. A. (1972) 'Suppliers' Credit in Ghana and Uganda – An Aspect of the Imperialist System', *The Journal of Modern African Studies* (vol. 10, no. 4).
CONVENTION PEOPLES PARTY (1962) *Programme for Work and Happiness* (Accra: Government Printer).
DASGUPTA, P., SEN, A. and MARGLIN, S. (1972) *Guidelines for Project Evaluation*, UNIDO Project Formulation and Evaluation Series no. 2 (New York: United Nations).
DE GRAFT-JOHNSON, J. C. (1967) 'Some Historical observations on Money and the West African Currency Board', *The Economic Bulletin of Ghana*, Accra (vol. XI, no. 2).
DICKINSON, H. (1982) 'The Volta Dam: Energy for Industry' in M. Fransman (ed.), *Industry and Accumulation in Africa* (London: Heinemann).
DICKSON, K. B. (1969) *A Historical Geography of Ghana* (London: Cambridge University Press).
DICKSON, K. B. and BENNEH, G. (1980) *A New Geography of Ghana* (London: Longman).
DOWSE, R. E. (1969) *Modernisation in Ghana and the USSR: A Comparative Study* (London: Routledge & Kegan Paul).
DUMONT, R. (1966) *False Start in Africa* (London: André Deutsch).
ESSEKS, J. D. (1975) 'Economic Policies' in D. Austin and R. Luckham (eds), *Politicians and Soldiers in Ghana* (London: Frank Cass).
EWUSI, K. (1973) *Economic Development Planning in Ghana* (New York: Exposition Press).
EWUSI, K. (1977) *Rural-Urban and Regional Migration in Ghana* (Legon: ISSER, University of Ghana).
EWUSI, K. (1978) *Planning for the Neglected Rural Poor in Ghana* (Legon: ISSER, University of Ghana).
FAGE, J. D. (1969) *A History of West Africa: An Introductory Survey* (London: Cambridge University Press).
FITCH, B. and OPPENHEIMER, M. (1966) *Ghana: End of an Illusion* (New York: Monthly Review Press).
FORSYTH, D. J. C. (1979) *The Choice of Manufacturing Technology in Sugar Manufacturing Technology in Less Developed Countries* (London: HMSO).
FRANCO, R. G. (1979) 'Domestic Credit and the Balance of Payments in Ghana', *Journal of Development Studies* (vol. 15).
GORDON, D. F. and PARKER, J. C. (1984) 'The World Bank and its Critics: The Case of Sub-Saharan Africa', Discussion Paper no. 108, Centre for Research on Economic Development (Ann Arbor: University of Michigan).
GRAYSON, L. E. (1971) 'Ghanaian Industrial Strategy: Some Problems for the 1970s', *The Economic Bulletin of Ghana* (vol. I, no. 3).
GRAYSON, L. E. (1973) 'The Role of Suppliers' Credits in the Industrialization of Ghana', *Economic Development and Cultural Change* (vol. 21, no. 3) April.
GREEN, R. H. (1971) 'Reflections on Economic Strategy, Structure, and Necessity: Ghana and the Ivory Coast, 1957-67', in P. Foster and A. R. Zolberg, *Ghana and the Ivory Coast: Perspectives of Modernization* (Chicago and London: University of Chicago Press).

GYEKYE, L. O. (1983) 'Ghana Cocoa Marketing Board at Cross Roads' Paper presented at International Seminar, Leiden, 19-23 September 1983, African Studies Centre, Leiden.

GYEKYE, L. O. (1984) *The Role of the Group Lending Scheme in Food Production in Ghana: An Appraisal.* Paper presented at the ISSER Conference on Food Self-Sufficiency in West Africa, ISSER, University of Ghana, Legon 1-3 May.

HAKAM, A. N. (1972) 'Impediments to the Growth of Indigenous Industrial Entrepreneurship in Ghana: 1946-68, *The Economic Bulletin of Ghana* (vol. 2, no. 2).

HART, D. (1980) *The Volta River Project. A Case Study in Politics and Technology* (Edinburgh: Edinburgh University Press).

HARVEY, C. (1983) *Analysis of Project Finance in Developing Countries* (London: Heinemann).

HAWRYLYSHYN, O. (1978) 'Capital-intensity Biases in Developing Country Technology Choice', *Journal of Development Economics* (vol. 5, no. 3) September.

HAZLEWOOD, A. (1979) *The Economy of Kenya: The Kenyatta Era* (Oxford: Oxford University Press).

HEGGIE, I. G. (1976) 'Practical Problems of Implementing Accounting Prices', in I. M. D. Little and M. Scott (eds), *Using Shadow Prices* (London: Heinemann).

HOGAN, W. (1968) 'Capacity Creation and Utilization in Pakistan's Manufacturing Industry', *Australian Economic Papers* (vol. 7) June.

HONNY, L. A. (1982) *Preliminary Report on Farm Management Data Collection System* (Cape Coast: Centre for Development Studies, University of Cape Coast).

HONNY, L. A. and MICAH, J. A. (1983) *Socio-Economic Baseline Survey of Twifu-Prasu Agricultural District* (Cape Coast: Centre for Development Studies, University of Cape Coast).

HUQ, M. M. and ARAGAW, H. (1981) *Choice of Technique in Leather Manufacturing* (Edinburgh: Scottish Academic Press).

HUQ, M. M. and PRENDERGAST, C. C. (1983) *Machine Tool Production in Developing Countries* (Edinburgh: Scottish Adademic Press).

HUQ, M. M. (1984a) 'Ghana: Economic Decline, and a Case for Recovery'. Paper presented in a seminar at the Institute of Statistical, Social and Economic Research, University of Ghana, Legon, January 1984.

HUQ, M. M. (1984b) *Investment Decisions and Technology Choice: A Case Study of Ghana,* Centre for Development Studies, University of Cape Coast, Cape Coast 1984.

HUQ, M. M. (1986) 'Use of Imported and Indigenous Technologies – Observations from East and West Africa, and a Case for Technology Policy', *Science, Technology and Development* (vol. 4, no. 1).

ISLAM, N. (1981) 'Aid Requirements and Donor Preferences' in Just Faaland (ed.), *Aid and Influence: The Case of Bangladesh* (London: Macmillan).

JEFFRIES, R. (1982) 'Rawlings and the Political Economy of Underdevelopment in Ghana', *African Affairs* (vol. 81).

KAFKA, A. (1956) 'The Brazilian Exchange Auction System', *Review of Economics and Statistics* (vol. 38).

KARKARI, E. B., et al (1983) *Choice of Technology in the Sawmilling Industry in Ghana*. Project Report Prepared in the Workshop on Choice of Technology and Project Evaluation, Centre for Development Studies, University of Cape Coast, Cape Coast, October 1983 (mimeo).

KEDDIE, J. and CLEGHORN, W. (1979) *Brewing in Developing Countries* (Edinburgh: Scottish Academic Press).

KHAN, M. R. and RAHIM, E. (1985) *Corrugated Board and Box Production* (Edinburgh: Scottish Academic Press).

KILLICK, T. (1978) *Development Economics in Action: A Study of Economic Policies in Ghana* (London: Heinemann).

KILLICK, T. (1966) 'External Trade', in Birmingham (1966), op. cit.

KILLICK, T. (1966) 'Labour: A General Survey' in Birmingham (1966).

KILLICK, T. (1966) 'Mining' in Birmingham (1966).

KILLICK, T. (1966) 'The Monetary and Financial System' in Birmingham (1966).

KILLICK, T. (1966) 'The Volta River Project' in Birmingham (1966).

KRASSOWSKI, A. (1974) *Development and the Debt Trap: External Borrowing in Ghana* (London: Croom Helm).

KUZNETS, S. (1955) 'Economic Growth and Income Inequality', *American Economic Review* (vol. 45, no. 1) March.

LA ANYANE, S. (1963) *Ghana Agriculture* (London: Oxford University Press).

LAWSON, R. M. (1966) 'Inflation in the Consumer Market in Ghana: Report of the Commission of Enquiry into Trade Malpractices in Ghana', *Economic Bulletin of Ghana*, Accra (no. 1).

LEITH, J. C. (1974) *Foreign Trade Regimes and Economic Development: Ghana* (National Bureau of Economic Research, Columbia University Press).

LEWIS, S. R. (Jr) and GUISINGER, S. E. (1971) 'The Structure of Protection in Pakistan' in B. Balassa (ed.) *The Structure of Protection in Developing Countries* (Baltimore: The Johns Hopkins Press).

LEWIS, W. A. (1965) *Politics in West Africa* (London: George Allen & Unwin Ltd).

LEWIS, W. A. (1953) *Report on Industrialization in the Gold Coast* (Accra: Government of Gold Coast).

LEWIS, W. A. (1955) *The Theory of Economic Growth* (London: George Allen & Unwin Ltd).

LITTLE, I. M. D. and MIRRLEES, J. A. (1968) *Manual of Industrial Project Analysis*, vol. II (Paris: OECD Development Centre).

LITTLE, I. M. D. and MIRRLEES, J. A. (1974) *Project Appraisal and Planning for Developing Countries* (London: Heinemann).

LITTLE, I., SCITOVSKY, T. and SCOTT, M. (1970) *Industry and Trade in Some Developing Countries: A Comparative Study* (London: Oxford University Press).

MANU, J. E. A. (1973) 'Cocoa in the Ghana Economy'. Paper presented to Cocoa Economics Research Conference 9–12 April, 1973 (Legon: ISSER, University of Ghana).

MCBAIN, N. S. (1977) *The Choice of Technique in Footwear Manufacture for Developing Countries* (London: HMSO, Ministry of Overseas Development).

MACGAFFEY, J. (1983) 'How to Become Rich Amidst Devastation: The Second Economy in Zaire', *African Affairs* (vol. 82, no. 328) July.

MASON, E. S. et al (1980) *The Economic and Social Modernisation of the Republic of Korea* (Cambridge, MA and London: Harvard University Press).

MINHAS, B. S. (1963) *An International Comparison of Factor Cost and Factor Use* (Amsterdam: North-Holland Publishing Co.).

MORAWETZ, D. (1974) 'Employment Implications of Industrialization in Developing Countries: A Survey', *Economic Journal* (vol. 85) September.

MYINT, H. (1967) *Economics of the Developing Countries* (London: Hutchinson) 3rd edition.

MYRDAL, G. (1964) *Economic Theory and Under-developed Regions* (London: Methuen).

NATHAN, R. R. and Associates (1970) *Sector Studies on Transportation, Agricultural and Water Resources*, various volumes, Accra (mimeo).

NEWLYN, W. T. (1967) *Money in an African Context* (Nairobi: Oxford University Press).

NKRUMAH, K. (1957) *Ghana: An Autobiography* (London: Nelson).

NKRUMAH, K. (1963) *Africa Must Unite* (London: Heinemann Educational Books).

NURKSE, R. (1953) *Problems of Capital Formation in Under-developed Countries* (Oxford: Blackwell).

NWANKWO, G. O. (1980) *The Nigerian Financial System* (London: Macmillan).

NYANTENG, V. K. (1980) *The Declining Ghana Cocoa Industry: An Analysis of Some Fundamental Problems* (Legon: ISSER).

OMABOE, E. N. (1963) 'Some Observations on the Statistical Requirements of Development Planning in the Less-Developed Countries', *Economic Bulletin of Ghana* (no. 2).

OMABOE, E. N. (1969) *Development in the Ghanaian Economy between 1960 and 1968* (Accra: Ghana Publishing Corporation).

OMABOE, E. N. (1966) 'An Introductory Survey' in Birmingham (1966).

OMABOE, E. N. (1966) 'The Process of Planning' in Birmingham (1966).

OMANE, I. A. (1980) *Ghana's External Indebtedness Problem: A Case Study with Special Reference to the Role of Corruption and Suppliers' Credit 1960-74* (Unpublished Ph.D. thesis, University of Strathclyde, Glasgow).

ORLEANS-LINDSAY, J. K. (1967) *Bank of Ghana 1957-67*. A pamphlet prepared to commemorate the tenth anniversary of the establishment of the Bank in Accra.

PICKETT, J. (1977) 'The Work of the Livingston Institute on "Appropriate Technology" - Editor's Introduction', *World Development* (vol. 5, nos. 9/10) September-October.

PICKETT, J., FORSYTH, D. J. C. and McBAIN, N. S. (1974) 'The Choice of Technology, Economic Efficiency and Employment in Developing Countries', in E. O. Edwards (ed.) *Employment in Developing Nations* (New York: Columbia University Press).

RADO, E. R. (1985) 'Ghana Revisited', Centre for Development Studies, University of Glasgow (Typescript).

RADO, E. R. (1986) 'Notes Towards a Political Economy of Ghana Today', *African Affairs* (vol. 85, no. 341) October.

Bibliography

RAHIM, E. (1979) 'Editorial Introduction' in J. Keddie and W. Cleghorn, *Brewing in Developing Countries* (Edinburgh: Scottish Academic Press).
RIMMER, D. (1966) 'The Crisis in the Ghana Economy', *Modern African Studies*, May.
RIMMER, D. (1983) 'Ghana Economy' in *Africa: South of the Sahara 1983-84* (London: Europa Publications Ltd.).
RIMMER, D. (1984) *The Economies of West Africa* (London: Weidenfeld and Nicolson).
ROEMER, M. (1984) 'Ghana, 1950-80: Missed Opportunities' in A. C. Harberger (ed.) *World Economic Growth: Case Studies of Developed and Developing Nations* (San Francisco: Institute for Contemporary Studies).
ROSTOW, W. (1960) *The Process of Economic Growth* (Oxford: Clarendon Press) second enlarged edition.
ROSTOW, W. (1971) *The Stages of Economic Growth* (Cambridge: Cambridge University Press) second edition.
SAFO, D. B. (1981) *The West African Dictionary of Modern Economics* (Cape Coast: University of Cape Coast).
SAFO, D. B. (1983) 'The Illusion of Price Controls in Ghana' (Cape Coast: University of Cape Coast).
SEIDMAN, A. W. (1978) *Ghana's Development Experience* (Nairobi: East African Publishing House).
SINGAL, M. S. and NARTEY, J. D. M. (1971) *Sources and Methods of Estimation of National Income at Current Prices in Ghana* (Accra: Central Bureau of Statistics).
SCHULTZ, T. W. (1960) 'Capital Formation by Education', *Journal of Political Economy*.
SCHULTZ, T. W. (1961) 'Investment in Human Capital', *American Economic Review*, March.
SCHULTZ, T. W. (1971) *Investment in Human Capital* (New York: The Free Press).
SMITH, A. (1950) *The Wealth of Nations* (vols. I and II), edited by Edwin Cannan (London: Methuen) sixth edition.
SOLIGO, R. and STERN, J. (1965) 'Tariff Protection, Import Substitution and Investment Efficiency', *Pakistan Development Review* (vol. 5, no. 2) Summer.
STEEL, W. F. (1972) 'Import Substitution and Excess Capacity in Ghana', *Oxford Economic Papers*, New Series (vol. 24, no. 2) July.
STEEL, W. F. (1977) *Small-Scale Employment and Production in Developing Countries - Evidence from Ghana* (New York: Prager Publications Inc.).
TANZI, V. (1983) 'The Underground Economy: The Causes and Consequences of this World-Wide Phenomenon', *Finance and Development* (vol. 20, no. 4) December.
TIMMER, C. P. (1975) 'The Choice of Technique in Indonesia' in C. P. Timmer et al, *The Choice of Technology in Developing Countries* (Harvard University: Centre for Industrial Affairs).
TRIBE, M. (1983) 'The National Investment Bank and the Choice of Technology - A Case Study'. Draft Discussion paper presented at CDS/DLI

Programme in Choice of Technology, Centre for Development Studies, University of Cape Coast.

WARREN, J. (1977) 'Savings and the Financing of Investment in Ghana, 1960 to 1969' in W. T. Newlyn (ed.) *The Financing of Economic Development* (Oxford: Clarendon Press).

WINSTON, G. C. (1971) 'Capacity Utilisation in Economic Development', *Economic Journal* (vol. 81).

WINSTON, G. C. (1979) 'The Appeal of Inappropriate Technologies: Self-inflicted Wages, Ethnic Pride and Corruption', *World Development* (vol. 7, no. 8/9) August–September.

Index

Abidjan, 198
Abrahams Commission, 199, 218
accounting prices, see shadow prices
Acheampong Government, 16-17
  misrule during, 26, 28
Adei, S., 318n, 331-2n
Adelman, I., 320n
Adjetey, S., 326n
African Development Bank, 318n
  loan for Ghana Railway, 66
  support for development banks, 188, 192-3
African Manganese Company, 156
Agbodeka, F., 319n
agricultural credit, 96-100
  see also Agricultural Development Bank; money-lenders; rural banks
Agricultural Development Bank, 97-100, 187-90, 323n, 327n, 331n
  loans to agricultural sector, 98-100
  loans to livestock sub-sector, 103
  study on money-lenders, 97
  technology choice, 271-2
  use of shadow prices, 146, 261-3
Agricultural Credit and Co-operative Bank, 187
Agricultural Indebtedness, Government Committee on (1958), 96
agriculture, 81-107
  extension services, 95-6
  fertilisers, 93-4
  food crops, 81-3
  inter-cropping, 87-8
  irrigation, 92-3
  labour, 89-91
  land tenure, 86-7
  land use, 83-8
  output, 81-3
  seeds, 93, 95
  share of GDP, 50-2
  technology, 88-9
Agriculture, Ministry of, 95, 103, 322-3n
Ahluwalia, M., 56, 320n
Ahmad, N., 13, 331n
aid, see foreign aid
air transport, 63, 68
Akim Concessions, 155
Akosombo dam, 36, 68-9
  see also Volta River Project

Akwatia Consolidated Diamonds Ltd, 155
Alajo Brick Factory, 263
alternative techniques/technologies, see technology
aluminium industries
  major world companies, 158
Ameyaw, S., 330n
Amonoo, E., 88, 93, 96, 322-3n
Anwia-Nkwanta Oil Mill, 263
Apaloo Report, 318n
Aragow, H., see Huq, M.
Argentina, 100
Arrow, K., 267, 331n
Ashanti kingdom, 35
Atabu, F., 316n
Atsu, S., 92, 322n
  with Owusu, P., 92-3, 322n
Attoh-Okine, G., 331n
Austin, D., 318n
Australia, 100

Baah-Nuakoh, A., 329n
balance of payments, 212-13
Bank for Housing and Construction, 182
Bank of Ghana, 167-70, 195, 283, 326-7n, 330
  cattle ranch owned by, 103
  and money supply, 174-7
  promotion of development banking, 181
  Research Department, 74
  technology choice, 271-2
  use of shadow prices, 261-3
banking structure, 167-73
banks, see commercial banks
Barclays Bank (Ghana Ltd), 167, 170
'barter' system, 223
bauxite, 157-8
  capital intensity, 159-61
  employment, 160
  output, 159-60
  value added, 160-1
Bangladesh, 197
Benin, 158
Benneh, G., see Dickson, K.
Bhagwati, J., 328n
Birmingham, W., 284, 325n
Bissue, I., 8, 315n

black market, *see kalabule* economy; parallel market
Black Star Line, 25, 67
Blomqvist, A., 323n
Bogosso Mills Ltd, 263
Bond, M., 322n
Botchwey, K., 331n
Brazil, 116, 118, 198
Britain, 67, 76, 117, 170, 207
  see also United Kingdom
Brooks, R., 322n
Brown, C., 95, 323n
Brown, T., 284
Buah, F., 319n
budget deficits
  by Central Government, 176
  recovery from, 278
  see also deficit financing
Burkina Faso, 35-6, 39, 54-5, 100, 159, 208
Busia Government, 28, 35
  import liberalisation, 199
  price controls, 218
  stabilisation attempts, 15-16
  trade liberalisation, 215

Cameroon, 116
Canada, 158
capacity utilisation
  in GIHOC, 242
  in manufacturing sector, 142-5
capital formation, 48-9, 295
  time series data, 289-90
  see also investment
capital intensity
  of medium and large firms, 135-9
  of small firms, 135-7
Capital Investment Board, 130-1
  see also Ghana Investments Centre
Capital Investments Act (1963), 130
Capital Investments Decree (1973), 131
capital-output ratio, 265-6
ceramic industry, 153
census, of population, *see* population
central bank, *see* Bank of Ghana
Central Bureau of Statistics (CBS), 74, 229, 283-4, 319-33n
  on exodus of professionals, 229
  on inflation, 216
Centre for Development Studies (CDS), 74, 87, 113, 136, 284, 321n, 325n, 327n
  on cocoa, 113
  on income distribution, 58
  on rural banking, 192
  on small scale industries, 136-8
CFA (monetary) zone, 36, 313
Chenery, H. B.
  and Bruno, M., 320n
  and MacEwan, A., 320n
  and Strout, A. M., 320n
China, 55, 208-9
cocoa, 109-118
  comparative advantage in, 113-14
  demand elasticity, 111-12
  inputs, 114-15
  input subsidy, 114-15
  output, 109-11
  problems and constraints, 115-16
  producers' price, 22, 112-14
  production, 109-11
  smuggling into neighbouring countries, 22, 114
cocoa butter, 110, 112, 117, 133, 201
Cocoa Co-operative Societies, 97-8, 172
Cocoa Marketing Board (CMB), 21, 112, 114, 116-18
  overmanning, 229, 279
cocoa paste, 110, 112, 117, 133, 201
Cocoa Producers' Alliance, 116
Cocoa Purchasing Company (CPC), 98
Cocoa Research Institute, 74
Cohen, D., 318n, 330n
command economy, 14
  see also planning
commercial banks, 170-3 (*see also* Barclays Bank; Ghana Commercial Bank; Standard Bank; credit supply)
  the 'big three', 171
  loans and advances by, 178-9
  loans to agricultural sector, 99
  primary banks, 167, 170
  secondary banks, 167
Commission on Salaries and Wages, 329n
comparative advantage
  in cocoa, 22
  in timber, 22
comprehensive economic planning, *see* planning
conditionality, *see* World Bank-IMF conditionality
Consolidated African Selection Trust, 155
Consolidated Plan, 7
consumer price index, time-series data, 304

Index

consumption, *see* private consumption; government consumption
controlled prices, 20-1, 219-20
Convention People's Party (CPP), 35, 139, 325n
Co-operative Bank, 98
corruption, 15, 26
CPP, *see* Convention People's Party
Council for Scientific and Industrial Research (CSIR), 73
Council of State, 196-7, 318, 333n
  Report by, 26-8, 228-9
credit supply, 177-80
  see also agricultural credit
Crystal Oil Mills, 140
current revenue, 3
  see also government revenue

Darko Poultry Farm, 103
David Livingstone Institute studies on technology choice, 268
debt
  burden of, 25
  see also external debts
debt service ratio
  for external borrowing, 25
  for project appraisal, 184, 188
decimal currency, 168-9, 195-6
decimal currency committee, 168
decision-making, *see* investment decisions
deficit financing
  and monetary expansion, 13-15
  see also budget deficits
deflators, implicit
  of components of GDP by industrial origin, 298-9
  of components of GDP by uses, 296
  of types of assets in capital formation, 297
demographic features, 38
depreciation
  in mining assets, 160
  time-series data, 295
devaluation, 15, 195-7, 277
  (1967), 195
  (1971), 15-16
  (1983), 131, 192, 277
  (1984), 13
development banking, 181-93
  need for, 181-2
development expenditure, *see* government expenditure

development planning, *see* planning
Development Service Institute (DSI), of the NIB, 184
development strategy, 7, 9-12
  see also industrialisation strategy
diamonds, 155-6
  capital intensity, 159-61
  employment, 159-60
  value added, 160-1
  output, 160
  see also mining
Diamond Marketing Corp., 156
Dickinson, H., 75
Dickson, K. B., 319n, 321n, 326n
direct taxes, 240
discounted cash flow method
  use in project appraisal, 184, 188
distorted price stucture, *see* price distortions
distributive trade
  sales by the 'big seven', 223
  see also trade
domestic capital formation,
  see capital formation
domestic resource costs (DRC), 147-9
domestic savings, *see* savings
Dorman Long, 140
Dowse, R., 14, 19, 316-17n
dualism, 104
  see also technological dualism
Dumont, R., 317-18

Economic Commission for Africa, 318n
Economic Community of West African States (ECOWAS), 133, 200, 208-9
economic decline, 1-3, 31-2
economic development since independence, 2-6
economic management, 12-25
  inefficiency in, 16-17
  see also management
economic recovery, in recent years, 32, 277, 280
Economic Recovery Programme, 11-12, 31-2, 234, 277, 280-1, 318, 321n
  see also Provisional National Defence Council
ECOWAS, *see* Economic Community of West African States
education, 71-2
  exodus of trained teachers, 72
  investment in, 71
  see also literacy
Education, Ministry of, 279

effective rate of protection, see protection
Egypt, 2, 46, 55, 207-8
electricity supply, 68-9
  from Akosombo dam,
  from different sources, 69
  generation, 63
Electricity Corporation of Ghana, 69, 158
employment
  distribution by industries, 42
  distribution by sex, 42-3
  in manufacturing, 120
  in private sector, 225-30
  in public sector, 225-30
  and technology choice, 225
  see also overmanning
Engel's law, 112, 323n
engineering industries
  dependence on imported materials, 127
  employment, 127
  medium and large scale, 126
  small scale, 127
engineering sector, 126
engineering workshops, 127
  see also Suame Magazine
Esseks, J. 15, 316
Ewusi, K., 8, 247, 315, 318, 319n, 330n
Exchange Control Act, 195
exchange rate
  adjustments of, 4
  effective rates of, 200-1
  multiple (rates), 198, 277
  policy, 195-8
  see also devaluation; foreign exchange
expatriate trading companies, 218, 222-3
exports
  bonus scheme, 199
  direction of, 207-9
  fall in, 49-50, 198
  increase in, 32
  major items, 201-4
  policy, 198-9
  recent growth in, 280
  time-series data of, 289-90, 305
  see also external trade
external capital, 246-50
  loans and grants, 249-50
  private investment, 247
  suppliers' credit, 247-8
external debts
  repayments, 280-1
  rescheduling, 248

external dependence, 24-5
external trade, 195-213
  see also balance of payments; exports; imports; terms of trade

factor substitution, 267
Fage, J., 319n
fertilisers, 93-5
Finance, Ministry of, 195
Finance and Economic Planning
  Ministry of, 8, 74, 232
  Secretary for, 197, 280, 317
financial institutions, 74-5
Financial Times, 332n
firms, size of, 134-8
First Ghana Building Society, 74
fiscal policy, 11
  see also government expenditure; government revenue
fishing, 103-4
Fitch, B., 117, 324n
Five Year Plan (1975-80), 10-11, 18-20, 134, 317n, 324n, 326n
  on large scale, 119
  view on foreign private capital, 163
food
  crops, 82-3
  imports, 19
Food Production Corp., 89
forced savings, 245
Fordwor, K. D., 260
foreign aid
  in economic recovery, 32, 279
foreign capital, see external capital
foreign debts, see external debts; debt
foreign exchange
  auctioning, 277-8
  constraint, 30, 77
  see also exchange rate
foreign exchange gap, 24
foreign loans and grants, 249-50
foreign indebtedness, see external debts
foreign investment, in mining, 163
Forsyth, D., 331n
forestry, 105-7
  employment, 107
  export, 106
France, 158
Francophone zones, 265
Frankfurt, 131

Gabon, 116
Geological Survey Department, 154
Ghana Airways, 68

Index

Ghana Airways Corp., 68, 243
Ghana Aluminium Products Ltd, 140
Ghana Cement Works, 140
Ghana Cocoa Board, see Cocoa Marketing Board
Ghana Commercial Bank, 31, 98-9, 167, 170-2
Farmers Association, 99
Ghana Empire, 35
Ghana Enterprises Development Commission, 74
Ghana Export Promotion Company, 169
Ghana Food Distribution Corp., 222
Ghana Highway Authority, 64, 321n
Ghana Industrial Holding Corp. (GIHOC), 126, 129, 139-40, 242-4, 330-1n
 joint venture enterprises, 140
 technology choice, 271-2
 use of shadow prices, 261-4
 various division of, 140
Ghana Investments Centre (GIC), 23, 74, 260-2, 269, 324n
 and technology choice, 271
 use of shadow prices, 23, 261-2
Ghana National Association of Garages, 325n
Ghana National Manganese Corp. (GNMC), 156
Ghana National Trading Corp. (GNTC), 25, 218, 222, 243
Ghana Pioneer Aluminium Factory, 140
Ghana Railway, 76
 government subventions, 66
 inability to transport timber, 106
 workshop at Takoradi, 126
Ghana Railway Corp., 321n
 see also Ghana Railway
Ghana Sanyo Electrical Manufacturing Company, 140
Ghana Sugar Estates Limited, 140, 242-3
Ghana Textiles Printing, 140
Ghana Tourist Development Company, 169
Ghana Water and Sewerage Corp., 70
gold, 153-5
 capital intensity, 159-61
 employment, 159-60
 output, 160
 mining, 153-5
 Tarkwa deposits, 154
 value added, 160-1
 see also mining
Gordon, D., 28-9, 317-18n

government consumption, 48-50
 time-series data, 289-90
government expenditure
 development, 239
 non-development, 239
 on roads and waterways, 65
 time series data, 300
government revenue, 239
 share of non-tax revenue, 239-41
 share of taxes, 238-40
 time series data, 300
Government's Economic Programme (1981), 321n
Grains Warehousing Company, 263
Grayson, L., 151, 325-6n, 330n
Green, R., 12, 316n
gross domestic product (GDP)
 distribution by industrial origin, 50-3, 291-4
 growth during 1890-1910, 45-6
 growth since independence, 2, 46-8
 per capita time-series data, 32, 47-8, 277, 288
 recovery from decline, 277
 time-series data, 47-8, 287, 289-94
gross domestic savings, 237-46
 private savings, 244-6
 public savings, 238-44
gross investment, 253
 time-series data, 302
Guggisberg, G., 6
Guinea, 55
Guisinger, S., see Lewis, S.
Gyekye, L., 323n

Hakam, A., 325n
harmattan, 36
Hart, D., 326n
Harvey, C., 275, 332n
Hawrylyshyn, O., 331n
Hazelwood, A., 174, 319
health, 71-3
 expenditures on, 73
 indicators, 72
Holland, see Netherlands
Hogan, W., 325n
Honny, L., 322-4n
human resources, see labour; education
Huq, M., 321-2n, 325, 327, 330n, 332n
 with Aragaw, 331n
 with Prendergast, 126, 324n
Hungary, 320n
hydro-electric power, see electricity supply

Import and Export Act (1980), 199
imports
 controls, 195-6
 decline in, 198
 increase in, 32
 liberalisation, 199
 licensing, 199-200
 major items, 204-7
 origin of, 209-10
 policy, 199-200
 time-series data of, 289-90, 306
income distribution, 56-8
India, 2, 30, 46, 53, 55
indirect taxes, 239-40
Indonesia, 55, 274
Industrial Development Corp. (IDC), 128-9
industrialisation strategy, 128-34
 export expansion, 133
 import-substituting, 31, 129-30
 incentives, 130-2
 location, 133-4
 projection, 132-3
Industries, Ministry of, 129, 242, 263
inflation, 3, 15, 215-17, 278
 rates of, 216-17
 small-scale manufacturing, 134-8
infrastructure, 61-77
 institutional, 73-5
 physical, 62-71
 social, 71-3
Institute of Statistical, Social and Economic Research (ISSER), 74, 284
interest rate
 nominal, 245-6
 real, 246
interest rate policy, 246, 278
internal trade, 221-4
 women's participation in, 221
International Bank for Reconstruction and Development, see World Bank
International Monetary Fund, 170, 326n
 advice from, 15, 17, 27-8
 assistance from, 279
 role in economic recovery, 279
 see also World Bank
international trade, see external trade; trade
investment
 finance for, 237-51
 low level of, 253
 by private sector, 254-5, 302
 by public sector, 254-5, 302

sectoral allocation, 256
 in selected development countries, 55
 time-series data, 302
 see also capital formation; external capital; gross investment
investment decisions, 256-66
Investment Holding Company, 173
Investment Policy Decree (1973), 185
investment-push strategy, under Nkrumah government, 254
irrigation, 92-3
Irrigation Authority, 92
Islam, N., 330n
Ivory coast, 35, 39, 48, 53, 55, 110, 114, 116, 118, 204

Japan, 207, 208
Jeffries, R., 14-16, 26, 316, 318
Jibowu Commission (1956), 98
Juapong Textiles, 140

Kafka, A., 328n
Kaiser Corporation, 68, 75, 158
 see also Volta River Project
kalabule
 economy, 5, 309-14
 practices, 3
 trading, 5
Kaldor, N., 27
Karkari, E., 105n
Keddie, J., 331n
Kenya, 30, 55
Kessels Committee, 168-9
Khan, R., 331n
Killick, T., 8-10, 13, 57, 259, 270, 284, 315-18n, 320n, 328n, 330-2n
 on income distribution, 57
 on investment decisions, 259-60
 on terms of trade, 210
Komenda Irrigation Project, 92
Korea, South, 30, 53
Korle Bu Hospital, 6
Kpong dam, 36, 69
Krassowski, A., 331n
Kumasi Brewery, 140
Kuznets, S., 320n
Kwahus, 222

La Anyane, S., 323n
labour
 agricultural, 89-91
 family, 90-1
 hired (wage), 90-1
 productivity, 123-6, 159-60

Index

shortage of, 90-1
supply, 38-41
  see also wages and salaries;
  population; redeployment
labour-intensive strategy, 225
labour-intensive technique, in carpentry workshop, 107
land productivity, 83-4
land tenure patterns, 86-7
Lange, O., 29
Lawson, R., 221, 328n
Leith, C., 132, 147-9, 200, 284, 315n, 320n, 326n, 328n
Lever Brothers, 140
Lewis, S., 320n, 325n
Lewis, W. A., 19, 27, 128, 317n
liberalisation of trade, 199, 215, 219
Liberia, 48, 55
Liberman reforms, 29
Libya, 200
license, 199-200
  open general, 199
  special, 199
  specific, 200
Limann Government, 11, 16-17, 26, 28
  indecisive rule, 26
linkages, inter-industry, 127-8
literacy, 59
  see also education
Little, J., 264
  with Mirrlees, J., 264, 331n
  with Scitovsky, T. and Scott, M., 325n
livestock, 100-3
  by population by regions, 101-2
  by population type, 100-1
Lomé Convention, 200
London-Rhodesia Company (LONRHO), 155

McBain, N., 332n
MacGaffey, J., 333n
machine tools, manufacture of, 126
Mali, 100
Malaysia, 30, 112, 116, 118
Managed Input Delivery and Agricultural Services (MIDAS), 188
management
  review of, 17-25
  inefficiency in, 16-17
  see also economic management
Management Development Productivity Institute (MDPI), 329n
manganese, 156
  capital intensity, 159-61
  output of, 159-60
  value added, 160-1
Mankoadze Fisheries, 104
Manpower Utilisation Committee, 234
Manu, J., 323n
manufacturing, 119-51
  capacity utilisation, 142-3, 280
  capital intensity, 124-6
  efficiency of production, 147-50
  employment, 120-5
  output, 119-23
  ownership, 138-42
  share of GDP, 51-2
  of simple machinery, 126
  sources of inputs, 144-5
  technology, 145-6, 150-1
  wages and salaries, 120
market
  mechanism, 14, 20, 29-30
  'mummies', 224
  see also price mechanism
Mason, E., 319n
Merchant Bank Ghana Ltd, 172-3
Mensah, J. H., 316n
migration of skilled personnel, 31, 229-30
military takeover, 9, 10-11, 16
Minhas, B., 331n
mineral deposits, see mining
mining, 153-63
  employment, 160
  factor intensity, 159-61
  'Jungle Rush', 153
  problems and constraints, 161-3
  see also bauxite; diamonds; gold; manganese; oil
mismanagement, economic, 281
monetary
  expansion, 13-14
  policy, 173-6
  stability, 169
money lenders, 96-7
  see also agricultural credit
money supply, 173-7
  annual growth rate of, 175
  fiduciary issue, 168, 176
  time series data of, 303
  see also deficit financing
monopoly rent, 220
Morawetz, D., 331n
Morris, C., see Adelman, I.
Myint, H., 53, 320n

NLC, see National Liberation Council

NRC, see National Redemption Council
Nasia Rice Company, 262
Nathan Consortium, 62, 66-7, 321-2n
  report, 62, 65-7, 75
National Investment Bank (NIB), 23, 98-100, 131, 135, 182-6, 192-3, 327n, 331n
  loans to agricultural sector, 98-100
  technology choice, 271
  use of shadow prices, 23, 261-2
National Liberation Council (NLC), 9, 15, 35
  rescheduling of external debts, 248
  trade liberalisation, 215
National Pig Multiplication Centre, 103
National Planning Commission, see Planning Commission
National Programme for Economic Development, 332-3n
National Redemption Council (NRC), 4, 9, 10, 16, 20
  price controls, 218-19
National Savings and Credit Bank (NSCB), 171
National Stock Broker Ltd, 173
National Trust Holding Company, 74-5
natural resources, 1
Neoplan Ltd, 140
net present value, 269
Netherlands, 111
New Match Factory, 140
Newlyn, W., 326n
Niculescu, 7
Nigeria, 2, 20, 46, 48, 55, 72, 116, 200, 207, 208, 274, 309
  expulsion of Ghanaians, 4-5, 277
  migration to, 39, 229-30
Nkrumah, K., 9, 19-20, 26-7, 35, 325n
  attitude towards foreign investment, 247
  desire for modernisation, 270
  liking for planning, 14
  socialist objective, 128-9
  view on industrialisation, 128-9
Nkrumah Government, 7, 13-14, 73
  and corruption, 26
  dependence on suppliers' credit, 14
  on manufacturing ownership, 139
  plan implementation, 7
  policy of industrialisation, 130
  refusal to devalue, 14-15
  view to diversify the economy, 269
non-price decisions, see investment decisions; resource allocation

non-tax revenue, see government revenue
Nurkse, R., 53, 320
Nwankwo, G., 327n
Nyanteng, V., 114-15, 323-4n

official prices, see controlled prices
oil, 153
  see also petroleum
Okyereko Irrigation Project, 92
Omaboe, 7, 45, 259, 315, 318, 320, 330-1n
  on investment decisions, 9, 259
  on politicians, 8
Omane, I., 317
Omnibus Service Authority, 243
One-Year Plan, 9-10
'Operation Feed Yourself', 82
Oppenheimer, M., see Fitch, B.
Organisation of African Unity, 318n
Orleans-Lindsay, J., 326n
overmanning, 17
  in cocoa marketing board, 229
  retrenchment of surplus labour, 279
  in state enterprises, 244
  of unskilled labour, 229
  see also employment; redeployment
overvaluation of the cedi (domestic currency), 21-3, 106, 132, 151, 162, 196-7
  fall in mining output, 162
  and misallocation of resources, 22
ownership, in manufacturing, 38-42
  private, 140
  state, 139-41
Owusu, P., see Atsu, S.

Pakistan, 197
PNDC, see Provisional National Defence Council programme for reconstruction
parallel market
  in hard currencies, 4
  rate of exchange, 4, 196, 278
Parker, J., see Gordon, D.
petroleum, 37
  see also oil
Philippines, 55
Pickett, J., 268, 331n
planning, 6-12
  and allocation through price incentives, 30
  non-implementation of, 18-21
  review of, 17-25

Index

Planning Commission, 8-9
Poland, 320n
Pomadze Poultry Farm, 103
population, 35-43
  census, 37-8
  distribution of, 39-40
  and labour force, 37-43
  migration to Nigeria, 39
  occupational distribution, 40-3
  by Regions, 40-2
  time-series data, 288
Ports Authority, 66
postal service, 70-1
  growth of, 63
  rates, 76
Post Office Savings Bank, 171
poultry
  Newcastle disease, 103
  production, 101-3
power, see electricity
Premier Bank (Ghana) Ltd, 172-3
Prendergast, C. C., see Huq, M.
present value cost (PVC), 268
price controls, 4, 217-20
price distortions, 20-1, 23, 150-1
  and low output growth, 21
price index, see consumer price index
price mechanism, 29-31
  see also market mechanism
'price rings', 221-2
prices, see controlled prices; price distortions; shadow prices
Prices and Incomes Board (PIB), 21, 218-19, 221, 223, 244
  in fixing wages and salaries, 232
pricing policy, 29-31 (see also resource allocation; price mechanism)
  need for realistic prices, 76-7
private consumption, 48-50
  time-series data, 289-90
private enterprise, 9
production function, 267
productivity, 123-6
  of labour, 159-60
  of land, 83-4
professional advice, 27-9, see also International Monetary Fund; World Bank
professional manpower, exodus of, 229-30
profitability
  private, 20
  social, 20

project appraisal, 173, 183-4
  absence of, 8, 262
protection, 132-3
  effective rate of (ERP), 132, 147-9
Provisional National Defence Council (PNDC), 11, 16-17, 27-8, 233
  interest rate policy, 246, 278
  policy of economic recovery, 11-12
public debt, increase in, 13-14
public savings
  fall in, 2-3, 238
  see also savings
public services
  decline of, 5

quasi-money, see money supply
Queen Mothers' role in market control, 221-2

Rado, E., 16, 27, 316-18, 329
Rahim, E., 319n
  see also Khan, R.
railway, 63, 65-6
Railway Corporation, see Ghana Railway Corp.
rate of return, 20, 147
  in cocoa sector, 113
  see also profitability
realistic pricing policy
  need for, 29-31
recovery, 277-81
  difficulties of, 29, 32
  signs of, 6
Recovery Programme, see Economic Recovery Programme
recurrent expenditure, see government expenditure
redeployment of labour, 17, 234
regional imbalance, 58-60
Regional Institute of Population Studies, 74
research, 73-4
resource allocation through administrative measures, 20, 29-30
retrenchment of labour
  in civil service, 279
  in cocoa board, 279
  Ghana Education Service, 279
return, see rate of return
revaluation of the cedi, 195-6, 216
Rimmer, D., 8, 259, 315, 328n, 330-1
road transport, 62-5
roads, 63-4
Roemer, M., 315

Rostow, W., 320
rural banks, 74, 93, 99-100, 190-2
  see also development banking

Safo, D., 281, 333n
salaries, *see* wages and salaries
salt, 158-9
savanna grassland, 101
savings, 53-5, 237-46
  domestic, 237-46
  foreign, 237, 246-50
  marginal rate of savings, 54
  private, 244-6, 301
  public, 238-44, 301
  for selected developing countries, 54-5
  time-series data, 301
  voluntary savings, 244-5
  see also external capital; forced savings; gross domestic savings
savings-investment gap, 247
Schultz, T., 71, 321n
Scitovsky, T., *see* Lewis, W. A.
Second (Five-Year) Development Plan (1959-64), 7, 12, 27, 128, 325n
Seidman, A., 24, 75, 316-18n, 322n
service industry, 53
Seven Year Plan, 7-11, 18, 20, 128-9, 247-8, 258, 269-70, 325n, 331-2
  criticisms of, 259
shadow prices, 20, 23, 113
  applications of, 260-5
  use by ADB, 146
  use by NIB, 184
Shai Hills Cattle Ranch, 263
Smith, A., 61, 77, 321n-2
smuggling, 26
  of cocoa, 22, 114
social opportunity costs, *see* shadow prices
social prices, *see* shadow prices
Social Security Bank (SSB), 171-2
Social Security and National Insurance trust, 74, 172
Soligo, R., 325
Soviet Union, *see* USSR
Special Unnumbered Licence (SUL), 199-200
Standard Bank, 167, 170-1
State Construction Corporation, 243
state enterprises, 241-4
State Enterprises Commission, 241, 330n
State Enterprises Secretariat, 129

State Farms Corp., 89, 242-3
State Fishing Corp., 104, 242-3
State Gold Mining Corp. (SGMC), 154-5, 243
State Hotels and Tourism Corp., 243
State Insurance Corp. (SIC), 25, 75, 243
  overmanning, 244
State Transport Corp., 243
Steel, W., 147-9, 284, 325-6n
Stern, J., *see* Soligo, R.
stocks
  changes in, 49-50
  time-series data, 290-1
structural inflation, 13
Suame Magazine (Kumasi), 127
suppliers' credits, 13-14, 247-8
  and import of inappropriate equipment, 275
  for manufacturing investment, 151
  see also external capital
Switzerland, 158

Tano Irrigation Project, 92
Tanzi, V., 333n
tariffs, rates on manufacturing imports, 132
taxes
  direct, 238-40
  indirect, 239-40
  revenue from, 238-40
  see also government revenue
technological dualism
  in fishing industry, 104
  in pig farming, 101
technology
  capital intensive, 268-9
  domestic, 147, 150
  labour-intensive, 268-9
  least-cost, 269
  less-developed country, 268-9
technology choice, 150, 266-9
  case for a policy, 274-5
  in Ghana, 269-75
  in leather manufacturing, 268-9
Technology Consultancy Centre (TCC), 74, 325n
telecommunications, 70-1
  see also telephone service
telephone service, 5, 63, 73
  see also telecommunications
Tema Food Complex, 332n
Tema Textiles Ltd, 140

Index

Ten Year Plan
  (of 1920-30), 6
  (of 1946-56), 6-7
  (of 1951-61), 7
terms of trade, 22, 202-3, 210-13
Thailand, 55
Three-Year Medium Term Plan
  (1984-86), 31
  see also Economic Recovery
    Programme
thrift societies, 170
timber
  export of, 106
  output, 106
Timmer, C., 274, 331-2n
Togo, 35, 110, 114, 116, 158, 159, 204, 208
tourist industry, effects of over-valued
  currency, 198
trade
  deflator, 299
  external, see external trade
  internal, see internal trade
  intra-ECWAS, 209
  intra-West African, 208
Trade, Ministry of, 195, 218
transportation, 62-8
  by sea, 63
  see also water transport
Tribe, M., 327n, 332n
  with Cohen, 318n, 330n
Tsetse fly, 101
Turkey, 55
Two-Year Plan (1968-70), 9, 260, 316, 331n
  on investment decisions, 260

Uganda, 309
under-utilisation, of capacity, see
  capacity utilisation
unemployment, 4
  rate of, 39
United Ghana Farmers Co-operative
  Council (UGFCC), 98, 172
United Kingdom, 111, 207, 209
United Nations Conference on Trade
  and Development (UNCTAD), 116
University
  of Cape Coast, 71
  of Ghana, 71, 73
  of Science and Technology (Kumasi), 71, 73, 325n
urbanisation, 37, 59
USA, 111-12, 117, 156-8
USAID, 95
USSR, 29, 111, 200, 208-9

VALCO, see Volta Aluminium
  Company
Vea Irrigation Project, 92
Volta Aluminium Company, 69, 157-8
Volta dam, see Akosombo dam
Volta Lake Company, 67
Volta River Authority (VRA), 68-9, 158, 241, 243
Volta River (Hydro-Electric) Project, 68-9, 75, 157
  see also Akosombo dam; Kaiser
    Corporation
voluntary savings, see savings

WAFF Ltd, 263
Wage Earners Scheme, experience of
  Bangladesh, 197, 328n
wages and salaries, 230-4, 278-9
  fall in real wages, 3, 231-2
  in manufacturing, 120
  minimum (statutory), 230
  and tax relief, 278-9
water supply, 59-60, 63-4, 69-70
water transport, 67-8
Weija Irrigation Project, 92
West African Currency Board, 168, 173-4
Western Castings (Takoradi), 126
West Germany, 111, 131
Whitman Report, 12, 316
Willowbrook Ltd, 140
Winston, G., 274, 325n, 331-2
World Bank, 4, 21, 24-5, 54, 170, 234
  advice from, 17
  advice on shadow wage rate, 262
  help for development banks, 182
  GDP estimate (1984), 48
  loan and assistance from, 17, 27-8, 66, 192-3, 279 (see also International
    Monetary Fund)
World Bank-IMF conditionality, 32
World Health Organisation, 69

Zaire, 309
Zambia, 55